Frederick J. Furnivall, Andrew Boorde, Milton Barnes

The Fyrst Boke of the Introduction of Knowledge Made by Andrew Borde

of physycke doctor - A compendyous regyment - A dyetary of helth made in Mountpyllier

Frederick J. Furnivall, Andrew Boorde, Milton Barnes

The Fyrst Boke of the Introduction of Knowledge Made by Andrew Borde
of physycke doctor - A compendyous regyment - A dyetary of helth made in Mountpyllier

ISBN/EAN: 9783337221171

Printed in Europe, USA, Canada, Australia, Japan

Cover: Foto ©berggeist007 / pixelio.de

More available books at **www.hansebooks.com**

The Fyrst Boke of the Introduction of Knowledge

made by

Andrew Borde

of Physycke Doctor.

A Compendyous Regyment

or

A Dyetary of Helth

made in Mountpyllier, compyled by

Andrewe Boorde

of Physycke Doctour.

Barnes

in the Defence of the Berde:

a Treatyse made, answerynge the Treatyse of Doctor Borde vpon Berdes.

EDITED, WITH A LIFE OF ANDREW BOORDE, AND LARGE EXTRACTS FROM HIS *BREUYARY*, BY

F. J. FURNIVALL, M.A., Trin. Hall, Camb.,

EDITOR OF *THE BABEES BOOK*, &c.

LONDON:

PUBLISHED FOR THE EARLY ENGLISH TEXT SOCIETY

By KEGAN PAUL, TRENCH, TRÜBNER & CO., LIMITED,

DRYDEN HOUSE, 43, GERRARD STREET, SOHO, W.

1870.

[*Reprinted 1893, 1906.*]

TO

Colonel James Halkett,

BARON HALKETT)

OF *GREAT FOSTERS*, NEAR EGHAM,

LATE OF THE COLDSTREAM GUARDS, ETC. ETC.

My dear Colonel,

You are our most widely-travelled friend here. Your steps have wandered far beyond Boorde's range. Asia, North and South, Africa, North and South too, the Indies, and America, have seen you; the Crimea has been stained by your blood; and there are few Courts and cities in Europe where you have not been. I may therefore well dedicate to you Boorde's records of his travels, more than 300 years ago, in his *Introduction of Knowledge*.

On the Elizabethan porch of your fine old Tudor house is the date of 1578, while Anne Boleyn's badge is the centre ornament of your dining-room ceiling, and Tudor badges are about it. I may therefore well dedicate to you Boorde's *Dyetary* of 1542, which starts with directions that may have been studied by the builder of your own house, or the early dwellers in it. As it was once my Father's too, and has been the scene of many a happy visit at different times of my life, I like to mix the thought of the old house with my old author, Andrew Boorde, and to fancy that he'd have enjoyed ordering where the moat was to be, the stables, and all the belongings, and lecturing the owner as to how to manage house and servants, wife and child, pocket and body.

That health and happiness may long be the lot of you and the charming sharer of your name, whose taste has beautified the old house that you have together so admirably restored, is the hope of

Yours very sincerely,

F. J. FURNIVALL.

Walnut Tree Cottage, Egham,
August 3, 1870.

CONTENTS.

FOREWORDS : PAGE

Part I. Andrew Boorde's Works and the Editions of them ... 11

A. Genuine Works 11

B. Doubtful Works 26

C. Works probably spurious 27

Part II. Andrew Boorde's Life and Opinions, with Extracts from his *Brevyary of Health* (p. 74-106) 36

The present Edition 106

THE FYRST BOKE OF THE INTRODUCTION OF KNOWLEDGE (being *A Handbook of Europe, Barbary, Egypt, and Judæa*, in 39 chapters, with 'Contents,' p. 112-115) 111

A COMPENDYOUS REGYMENT OR A DYETARY OF HELTH ... 223
(The Table of the Chapytres, p. 229-231.)

THE TREATYSE ANSWERYNGE THE BOKE OF BERDES, compyled by Collyn Clowte, dedycatyd to Barnarde Barber, dwellyng in Banbery; *A Treatise otherwise called*, BARNES IN THE DEFENCE OF THE BERDE 305

HINDWORDS (including accounts of Boorde's *Introduction* and *Dyetary*; to be read after p. 104 of the Forewords) ... 317

NOTES 325

INDEX 352

FULLER'S ACCOUNT OF ANDREW BOORDE 384

SUPPLEMENT 385

X. ANDREW BOORDE'S INTRODUCTION, &c.

p. 18, note 7, after *day*, insert [of August]

p. 44, l. 4. The 'old writer' referred to was Roy, in his *Rede me and be not wroth*, p. 104-5 of Pickering's Reprint. The passage is quoted in my "Ballads from MSS," illustrating the Condition of Tudor-England, p. 82.

p. 57, note 3. 'my lord of chester' means 'the Abbot of St. Werburgh's.' E. A. Freeman in the *Saturday Review*, 10 Feb. 1872, p. 189, col. 1.

p. 116-17. On English changes of fashion, see the Society's *Four Supplications*, 1871, p. 51.

p. 156, l. 18. "Argentyne, we suppose, is Argentoratum or Strassburg." E. A. Freeman.

p. 165, note 1. "Andrew Borde does not at all speak as a Saxon heretic, but as a dutiful subject of King Henry the Eighth, who dedicates his book to that King's daughter. In the eyes of such a one the Saxons were praiseworthy in so far as they had cast off the usurped authority of the Bishop of Rome, blameworthy in so far as they had fallen into the heretical innovations of Martin Luther." E. A. Freeman, *Saturday Review*, 10 Feb. 1872, p. 189, col. 2.

p. 194, last side-note; p. 362, col. 1, Emperor; *for* Austria *read* Germany (Charles V.).

p. 287, l. 6-7. The Hebrecyon, and Cynomome. This saying is quoted in Cogan's *Haven of Health*, 1596, p. 109 (*N. & Q.*), and is not in the *Regimen Sanitatis Salerni* (as saith Riley's Dict. of Latin Quotations), in which however is a similar and well-known line, "Cur moriatur homo cui *salvia* cressit in horto?" Villanova, c. 60. Crokes, Sir Alex. 1830.—C. Innes Pocock.

p. 308, note 1, line 1, for *Ovium* read *Ædium*.

The short review of *Boorde* in the *North British Review*, No. 106, p. 559-61, notes that "his letters of the alphabet representing Hebrew numerals are given instead of the numerals themselves. . . . His Italian geography is full of confusion. He intimates that Jerusalem is out of Asia, and places Salerno [in Italy] in the neighbourhood of Constantinople. Writing in 1542, he describes the mosque of St. Sophia as a Christian Church. Then again, his statements, pp. 77, 178, respecting St. Peter's at Rome, will not bear comparison with the graphic account left by his contemporary, Thomas, of the basilica, as it stood in the 16th century, grand and magnificent, though uncompleted. (*Historie of Italie*, ed. 1549, fol. 40.) Every detail supplied by Thomas, from the '30 steppes of square stone, the solemnest that I have seene,' to 'the newe buildyng [which] if it were finished, wolde be the goodliest thyng of this worlde,' stamps his description as authentic."

FOREWORDS.

PART I. ANDREW BOORDE'S WORKS, AND THE EDITIONS OF THEM.

A. GENUINE WORKS.

§ 1. The Dyetary *of* 1542 (p. 11)

§ 2. *Is Powell's edition* 1547 *or* 1567? *Wyer's undated edition; Colwel's of* 1562 (p. 13)

§ 3. The Fyrst Boke of the Introduction of Knowledge, *written by* 1542, *not published till* 1547 *or after; meant mainly to be a book of Medicine, though Book I. is one of Travels. Two editions of it* (p. 14). *Is mentioned in the* Dyetary, Pryncyples of Astronamye, *and the* Brevyary (p. 15), *and* Barnes in the Defence of the Berde. *The Lothbury edition is of* 1562 *or* -3 A.D. (p. 18).

§ 4. Barnes in the Defence of the Berde *must be dated* 1542 *or* -3 ... ... (p. 19)

§ 5. The Breuyary of Health, *written also by* 1542, *though no edition is known till* 1547. *Boorde's account of its name* (p. 21). *His motives in writing it* (p. 21). *It's a companion to the* Dyetary ... ... (p. 20)

§ 6. The Pryncyples of Astronamye. *Its contents*, p. 22 (*an extract from it at* p. 16)

§ 7. The Peregrination, *or Itinerary of England, with its notice of "Boord's Hill, the authour's birth place"* (p. 23)

§ 8. *The lost* Itinerary of Europe, *which Boorde lent to Thomas Cromwell* (p. 24)

§ 9. A Boke of Sermons *lost by old noodle Time* ... (p. 24)

§ 10. The Pronostycacyon for 1545 ... ... (p. 25)

§ 11. *The lost* Treatyse vpon Berdes ... ... (p. 26)

B. DOUBTFUL WORKS.

§ 12. Almanac & Prognostication ... ... (p. 26)

C. WORKS PROBABLY SPURIOUS.

§ 13. Merie Tales of the Mad Men of Gotam ... (p. 27)

§ 14. Scogin's Jests (p. 31). *The Prologue and First Jest use Boorde's phrases* (p. 32).

§ 15. The Mylner of Abyngton (p. 32)

§ 16. *The* Promptuarium Physices *and* De iudiciis Urinarum, *of Bale's list* ... (p. 33)

§ 17. "Nos Vagabunduli" *and the Friar "Hindrance"* (p. 34)

PART II. ANDREW BOORDE'S LIFE.

§ 18. *Table of the known events of Boorde's Life* ... (p. 36)
§ 19. *His Birth at Boord's Hill* (p. 38)
§ 20. *His Bringing-up, probably at Oxford* ... ... (p. 40)
§ 21. *Mr Lower's identification of our Andrew with Lord Bergevenny's* nativus *or bondman in* 1510, *not to be trusted* (p. 41), *as Boorde was a Carthusian Monk before this date* (p. 43)
§ 22. *Boorde accused of being "conversant with wonnen"* (p. 44)
§ 23. *Boorde appointed "Suffrygan off Chychester"* (p. 44)
§ 24. *Boorde's First Letter, to Prior Hinton* (p. 45-7); *and, here anent, of the Carthusians* (p. 46)
§ 25. *Boorde has "lycence to departe from the Relygyon," goes abroad, and studies medicine* (p. 47)
§ 26. *He returns to England, and attends the Duke of Norfolk* ... ... (p. 48)
§ 27. *Boorde's second visit to the Continent to study Medicine, with notice of his later Travels, especially to Compostella* (p. 49)
§ 28. *Boorde again in the London Charter-house. He takes the oath to Henry VIII's Supremacy* ... ... (p. 51)
§ 29. *Boorde is in thraldom in the Charterhouse. He writes to Prior Howghton in the Tower* (p. 52)
§ 30. *Boorde is freed by Cromwell, whom he visits* (p. 52)
§ 31. *Boorde's third visit to the Continent, and Second Letter, from Bourdeaux,* 20 *June,* 1535, *to Cromwell* (p. 53)
§ 32. *Boorde in Spain, &c., sick. His Third Letter, to Cromwell* ... ... (p. 54)
§ 33. *Boorde at the Grande Chartreux. His Fourth Letter, to his Order in England* (p. 56)
§ 34. *Boorde in London again. His Fifth Letter, to Cromwell* (p. 58)
§ 35. *Boorde studies and practises medicine in Scotland. His Sixth Letter, to Cromwell* (p. 59). *"Trust yow no Skott"* (p. 59, last line but one).
§ 36. *Boorde in Cambridge. His Seventh Letter, to Cromwell,* 17 *Aug.*, 1537 (?) (p. 61)
§ 37. *Boorde's fourth visit to the Continent. The range of his Travels. He settles at Montpelier, and by* 1542 *has written the* Introduction, Dyetary, Breuyary, *and* (?) Treatyse vpon Berdes ... ... (p. 63)
§ 38. *Boorde at Winchester before or by* 1547, *and probably in London in* 1547 *to bring out his* Dyetary II, Introduction, Breuyary, *and* Astronamye (p. 64)
§ 39. *Bp Ponet's charge against Boorde of keeping three whores for himself and other Papist priests* ... ... (p. 65)
§ 40. *Guilty or not guilty?* (p. 67)
§ 41. *Boorde in the Fleet Prison* (Mr J. Payne Collier, p. 71-2). *Boorde's Will* (p. 73) *and the Proving of it* ... ... (p. 70)
§ 42. *Portraits of Andrew Boorde* (p. 74)
§ 43. *Characteristic Extracts from Boorde's* Breuyary :

α *Where he speaks of himself or his tastes, &c.* (p. 74)
β *His remarks on England, his Contemporaries, and the Poor* ... ... (p. 82)
γ *Some of Boorde's opinions* (*on Mirth*, p. 88) (p. 87)
δ *Boorde's Treatment of certain Diseases* (p. 96), *and herein of Chaucer's Somonour's* Saucefleem Face (p. 101)
ε *Boorde serious* ... (p. 102)
[*See the* Hindwords, p. 317.]
§ 44. *Boorde's character* (p. 105)
§ 45. *Esteem in which he and his books were held* (p. 105)
§ 46. *The present Reprints; some Cuts used indiscriminately* (p. 106)
§ 47. *The Editor's task* (p. 109)

---

§ 1. Among the many quaint books from which I quoted in my notes to Russell's Book of Nurture in the *Babees Book* (E. E. T. Soc. 1868), one of the quaintest was Andrew Boorde's *Dyetary*, as readers, no doubt, convinced themselves by the long extract on pages 244-8, and the shorter ones on p. 205, 207, &c. Since then I have always wished to reprint the book, and the securing, for 32*s*. at Mr Corser's sale last February, of a copy of the 1562 edition not in the British Museum,[1] made me resolve to bring out the book this year. Wishing, of course, to print from the first known edition, I turned to Mr W. C. Hazlitt's *Handbook* to find what that was, and where a copy of it could be got at, and saw, after the title of the *Dyetary*, the following statement:

"Wyer printed *at least* 3 editions without date, but in or about 1542. Two editions, both differing, are in the British Museum; a third is before me; and a fourth is in the public library at Cambridge.[2] All these vary typographically and literally."

[1] It is in the Cambridge University Library, perfect. Mr Bradshaw's description of it is as follows:

"Boorde (Andr.)

A compendious regiment or dietary of health.

London, Tho. Colwel, 12. Jan. 1562. 8°.

(b) *Title* (*within a single line*): Here Folo-/weth a Compendyous Re-/gimente or Dyetary of health, / made in Mount pyllor: Com-/pyled by Andrewe Boorde, of Phy-/sycke Doctor / Anno Domini. M. D. LXII. / XII. Die Mensis / Januarij./ [*woodcut of an astronomer.*] *Imprint:* Imprinted by / me Thomas Colwel. Dwel-/lynge in the house of Robert Wyer, / at the Signe of S. Johñ Euan-/gelyst besyde Charynge / Crosse./ ☞ /

Collation: ABCDEFGH^{8}; 64 leaves (1—64). Leaf 1^{a} title (as above); 1^{b}—4^{a} Table of chapters; 4^{a}—64^{b} Text; 64^{b} Imprint (as above)."

[2] This is the same book as the one undated Wyer edition in the Museum. Mr Bradshaw's description of it is:

A visit to the British Museum soon showed that one of these 'editions'[1] in the British Museum was only a title-page stuck before a titleless copy of Moulton's *Glasse of Health*, on to which had been stuck a colophon from some other book of Wyer's printing. The other Museum edition, in big black-letter, had not, on the front and back of its title, the dedication to the Duke of Norfolk that the other title-page had, and I therefore wrote to Mr Hazlitt to know where was the "third" copy that was "before" him when he wrote his Boorde entries. He answered that he had sold it to Mr F. S. Ellis of King St., Covent Garden, in one of whose Catalogues he had afterwards seen it on sale for four guineas. I then applied to Mr Ellis for this copy, and he very kindly had search made for it through his daybooks of several years, and found that it had been sold to our friend and member, Mr Henry Hucks Gibbs. Mr Gibbs at once lent me his copy, and it proved to be a complete one of the edition of which the Museum had only a title-page. It had a dedication to the Duke of Norfolk,—whom Boorde had attended in 1530,—dated 5 May, 1542, which was not in the undated edition in the Museum, and Mr J. Brenchley Rye of the Printed-Book Department was clearly of opinion that the type of the 1542 copy was earlier than that of the bigger black-letter of the undated one, though it too was printed by Robert Wyer, or said so to be.

Further, Mr Gibbs's copy was printed by Robert Wyer for John Gowghe; and the latest date in Herbert's *Ames* for Robert Wyer is 1542, while the latest for John Gough is 1543. One felt, therefore, tolerably safe in concluding that the 1542 copy was the first edition

"BOORDE (Andr.)
A compendious regiment or dietary of health.
London, Robert Wyer, *no date*. 8°.
(a) *Title* (*within a border of ornaments*): ¶ Here Folo-/weth a Compendyous Re-/gyment or a Dyetary of / helth, made in Mount-/pyllor: Compyled / by Andrewe / Boorde, of / Physicke / Doctor./ [*woodcut of an astronomer.*] *Imprint:* ¶ Imprynted by me Robert / Wyer: Dwellynge at the / sygne of seynt Iohñ E-/uangelyst, in S. Mar-/tyns Paryshe, besy-/de Charynge / Crosse./ ¶ Cum priuilegio ad imprimen-/dum solum.
Collation: ABCDEFGHIKLMNOPQ⁴; 64 leaves (1—64) in octavo. Leaf 1ᵃ title (as above); 1ᵇ—4ᵃ Table of chapters; 4ᵃ—64ᵇ Text; 64ᵇ imprint (as above).
The copy in the Cambridge University Library is perfect."

[1] Some bibliographers (if not most) are sadly careless dogs.

of *The Dyetary*, and that it was publisht in 1542, the year in which its Dedication bears date.

§ 2. But, this granted, came the question, When was the undated edition, printed by Robert Wyer, publisht? Before trying to answer this question, I must say that the Museum possesses a copy of another edition of the *Dyetary*, with a Dedication to the Duke of Norfolk, dated 5 May, 1547 (MDXLVII), altered from the Dedication of 5 May, 1542, while, as I have said before, the undated edition has no Dedication. But the colophon of this 1547 edition says that it was printed by Wyllyam Powell in 1567 (MDLXVII), the X and L having changed places in the two dates. Was then 1547 or 1567 the real date of this edition by William Powell? 1547, I think; for, 1. Boorde died in 1549, and the Dedication is altered in a way that no one but an author could have altered it; 2. the dates we have for William Powell's books are 1547-1566,[1] so that he could have printed the *Dyetary* in 1547; though we can't say he couldn't have printed it in 1567 too, as all his books are not dated.

If then we settle on 1547 for the date of Powell's double-dated edition, the question is, What is the date of Robert Wyer's undated one? Are we to put Wyer's date down from 1542 to 1549 or later, and explain the absence of the Dedication by the fact of Andrew Boorde's death in 1549?[2] or are we to explain it by the Duke of Norfolk's arrest on Dec. 12, 1546, and suppose Wyer to have issued his edition before Henry VIII's death on the night of Jan. 27, 1546-7, saved the Duke from following his accomplished son, Surrey, to the scaffold,[3] while Powell, who issued his edition in the summer of the same year, could safely restore Boorde's Dedication, since Norfolk, though excepted from the general pardon proclaimed on Edward VI's accession, was looked on as safe? The latter alternative is countenanced by Wyer's undated edition being printed from his first of 1542, rather than Powell's of 1547, as the collation shows; but I cannot decide whether the second Wyer, or Powell, was issued first.

[1] The last license to him in Collier's *Extracts*, i. 137, is about midway between July 1565 and July 1566.

[2] The Duke of Norfolk did not die till 1554.

[3] Surrey was beheaded on Jan. 19, 1546-7.

The possibility that the undated dedicationless Wyer was issued before 1542, and that the 1542 edition was the second, is negatived by Mr Rye's opinion on the types of the two editions, and perhaps by the omission of two of the woodcuts, the change of the third, and the results of the collation. Of later editions I know only that of 1562, 'imprinted by me Thomas Colwel in the house of Robert Wyer': see page 11. By that fatality which usually attends the most unsatisfactory plan of "Extracts," Mr Collier has in his "Extracts" missed the only two entries in the Stationers' Registers relating to Boorde's books that I wanted, namely, that of this 1562 edition of the *Dyetary*, and the Lothbury edition of the *Introduction*. The entry as to Colwel's print of the *Dyetary* is:

T. colwell *Recevyd* of Thomas Colwell,[1] for his fyne, for that he prented *the Deatory of helthe* / the Assyce of breade And Ale, *with* arra pater,[2] *with*out lycense. } xij*d*.

Company of Stationers' First Register, leaf 77, in the list of Fines, 22 July 1561, to 22 July 1562.

Lowndes enters other editions of "1564 (White Knights 507, mor. 9*s*. 1567 Perry pt. i, 468, 9*s*. Bindley pt. i. 460, 11*s*.) 1576."

As the date of the Dedication to the *Dyetary* is 5 May, 1542, while that of the *Introduction* is 3 May, 1542, I have put the former after the latter, though it (the Dyetary) was published five years before the *Introduction*. Still, the *Introduction*, the *Dyetary*, and the *Breuyary* (examined 1546, published 1547) were all written by Boorde by the year 1542.

§ 3. *The fyrst Boke of the Introduction of Knowledge.* This book was dedicated to the Princess Mary, afterwards Queen, daughter of Henry VIII, on May 3, 1542. It was intended to have a second book, in which the vices of Rome were mentioned,[3] and which second book *may* therefore[4] have been the *Breuyary*, as the vices of Rome are mentioned in its 2nd part, the *Extravagantes*, fol. v, back. It

[1] Colwell was admitted a freeman of the Stationers' Company on the 30th of August, 1560.

[2] An Almanack. See entries in Stat. Reg., and Hazlitt's *Handbook*.

[3] In the *Introduction*, chap. xxiii (repr. sign. R), Boorde says, "Who so wyl see more of Rome & Italy, let him loke in the second boke, the .lxvii. chapter" (p. 178 below).

[4] I don't think it was so.

was also intended to have been mainly a book on physic, for, besides the four quotations given under (*a*) below, Boorde says in his *Breuyary*, "no man shulde enterpryse to medle with Phisicke but they which be learned and admytted, as it doth appeare more largelyer in the *Introduction of knoweledge*" (Fol. iii, at foot); and again, Fol. v, and lxxvi back:

"I had rather not to meddle with Physicions and Chyerurgions then to haue them, yf I shulde dysplease them: for yf they be dyspleased, there is neither Lorde nor Lady nor no other person can haue any seruyce or pleasure of theym, for this matter loke forther in the *Introduction of knowledge*, and there shall you see what is good both for the soule and body in god. Amen."

The *Introduction* was also intended to have a book on Anatomy in it,—see the next quotation;—but it appeared as a book of Travels, with only a "fyrst Boke" in or after 1547, after both the *Dyetary* and *Breuyary*, and the *Astronamye* also, had been published. In each of these books the *Introduction* is mentioned as in the press. Take (*a*) the *Breuyary:*

"Euery man the which hath all his whole lymmes, hath ii.c. xlviii. bones, as it doth more playnely appeare in my Anothomy in *the Introduction of knowledge, whiche hath bene longe a pryntynge, for lacke of money and paper; and it is in pryntynge*, with pyctures, *at Roberte Coplande, prynter.*" (Breuyary, Pt I. fol. lxxxviii.)

"For kynges, and kynges sones, and other noble men, hath ben eximious Phisicions, as it appereth more largely in *the Introduction of knowlege, a boke of my makynge, beynge a pryntyng with Ro. Coplande* (*ib.* Fol. lxx, back). See p. 93 below.

"wherfore this science of medecines is a science for whole men, for sick men, and for neuters, which be neyther whole men nor sycke men; wherfore I do aduertyse euery man not to set lytle by this excellent science of medecines, consyderynge the vtilitie of it, as it appereth more largelier in *the introduction of knowlege.*" Fol. lxxvi, back.

"the kynges actes and lawes . . wylleth and commaundeth, with greate penalytie, that no man shulde enterpryse to medle with Phisicke, but they which be learned and admytted, as it doth appeare more largelyer in the *Introduction of knoweledge.*" Breuyary, Fol. iii, at foot.

(*β*) The *Dyetary*. Boorde says in his Dedication to the Duke of Norfolk:

"But yf it shall please your grace to loke on a boke the which I dyd make in Mountpyller, named *the Introductory of knowlege*, there shall you se many new matters / the whiche I have no doubte but that your grace wyl accept and lyke the boke, *the whiche is a pryntynge besyde saynt Dunstons churche within Temple barre, ouer agaynst* the *Temple.*" (p. 227, col. 1, below.)

(γ) *The Astronamye.* The full title of this book, the only known copy of which is in the Cambridge University Library, is :

"The pryncyples / of Astronamye / the whiche / diligently perscrutyd is in maner a / pronosticacyon to the worldes / end compylyd by Andrew / Boord of phisick / Doctor /,"

and the last words of the Preface are :

"And wher I haue ometted & lefft out mani matters apertayn-[yn]g to this boke, latt them loke in a book namyd the *Introduction of knowleg, a boke of my makyng, the which ys aprintyng at old Robert Coplands, the eldist printer of Ingland*, the which doth print thes yere [1] mi pronosticacions."

Accordingly, the colophon is, "Enprynted at London in ye Fletostrete / at the sygne of the Rose garland by / Robert Coplande."

The other references in this volume to Boorde's other works are on B vii (not signed) : "for this matter, looke in the *Breuyary of helth* and in the *Introduccyon of knowleg.*"

C. ii. (not signed) "And he that wyll haue the knowleg of all maner of sicknesses & dysesys, let them looke in the *breuyary of helth*, whiche is pryntyd at Wyllyam Mydyltons in flet stret."

The last paragraph of the *Astronamye* is :

"¶ Now to conclud, I desier euere man to tak this lytil wark for a pasttime.[2] for I dyd wrett & make this bok in .iiii. dayes, and wretten with one old pene with out mendyng. and wher I do wret ye sygnes in Aries, in Taurous, & in Leo, is, for my purpose it stondyth best for our maternal tonge."

A further and earlier[3] notice of the *Introduction* is found in the chaffy answer to Boorde's lost attack on beards,[4] which answer is

[1] A friend reads *thes yere* as 'these here;' but the words no doubt mean 'this year,' and the *pronosticacions* may be one of those of which a title of one, and a fragment of another—or a supposed other—are in the British Museum. See below, p. 25, 26-27.

[2] past time, *orig.* [3] I take Barnes's book to be of the year 1542 or 1543.

[4] As a substitute, take parson Harrison's: "Neither will I meddle with our varietie of beards, of which some are shauen from the chin like those of Turks,

called at the end 'Barnes in the defence of the Berde,' and is, on account of its connection with Boorde, reprinted at the end of this volume. The book opens thus:

"It was so, worshypful syr, that at my last beynge in Mountpyllour, I chaunsed to be assocyat with a doctor of Physyke / which at his retorne had set forth *.iij. Bokes to be prynted in Fleet strete, within Temple Barre, the whiche bokes were compyled togyther in one volume named the* INTRODUCTORIE OF KNOWLEDGE / wheryvpon there dyd not resort only vnto hym, marchauntes, gentylmen, and wymmen / but also knyghtes, and other great men, whiche were desyrouse to knowe the effycacyte and the effecte of his aforesayd bokes."

Now this looks certainly as if the *Introduction* was at first believed by Boorde's acquaintances to have been intended to contain his other two books written in or before 1542, namely, the *Dyetary* and *Breuyary;* but as Boorde himself says he meant to have an *Anatomy* in his *Introduction*, and evidently much other matter on physic (p. 14-15 above), we need not speculate further on Barnes's words. What we know is, that the *Introduction* must have been published after the *Breuyary* of 1547, and the *Astronamye* doubtless of the same year. I say the same year, for the Preface of the *Breuyary* shows that a treatise on Astronomy was wanted to ac-

not a few cut short like to the beard of marques Otto, some made round like a rubbing brush, other with a *pique de vant* (O fine fashion !) or now and then suffered to grow long, the barbers being growen to be so cunning in this behalfe as the tailors. And therfore if a man haue a leane and streight face, a marquesse Ottons cut will make it broad and large; if it be platter like, a long slender beard will make it seeme the narrower; if he be wesell-becked, then much heare left on the cheekes will make the owner looke big like a bowdled hen, and so grim as a goose; if Cornelis of Chelmeresford saie true, manie old men doo weare no beards at all."—*Harrison's Description of England*, ed. 1586, p. 172, col. 2.

See on this Beard question the curious and rare poem,—by Wey? see the Roxb. Club print of it—"The Pilgrymage and the wayes of Ierusalem," in a paper MS of Mr Henry Huth's, about 1500 A.D., quoted below, p. 182.

Prestes of the New lawe:

The thyrd Seyte beyn prestis of oure lawe,
That synge masse at þe Sepulcore;
At þe same graue there oure lorde laye,
They synge þe leteny euery daye.
In oure maner is her songe,
Saffe, here *berdys be ryght longe;*
That is þe geyse of þat contre,
The lenger þe berde, the bettyr is he;
The ordere of hem be barfote freeres . . .

company it[1]; Boorde tells us that he wrote his *Astronamye* in four days with one old pen without mending[2]; and this *Astronamye* was printed by Robert Coplande, who, so far as we know, printed no book after 1547. The cutting of the 'pyctures' must have taken so much time[3], and the 'lacke of money and paper'[4] continued so long, that old Robert Coplande did not finish the book, but left his successor, William Coplande, to bring it out in Robert's old house,[5] in Flete strete, at the sygne of the Rose Garland,[6] no doubt late in 1547, or in 1548. This delay in the appearance of the *Introduction* accounts for a few words in it relating to Boulogne, which could not have been written till 1544, when Henry VIII took that city: "Boleyn is now ours by conquest of Ryall kyng He*n*ry the eyght.[7]"

Now, besides William Coplande's undated "Rose-Garland" edition of the *Introduction*, we know of another undated edition by him printed at Lothbury. In this "Lothbury" edition we do not find the above-quoted words of the "Rose-Garland" edition relating to Boulogne; and as we know that Edward VI restored Boulogne to the French in 1550, the Lothbury edition must have been after that date. It must also have been after the deaths of Henry VIII and Edward VI, when there was no king in England, as the Lothbury edition leaves out the Rose-Garland's "But euer to be trew to God and my kynge" (p. 117, l. 24). The Lothbury edition must also

[1] "but aboue al thinges next to grammer a Physicion muste haue surely his Astronomye, to know how, whe*n*, & at what time, euery medecine ought to be ministred."—*Breuyary*, The preface, A Prologe to Phisicions, Fol. ii, back. See also the 'Proheme to Chierurgions,' Fol. iiii.

[2] See p. 16, above.

[3] That is, if any but the Englishman and Frenchman were cut for it, which I doubt. But Boorde might have waited for money for more original cuts.

[4] See p. 15, above.

[5] Herbert remarks in his MS memoranda, 'though the book was printed by R. Copland, it was licensed to W. Copland.'—*Ames* (ed. Dibdin, 1816). I don't believe there is any authority for this "licensed." The Charter of the Stationers' Company was not granted till 1556.

[6] If the reader will turn to the Rose-Garland device at the end of the *Introduction*, he will see how William Coplande has used his predecessor's block: he has left R. C. in the middle, but has cut out the black-letter 'Robert' in the legend, and put his own 'William,' in thinner letters, in the stead of his predecessor's thicker 'Robert,' which matcht the 'Coplande.'

[7] The .xviii. day, the kinges highnes, hauyng the sworde borne naked before him by the Lorde Marques Dorset, like a noble and valyaunt conqueror rode into Bulleyn.—*Hall's Chronicle*, p. 862, ed. 1809.

have been after 1558, for the change of Boorde's description of the Icelander, "Lytle I do care for *matyns or masse*" (chap. vi. line 9, p. 141) into "Lytle do I care for *anye of gods seruasse*," shows that Mary's reign was over; besides being a specimen of William Coplande's notion of rimes. As we know further that William Coplande printed one book at least at the Three Cranes in the Vintry in 1561—Tyndale's Parable of the Wicked Mammon—we may at once identify the Lothbury edition with that which was licensed to William Coplande in 1562-3,[1] as appears by the following entry (omitted by Mr Collier[2]) on leaf 90 of the first Register of the Stationers' Company:

W. Coplande **Recevyd** of William Coplande, for his lycense for pryntinge of [a] boke intituled "the introduction to knowlege" } iiij[d]

Of Coplande's first, or Rose-Garland, edition, a unique copy was known in Mr Heber's library; but I could not hear of it, when first preparing the present volume, and was obliged to apply to the Committee of the Chetham Library for the loan of their copy of the 2nd, or Lothbury, edition. This they most kindly granted me; and Mr W. H. Hooper had copied and cut all the 'pyctures' in it, and the reprint was partly set-up, when a letter to that great possessor of old-book treasures, Mr S. Christie-Miller of Britwell House, brought me a courteous answer that he had the first edition, that I might correct the reprint of the second by it, and that Mr Hooper might copy the cuts—nine in number—that differed from those in the 2nd edition. These things have accordingly been done, and the varying cuts of the 2nd edition put into, or referred to in, the notes. The differences in the texts of the two editions are very slight, barring the Boulogne, King, and Mass passages noticed on this page and the foregoing one.

§ 4. The Dedications to the *Introduction* and the *Dyetary*, and the publication of the latter in 1542 (or 1543), coupled with the opening words of *Barnes in Defence of the Berde* which we quoted above, p. 17, leave no doubt in my mind that this last tract was written and

[1] This enables us too to settle that the other Lothbury books were printed after the Three-Cranes books. (One Lothbury book is dated 1566.)

[2] See p. 14, above.

published in 1543, and that Boorde returned to England from Montpelier to see his *Dyetary* through the press.

§ 5. *The Breuyary of Health.* Having thus discussed the dates of the three little books in the present volume, we have next to notice shortly Boorde's other books. The principal of these is the *Breuyary.* There is no copy of the first edition of it (A.D. 1547?) in the British Museum, Bodleian, or Cambridge University Library. Lowndes says that it was reprinted in 1548, 1552, &c., and I have seen a statement that the edition of 1552 is an exact reprint of that in 1547. A colophon at the end of the first book of the 1552 edition says, "Here endeth the first boke examined in Oxford, in June, the yere of our lord .M. CCCCC. xlvi. And in the reigne of our souerayne Lorde kynge Henry the .viii. kynge of Englande, Fraunce, and Irelande the .xxxviii. yere. ·. And newly Imprinted and corrected, the yere of our Lorde God .M. CCCCC. L. II." As I mean to give several extracts from the *Breuyary* further on, page 74 *et seq.*, in Boorde's Life, I shall only quote here his "Preface to the Readers of this Boke," of which the end will commend itself to my fellow-workers in the Society, who, too, "wryte for a common welth[1]," and "neuer loke for no reward, neyther of Lorde, nor of Prynter, nor of no man lyuing."

"Gentyll readers, I haue taken some peyne in makyng this boke, to do sycke men pleasure, and whole men profyte, that sycke men may recuperate theyr health, and whole men may preserue theym selfe frome syckenes (with goddes helpe) as well in Phisicke as in Chierurgy. But for as much as olde, auncyent, and autentyke auctours or doctours of Physicke, in theyr bokes doth wryte many obscure termes, geuyng also to many and dyuerse infirmyties, darke and harde names, dyffycyle to vnderstande,—some and mooste of all beynge Greeke wordes, some and fewe beynge Araby wordes, some beynge Latyn wordes, and some beynge Barbarus wordes,—Therefore I haue translated all suche obscure wordes and names into Englyshe, that euery man openlye and apartly maye vnderstande them. Furthermore all the aforesayde names of the sayde infirmites be set togyther in order, accordynge to the letters of the Alphabete, or the .A. B. C. So that as many names as doth begyn with A. be set together, and so forth all other letters as they be in order. Also there is no sickenes in man or woman, the whiche maye be frome the crowne of the head to the sole of the fote, but you shall fynde it in this booke,—as well the syckenesses the which doth parteyne to

[1] profit, good.

Chierurgy as to phisicke,—and what the sickenes is, and howe it doth come, and medecynes for the selfe same. And for as much as euery man now a dayes is desyrous to rede briefe and compendious matters, I, therefore, in this matter pretende to satisfye mens myndes as much as I can, namynge this booke accordyng to the matter, which is, 'The Breuiary of health:' and where that I am very briefe in shewynge briefe medecines for one sicknes, I do it for two causes: The fyrst cause is, that the Archane science of physycke shulde not be to manifest and open, for then the Eximyous science shulde fal into greate detrimente, and doctours the whiche hath studied the facultie shulde not be regarded so well as they are. Secondaryly, if I shulde wryte all my mynde, euery bongler wolde practyse phisycke vpon my booke; wherfore I do omyt and leue out many thynges, relynquyshynge that I haue omytted, to doctours of hygh iudgement, of whom I shalbe she*n*t for parte of these thynges that I haue wrytten in this booke: howe be it, in this matter I do sette God before mine eyes, and charitie, consyderynge that I do wryte this boke for a common welth, as god knoweth my pretence, not onely in making this boke, but al other bokes that I haue made, that I dyd neuer loke for no reward neyther of Lorde, nor of Prynter, nor of no man lyuing, nor I had neuer no reward, nor I wyl neuer haue none as longe as I do lyue, God helpynge me, whose perpetuall and fatherly blessynge lyght on vs all. Amen."

In his Preface to "The Seconde Boke of the Breuyary of Health, named the Extrauagantes," as in its colophon,[1] Boorde re-states his chief motive for writing the book:

"I do nat wryte these bokes for lerned men, but for symple and vnlerned men, that they may have some knowledge to ease them selfe in their dyseyses and infirmities. And bycause that I dyd omyt and leaue out many thynges in the fyrste boke named the Breuiary of Health,—In this boke named 'the Extrauagantes' I haue supplied those matters the whiche shulde be rehersed in the fyrst boke."

The *Breuyary* was intended by Boorde as a kind of companion to his *Dyetary;* for when treating 'of the inflacion of the eyes' and his remedies for it, he says:

"Aboue all other thynges, lette euery man beware of the premisses rehersed, in the tyme whan the pestilence, or the sweatyng syckenes, or feuers, or agues, doth reigne in a cou*n*tre. For these syckenesses be infectiouse, and one man may infecte an other, as it dothe appere in the Chapiters named Scabies, morbus Gallicus. And specially in *the dyatary of health.* wherfore I wolde that euery man hauynge

[1] Thus endeth these bokes, to the honour of the father, and the sonne, and the holy ghost, to the profyte of all poore men and women. &c. Amen.

this boke, shulde haue the sayd *dyetary of health* with this boke, consideryng that the one booke is concurrant with the other." '

Again, in his *Dyetary*, Boorde refers also frequently to the *Breuyary*,[1] and says, in his Dedication to the Duke of Norfolk:

"And where that I do speake in this boke but of dietes, and other thynges concernyng the same, If any man therfore wolde haue remedy for any syckenes or diseases, let hym loke in a boke of my makynge named *the Breuyare of helth.*"

The two books were, as Boorde says, concurrent in subject (l. 2, above), and probably also in date of writing, if not publication.

The *Breuyary* is an alphabetical list of diseases, by their Latin names, with their remedies, and the way of treating them. Other subjects are introduced, as *Mulier*, a woman—for which, see the extract p. 68, below,—*Nares*, nosethrilles, &c. Except for the many interesting passages and touches showing Boorde's character and opinions, the *Breuyary* is a book for a Medical Antiquarian Society, rather than ourselves, to reprint.

6. *The Pryncyples of Astronamye.* The second companion to the *Breuyary*—the *Dyetary* being the first—is the *Astronamye*, of which the title and an extract are printed above, p. 16. It is too astrological for us to reprint, though one or two chapters are generally interesting.

The following is its Table of contents:

¶ The Capytles of contentes[2]
of thys boock folowth.

The fyrst Capytle doth shew the names of the .xii. synes and of the .vii. planetes. And what the zodiack, and how many minutes a degre doth containe.

¶ The seconde Capytle doth shew what sygnes be mouable, and what sygnes be not mouable, and which be commone, and which be masculyn signes, and which be femynyne, and of the tryplycyte of them.

¶ The .iii. capytle dothe shewe in what members or places in man yᵉ sygnes hath theyr domynion, and how no man owt to be let

[1] "The Breviarie of health" was licensed to Tho. Easte on March 12, 1581-2. (*Collier's Extracts from the Registers of the Stationers' Company*, ii. 161.)

[2] *orig.* contences.

blod whan the moone is in ye sygne wher the sygne hath domynyon; and also what operacion the sygnes be of whan ye moone is in ther

¶ The .iii[i]. capytle doth shew of the fortitudes of the planetes, and what influens they doth geue to vs.

¶ The .v. Capitle doth shew the natural dyspocycyon of the mone whan she is in any of the .xii. sygnes.

¶ The .vi. capytle doth shew of ye nature of al y^e^ .xii. sygnes, And what influe*n*ce thei hath in ma*n*, And what fortitudes ye planetes hath in ye signes, *with* the names of the Aspects.

¶ The .vii. capytle doth shew ye natural dyspo[s]ycions of the planetes, And what operacyon they hath in mans body.

¶ The .viii. Capitle doth shew of the .v Aspectus, and of theyr operacyon

¶ The ix capitle doth shew of ye mutacio*n* of ye Ayer whan any rayne, wind, wedder, froste, and cold, shold be by the course of ye sygnes and planetes.

¶ The .x. capytle doth shew ye pedyciall of the aspectus of the mone and other planets, and what dayes[1] be good, and what dayes be not. &c.

¶ The .xi capytle doth shew of fleubothomy[2] or lettyng of blod[3]

¶ The xii capitle doth shew how, whan, & what tyme, a phisicion sholde minister medycynes

¶ The .xiii. Capitle doth shew of sowi*n*g of seedes, & plantynge of trees, and setyng of herbe.

Thus endyth the table.

As I have said before (p. 15, 17), I believe the *Astronamye* to have been published with the *Breuyary* in 1547.

§ 7. *The Peregrination.* The Itinerary of England, or '*The Peregrination of Doctor Boarde*,' which is the title in Hearne, may perhaps be taken as part of his lost *Itinerary of Europe*, and was printed by Hearne in 1735, in his *Benedictus Abbas Petroburgensis, de Vita et Gestis Henrici* III *et Ricardi* I, &c., vol. ii. p. 764—804. It is a list of

"Market townes in England, p. 764-771.

Castelles in England [& Wales], p. 771-775 (168 of them; whereof 7 were new, and 5 newly repaired).

In England be 24 suffragane bishops, p. 775.

Iles adjacent to England, p. 775-6.

The havens of England, p. 776-7.

Downes, mountaynes, hilles (including 'Boord's Hill, the authours birthplace'), dayles, playnes, & valleyes of England, p. 777-782.

[1] *orig.* dayer. [2] *orig.* flenbothomy. [3] *orig.* bold.

Fayre stone bridges in England, p. 782-3.
Rivers and pooles, p. 783-9.
Forestes and parkes in England, p. 789-797.
The high wayes of England, from London to Colchester, & Orford, p. 797-9.
The compasse of England round about by the townes on the sea coste, p. 800-4."

§ 8. *The Itinerary of Europe.* This, though lost to us now, may yet, I hope, turn up some day among some hidden collection of Secretary Cromwell's papers. Boorde gives the following account of it in the Seventh chapter of his *Introduction*, p. 145, below:

"for my trauellyng in, thorow, and round about Europ, whiche is all chrystendom, I dyd wryte a booke of euery region, countre, and prouynce, shewynge the myles, the leeges, and the dystaunce from citye to cytie, and from towne to towne; And the cyties & townes names, wyth notable thynges within the precyncte [of], or about, the sayd cytyes or townes, wyth many other thynges longe to reherse at this tyme, the whiche boke at Byshops-Waltam—.viii. myle from Wynchester in Ha*m*pshyre,—one Thomas Cromwell had it of me. And bycause he had many matters of [state] to dyspache for al England, my boke was loste, *th*e which myght at this presente tyme haue holpen me, and set me forward in this matter." (See p. 33.)

§ 9. *A Boke of Sermons.* This is not known to us, except by Boorde's own mention of it in *The Extrauagantes,* Fol. vi. (See p. 78.)

"shortly to co*n*clude, I dyd neuer se no vertue nor goodnes in Rome but in Byshop Adrians days, which wold haue reformed dyuers enormities, & for his good wyl & prete*n*ce he was poysoned within .iii. quarters of a yere after he did come to Rome, as this mater, with many other matters mo, be expressed in *a boke of my sermons.*"

This book one would at first assume to have been written before 1529-30, when Boorde was first 'dispensed of religion' in Prior Batmanson's days—as he says in his 5th Letter, p. 58 below,—especially as Pope Adrian VI died Sept. 24, 1523; but as we have no evidence that Boorde went abroad before 1529-30, and then to school to study medicine, we shall be safer in putting the probable date of the Sermons at between 1530 and 1534, when Boorde finally gave up his 'religion' or monkery; though it may have been later, as he was both monk and priest, and signed himself 'prest' in 1537. The loss of the book is assuredly a great one to us—one of the many losses for which that blind old noodle Time is to blame,—as we may be sure that the Sermons of a man like Boorde would have pictured his time for us better than almost any book we have.

§ 10. *A Pronostycacyon for the yere* 1545. Among Bagford's collection of Almanack-titles in the Harleian MS 5937, I have been lucky enough to notice the title-page of a hitherto uncatalogued work of Andrew Boorde's, which is, I suppose, unique:

"A Pronosty-/cacyon or an Almanacke for / the yere of our lorde .M. CCCCC. / xlv. made by Andrewe Boorde / of Physycke doctor an En-/glyshe man of the vni-/versite of Ox-/forde." Over a rose-shaped cut with a castle in the centre, used in the titleless edition of the Shepherd's Calendar in the British Museum, formerly entered as (?) Pynson's, but which, I am persuaded, is W. Coplande's.

On the back is "The Prologe to the reder.

I Were nat wyse, but inscipient, if I shulde enterpryse to wryte or to make any boke of p*ro*phesye, or els to pronostycate any mater of the occulte iugement of god, or to defyne or determyne any supe*r*natural mater aboue reson, or to presume to medle with the bountyfull goodnes of god, who doth dispose euery thing graciously. All such occulte and secrete maters, for any ma*n* to medle with-all, it is p*ro*hibited both by goddes lawe & the lawe of kynge Henry the eyght[1]. But for as muche as the excelle*n*t scyence of Astronomy is amytted dayly to be studyed & exercysed in al vniuersities, & so approued to be ye chiefe scie*n*ce amonge all the other lyberal sciences, lyke to the son, the which is in the medle of the other planetes illumynatynge as wel the inferyal planetes as ye superyal planetes, So in lyke maner Astronomy doth illucydat all the other lyberal sciences, indusing them to celestyall & terrestyall knowlege. D[o]the nat the planetes, sygnes, and other st[ers i]nduce vs to the knowlege of a c[reator of] the*m*, doth nat ye Mone gyue moyster to the[2]"

Coupling this with the fact already noticed, p. 16, l. 16, above, that Boorde in his *Astronamye* refers to Robert Coplande who prints 'thes yere my pronostycayons,' we must either conclude (as I do myself) that Boorde, like the Laets of Antwerp—grandfather, father, and son[3]—issued Prognostications yearly for some time, or that, if he

[1] Stat. 33, Hen. VIII, cap. 8, A.D. 1541-2. See *Queene Elizabethes Achademy*, notes.

[2] 'to the' are the catchwords.

[3] See my *Captain Cox*, or *Laneham's Letter*, for the Ballad Society, 1870.

only issued one, the date of his *Astronamye* is 1545, and not 1547, as I before supposed.

§ 11. *A Treatyse vpon Berdes.* All that we know of this book is got from the third tract in the present volume, called on its title-page, "The treatyse answerynge the boke of Berdes," and on its last page "Barnes in the defence of the Berde." The writer first speaks of Boorde's *spoken* answer to those who "desyred to knowe his fansye concernynge the werynge of Berdes" (p. 307), then says that Boorde "was anymatyd to *wryte his boke* to thende that great men may laugh thereat," as if he referred to the end of Boorde's Dedication of his *Dyetary* to the Duke of Norfolk (p. 225 below), and lastly heads his answer to Boorde "Here foloweth a treatyse, made, answerynge *the treatyse of doctor Borde vpon Berdes*" (p. 308). This makes it impossible to doubt the existence of such a book by Boorde; and the different charges which the writer (Barnes, whoever he may be) in his subsequent verses quotes from Boorde against the wearing of beards[1] are hardly consistent with a mere report of Boorde's sayings. Further, Wilson's allusion in 1553 to one who should 'dispraise beardes or commende shauen hiddes' (p. 307, note), probably points to this lost tract of Boorde's on Beards, as another passage of Wilson's does to Boorde's *Dyetary*, and *Introduction*, note on pages 116, 117, below. The reader can see for himself, in Barnes's lame verses, what arguments Boorde used against beards. Of Barnes's answers I can't always see the point; but that Boorde was a noodle for condemning beards, and advocating shaving, I am sure. Shaving is one of the bits of foolery that this age is now getting out of; but any one who, as a young man, left off the absurdity some three years before his neighbours, as I did, will recollect the delightfully cool way in which he was set down as a coxcomb and a fool, for following his own sense instead of other persons' reasonless customs.

§ 12. *Almanac and Prognostication.* In the British Museum (Case 18. c. 2, leaves 51, 52) are two bits of two leaves, belonging to

[1] Yet contrast Boorde's saying in his *Breuyary*, "The face may haue many impedimentes. The fyrst impedyment is to se a man hauyng no berde, and a woman to haue a berde." p. 95, below.

two separate Almanacs or Prognostications. The first bit is for the months of September, October, November, and December M. LLLLL. and xxxvii[. .],[1] signed at the foot . . . . . "e: Doctor of phisik." This *e* is supposed to be the last letter of *Boorde.* The second bit is of a Prognostication, with a date which is supposed to be 1540, "made by Maister" [no more in that line[2]] . . . . "cian and Preste." Put "Andrew Boorde physi" in the bit torn off the left edge, and you have one of the Pronosticacions which Robert Coplande in his day may have printed for our author (p. 16, above).

§ 13. *Jest-books.* I. *Merie Tales.* We come now to those books that tradition only assigns to Boorde: *The Merie Tales of the Mad Men of Gotam.* and *Scogin's Jests.* Though the earliest authority known to us for the former is above 80 years after Boorde's death, namely, the earliest edition of the book now accessible, that of 1630 in the Bodleian: "gathered together by A.B., of Physick, doctour:" yet Warton says: "There is an edition in duodecimo by Henry Wikes, without date, but about 1568, entitled Merie Tales of the madmen of Gotam, gathered together by A.B. of physicke doctour," *Hist. Engl. Poetry,* iii. 74, note *f.* ed. 1840; however, Warton had never seen it. Mr Halliwell, in his *Notices of Popular English Histories,* 1848, quotes an earlier edition still, by Colwell, who printed the 1562 edition of Boorde's *Dyetary,* "Merie Tales of the Mad Men of Gotam, gathered together by A.B. of Phisike Doctour. [Colophon] Imprinted at London in Flet-Stret, beneath the Conduit, at the signe of S. John Evangelist, by Thomas Colwell. n. d. 12°. black letter." Mr Hazlitt puts Colwell's edition before Wikes's, and quotes another edition of 1613 from the Harleian Catalogue.[3]

In a book of 1572, "the fooles of Gotham" is mentioned as a book: see p. 30, below. Mr Horsfield, the historian of Lewes,

[1] Boorde was in Scotland in 1536, in Cambridge in 1537; see p. 59-62 below.

[2] The blank looks to me like an intentional one, so that a different name might be inserted in each district the Prognostication was issued in.

[3] The chapbook copy in Mr Corser's 5th sale, of The Merry Tales of the Wise Men of Gotham (over a cut of the hedging-in of the cuckoo—a countryman crying 'Coocou,' and a cuckoo crying 'Gotam,' both in a circular paling—), Printed and Sold in Aldermary Church Yard, Bow Lane, London, contains 20 Tales, and six woodcuts.

affects to find the cause of these tales in a meeting of certain Commissioners appointed by Henry VIII.

"At a *last*[1] holden at Westham, October 3rd, 24 Henry VIII, for the purpose of preventing unauthorized persons 'from setting nettes, pottes, or innyances,' or any wise taking fish within the privileges of the marsh of Pevensey, the king's commission was directed to John, prior of Lewes; Richard, abbot of Begeham; John, prior of Mychillym; Thomas, Lord Dacre; and others.

"Dr Borde (the original Merry Andrew) founds his Tales of the Wise Men of Gotham upon the proceedings of this meeting—Gotham[2] being the property of Lord Dacre, and near his residence [at Herstmonceux Castle.]—Horsfield's *History of Lewes*, vol. i, p. 239, note; no authority cited:"—quoted by M. A. Lower, in *Sussex Arch. Coll.* vi. 207.

Anthony a Wood in his *Athenæ Oxonienses*, of which the first edition was published in 1691-2, over 140 years after Boorde's death, says at p. 172, vol. i., ed. Bliss, that Boorde wrote the Merie Tales:

"*The merry Tales of the mad Men of Gotham.* Printed at London in the time of K. Hen. 8; in whose reign and after, it was accounted a book full of wit and mirth by scholars and gentlemen. Afterwards, being often printed, is now sold only on the stalls of ballad singers. (An edition printed in 12mo. Lond. 1630, in the Bodleian, 8vo. L. 79. Art. 'Gathered together by A. B. of physicke doctor.')"

Those who contend for Boorde's authorship of this book are obliged to admit that the greater part of its allusions do not suit the Gotham in Sussex,[3] but do suit the Gotham in Nottinghamshire, except in three cases, where a Mayor, nearness to the sea, and putting

[1] "*Last*, in the marshes of Kent [and Sussex] is a court held by the twenty-four jurats, and summoned by the bailiff; wherein orders are made to lay and levy taxes, impose penalties, &c., for the preservation of the said marshes." *Jacob's Law Dict.*—Lower, *ib.*

[2] Gotham still possesses manorial rights. Gotham marsh is a well-known spot in the parish of Westham, adjacent to Pevensey; but the Manor-house lies near Magham Down in the parish of Hailsham.—Lower, *ib.*

[3] The manor of Gotham is the property of Lord Dacre, and near his residence, Herstmonceux Castle. The manor-house lies near Magham Down, in the parish of Hailsham.—*Sussex Arch. Coll.* vi. 206-7.

Lower. *Sussex Arch. Coll.* vi. 208. "In the edition of Mr Halliwell (which exhibits satisfactory evidence of some interpolating hand having introduced local names and circumstances, for the purpose of accommodating the anecdotes to the Nottinghamshire village) there are several jests which are still current as belonging to Sussex."

an eel in a pond to drown him, are alluded to[1]; but they argue that all the Nottinghamshire allusions have been introduced into the book since Boorde wrote it, and John Taylor the Water-Poet alluded to it. One may start with the intention to make the book Boorde's, and make it fit Sussex, by hook or by crook, or, from reading the book, turn cranky oneself, and write mad nonsense about it. There is no good external evidence that the book was written by Boorde, while the internal evidence is against his authorship.

The earliest collection known to us, of stories ridiculing the stupidity of the natives of any English county, is in Latin, probably of the 12th century, and relates to Norfolk. It was printed by Mr Thomas Wright in his *Early Mysteries and other Latin Poems of the Twelfth and Thirteenth Centuries*, 1838, p. 93-8, from 2 MSS of the 13th and 15th centuries in Trinity College Cambridge. In his Preface, Mr T. Wright says of this satire:

"The *Descriptio Norfolciensium* is said, in the answer by John of St Omer (p. 99-106), to have been written by a monk of Peterborough, and is, in all probability, a composition of the latter part of the twelfth century. It is exceedingly curious, as being the earliest known specimen of a collection of what we now call *Men-of-Gotham* stories; in Germany attributed to the inhabitants of Schildburg, but here, in the twelfth century, laid to the account of the people of Norfolk. The date of the German Schildburger stories is the sixteenth century[2]; the wise men of Gotham are not, I think, alluded to before the same century. Why the people of Norfolk had at this early period obtained the character of simpletons, it is impossible to say; but the stories which compose the poem were popular jests, that from time to time appearing under different forms, lived until many of them became established Joe Millers or Irish Bulls. The horseman (p. 95, l. 122-4) who carries his sack of corn on his own shoulders to save the back of his horse, is but another version of the Irish exciseman, who, when carried over a bog on his companion's shoulders, hoisted his cask of brandy on his own shoulders, that his porter's burden might be lessened. The story of the honey which was carried to market after having been eaten by the dog (p. 99-7, l. 147-172) re-appears in a jest-book of the seventeenth century."[3]

[1] Mr Lower thinks this clearly refers to the Pevensey practice of drowning criminals.—*Suss. Arch. Coll.* vi. 208; iv. 210.

[2] "For further information on this subject see an admirable paper on the Early German Comic Romances, by my friend Mr Thoms, in the 40th number of the Foreign Quarterly Review."—T. Wright.

[3] *Coffee House Jests*, Fifth Edition, London, 1688.—T. Wright.

The story of the sack of corn and the horse which Mr T. Wright instances from the 13th century, is, in fact, the Second Tale in the Gotham collection attributed to Boorde:

There was a man of Gottam did ride to the market with two bushells of wheate; and because his horse should not beare heauy, he carried his corne vpon his owne necke, & did ride vpon his horse, because his horse should not cary to heauy a burthen. Judge you which was the wisest, his horse or himselfe.

The Gothamites too were known before *The Merie Tales*, and if we may trust Mr Collier, the subject was open to any one. Mr J. P. Collier says:

"'The foles of Gotham' must have been celebrated long before Borde made them more ridiculous, for we find them laughed at in the Widkirk Miracle-plays, the only existing MS. of which was written about the reign of Henry VI. The mention of 'the wise men of Gotum' in the MS. play of 'Misogonus' was later than the time of the collector, or author, of the tales as they have come down to us, because that comedy must have been written about 1560: the MS. copy of it, however, bears the date of 1577. In 'A Briefe and necessary Instruction,' &c. by E. D., 8vo. 1572, we find the 'fools of Gotham' in the following curious and amusing company:—'Bevis of Hampton, Guy of Warwicke, Arthur of the round table, Huon of Bourdeaux, Oliver of the castle, the foure Sonnes of Amond, the witles devices of Gargantua, Howleglas Esop, Robyn Hoode, Adam Bell, Frier Rushe, the Fooles of Gotham, and a thousand such other.' Among the 'such other,' are mentioned 'tales of Robyn Goodfellow,' 'Songes and Sonets,' 'Pallaces of Pleasure,' 'unchast fables and Tragedies, and such like Sorceries,' 'The Courte of Venus,' 'The Castle of Love.'—This is nearly as singular and interesting an enumeration as that of Capt. Cox's library in Laneham's Letter from Kenilworth, printed three years later, although the former has never been noticed on account of the rarity of E. D.'s [possibly Sir Edward Dyer's] strange little volume.—William Kempe's 'applauded merriments,' of the men of Gotham, in the remarkable old comedy 'A Knack to know a Knave,' 1594, consists only of one scene of vulgar blundering; but it was so popular as to be pointed out on the title-page in large type, as one of the great recommendations of the drama."—Collier's *Bibliographical Account*, vol. i. p. 327.

I can see nothing in the *Merie Tales* that is like Boorde's hand; and if Colwell printed the book after Boorde's death, why shouldn't he have put Boorde's name on its title-page, as he did on the title-page of Boorde's *Dyetary* that he printed? So too with Wikes.

§ 14. "*Scogin's Jests,* an idle thing unjustly fathered upon Dr Boorde, have been often printed in Duck Lane," says Anthony a Wood, *Ath. Oxon.* i. 172, ed. Bliss. A copy of the first edition known to us is in the British Museum: "The first and best parts of Scoggins Iests: full of witty Mirth and pleasant Shifts done by him in France and other Places; being a Preseruatiue against Melancholy. Gathered by An. Boord, Dr of Physicke." London, F. Williams, 1626. Lowndes names an earlier edition in 1613, and an earlier still in black letter, undated. The work was licensed to Colwell in 1566.

Colwell **Recevyd** of Thomas colwell, for his lycense for pryntinge of the geyst*es* of skoggan, gathered together in this volume iiij[d].
MS Register A, *leaf* 134; (*Collier's Stat. Reg.* i. 120.)

The 'gathered together in this volume' looks as if this were the first collected edition of some old jests known in print or talk before. Anthony a Wood did not believe that Andrew Boorde ever had anything to do with this book. A modern follower of his might argue: "The way in which these attributions are got up, is well illustrated by a passage in Mr W. C. Hazlitt's *Early Popular Poetry*, vol. iii, p. 99:

'It is not unlikely that, besides the *Merie Tales of the Mad Men of Gotam,* and *Scogin's Jests,* Borde was the real compiler of the *Merie Tales of Skelton,* of which there was surely an impression anterior to Colwell's in 1567.'

"'Boorde recommends mirth in his books, says he has put jokes into one to amuse his patron, *therefore* he wrote all the jest-books issued during his life, and *à fortiori* those printed twenty years after his death.' Surely the more reasonable line to take is, 'In all his authentic books, Andrew Boorde declares himself, and otherwise enables us to identify him. In all, he writes about himself and his own work. If in any other books nothing of this kind is present, the odds are that Boorde did not write them. *Merie Tales* were put down to Skelton that he never wrote; may not those and the *Jests* put down to Boorde be in like case?'" A supporter of the authenticity of *Scogin's Jests* might answer, "I grant all this, and yet contend, 1. that the *Jests* do show evidence of being written by a Doctor, and, 2. that that Doctor is Boorde. In proof of 1. note how many of the Jests turn on doctors and medicine; in proof of 2. note how many are concerned with Oxford life, which we assume Boorde to have passed through. Also read the Prologue to the *Jests:*

'There is nothing beside the goodness of God, that preserves health so much as honest mirth used at dinner and supper, and mirth towards bed, as it doth plainly appear in the Directions for Health: therefore considering this matter, that mirth is so necessary for man, I published this Book, named *The Jests of Scogin,* to make men merry: for amongst divers other Books of grave matters I have made, my delight had been to recreate my mind in making something merry, wherefore I do advertise every man in avoiding pensiveness, or too much study or melancholy, to be merry with honesty in God, and for God, whom I humbly beseech to send us the mirth of Heaven, Amen.'

and then compare it with the extracts from Boorde's *Breuyary* on Mirth and honest Company, p. 88-9, below; lastly, compare the first *Jest* with Boorde's chapters on Urines in his *Extrauagantes,* and remark the striking coincidence between the *Jest's* physician saying, 'Ah . . . a water or urine is but *a strumpet;* a man may be deceived in a water,' and Boorde's declaring that urine '*is a strumpet* or an harlot, for it wyl lye; and the best doctour of Phisicke of them all maye be deceyued in an vryne' (*Extrav.* fol. xxi. back: see extract, page 34). If Boorde did not write the book, the man who fathered it on him made at least one designed coincidence look like an undesigned one." Still, I doubt the book being Boorde's. If it had been attributed to him in Laneham's time (1575), I should think that merry man would have told us that Captain Cox's "Skogan" was by "doctor Boord" as well as the "breuiary of health." (*Captain Cox,* or *Laneham's Letter,* p. 30, ed. F. J. F., 1870.)

§ 15. *The Mylner of Abyngton.* "Here is a mery Iest of the Mylner of Abyngton with his Wyfe and his Doughter, and the two poore scholers of Cambridge" [London, imprinted by Wynkyn de Worde] 4to, black letter.[1] Anthony a Wood says that a T. Newton of Chester wrote Boorde's name in a copy of this book as the author of it:

[1] Hazlitt's *Early Popular Poetry,* iii. 98.

"*A right pleasant and merry History of the Mylner of Abington, with his Wife, and his fair Daughter, and of two poor Scholars of Cambridge.* Pr. at Lond. by Rich. Jones in qu[arto]. And. Borde's name is not to it, but the copy of the book which I saw did belong to Tho. Newton of Cheshire, [Bodl. 4to. C. 39. Art. Seld.] whom I shall hereafter mention, and by him 'tis written in the title that Dr. Borde was the author. He hath also written *a Book of Prognosticks*, another *Of Urines*, and a third *Of every Region, Country and Province, which shews the Miles, Leeges, distance from City to City, and from Town to Town, with the noted Things in the said Cities and Towns.*"[1]—Wood's Athen. Oxon. i. 172.

This tale of *The Mylner of Abyngton* has been reprinted lately by Mr Thomas Wright in his *Anecdota Literaria*, p. 105-116, and by Mr Hazlitt in his *Early Popular Poetry*, iii. 100-118. It is a story like Chaucer's *Reeves Tale*[2], about the swiving of the Miller's wife and daughter by two Cambridge students, in revenge for his stealing their flour, and letting their horse loose. If any one will read Andrew Boorde's poetry, that is, doggrel, in his *Introduction of Knowledge*, and then turn to the *Mylner*, he will not need any further evidence to convince him that Boorde did not write the latter Tale.

§ 16. *Other Works.* The authority on which Wood assigns to Boorde his Books of Prognosticks and Urines, is doubtless that on which Warton (iii. 77, ed. 1840) also assigns to him the *Promptuarie of Medicine* and the *Doctrine of Urines*, namely, Bishop Bale, who in the 2nd edition of his *Scriptores* says:

"Andreas Boorde, ex Carthusianæ superstitionis monacho, malus medicus factus, in monte Pessulano in Gallijs eius artis professionem ac doctoratum, spreto diuini uerbi ministerio, suscepit. Congessit mœchus in sacerdotalis matrimonij contemptum. *Prognostica quædam, Lib.* 1. *Promptuarium Physices, Lib.* 1. *De iudicijs urinarum, Lib.* 1. Et alia."

Neither of the other books do I know by Bale's titles, though I suppose the *Promptuarium* to be Boorde's *Breuyary.* Of one of the *Prognostica* a leaf is printed above, § 10, p. 25. I should doubt Boorde's having written a separate treatise on Urine, as he has given more than six leaves to it in his *Extrauagantes*, Fol. xx-xxvi back, and had but a bad opinion of it:

[1] See above, p. 23-24.

[2] Not *Milleres Tale*, Mr Hazlitt.

"I do say that an vryne is a strumpet, or an harlot, for it wyl lye; and the best doctour of Phisicke of them all maye be deceyued in an vryne, and his cunnyng and learning not a iote the worse. I had rather to se the egestyon of a sycke person, then the vryne: bothe be good to loke on, as it doth appere in the Chapitre named *Egestio* in the fyrst boke named the Breuiary of health, &c."

§ 17. *A Latin Poem:* "Nos Vagabunduli." This was found in a book by Dr E. F. Rimbault, with Boorde's name to it, was printed by the Doctor in *Notes and Queries*, vol. v. p. 482-3, and reprinted by Mr M. A. Lower in his *Worthies of Sussex*, p. 34-5, with an English translation. Both Latin and English follow here, though it is clear to me that the poem is entirely alien to Boorde's known opinions on religion, and to his way of reproving vices in men professedly religious, even though he, as a monk and priest, may have hated friars as much as the rest of the regular clergy and monks did. The latter found the friars a 'hindrance,' like Bp Wilberforce's high-church clergy find the modern Dissenters; but I doubt Boorde's chaffing his 'hindrance' in this style:—

Latin Poem attributed to Andrew Boorde.	*Mr M. A. Lower's English paraphrase.*
"Nos vagabunduli,	"We're of wandering friars a pair,
Læti, jucunduli,	And jolly ones we truly are,
Tara, tantara teino	Down, derry down!
Edimus libere,	Freely we eat anything,
Canimus lepide,	And right merrily we sing,
Tara, &c.	Down, &c.
Risu dissolvimur,	With laughter oft our sides do crack,
Pannis obvolvimur,	And we've good cloth upon our back,
Tara, &c.	Down, &c.
Multum in joculis,	Much we deal in merry quips,
Crebro in poculis,	And full frequent are our sips,
Tara, &c.	Down, &c.
Dolo consuimus,	We are up to many a trick,
Nihil metuimus,	And at nothing do we stick,
Tara, &c.	Down, &c.
Pennus non deficit,	Our pouch is all unfailing still,
Præda nos reficit,	We pick up booty when we will,
Tara, &c.	Down, &c.
Frater Catholice,	Now, most holy catholic brother,
Vir apostolice,	Man apostolic (I'm another),
Tara, &c.	Down, &c.

Dic quæ volueris,	Call for anything that's nice,
Fient quæ jusseris,	It shall be served you in a trice,
Tara, &c.	Down, &c.
Omnes metuite	But let me humbly you beseech,
Partes gramaticæ,	Be careful of your parts of speech,
Tara, &c.	Down, &c.
Quadruplex nebulo	A fourfold rascal here have we,
Adest, et spolio,	All intent on booty he,
Tara, &c.	Down, &c.
Data licencia,	When there's too much license given,
Crescit amentia,	To what length is madness driven!
Tara, &c.	Down, &c.
Papa sic præcipit,	Thus commands our Holy Pope,
Frater non decipit.	A friar won't deceive his hope,
Tara, &c.	Down, &c.
Chare fratercule,	Now farewell, my brother dear!
Vale et tempore,	'Tis time that we were gone, I fear,
Tara, &c.	Down, &c.
Quando revititur,	When we meet again, my boy,
Congratulabimur,	We will wish each other joy,
Tara, &c.	Down, &c.
Nosmet respicimus,	Now we look upon each other,
Et vale dicimus,	And farewell, we say, dear brother,
Tara, &c.	Down, &c.
Corporum noxibus,	With right friendly hug we part,
Cordium amplexibus	And embraces of the heart,
Tara tantara teino."	Down, derry down!"
—*Notes & Queries*, vol. v. pp. 482, 483.	—M. A. Lower's *Worthies of Sussex*, pp. 34, 39.

Having thus run through the works written by Boorde, or attributed to him, I pass on to Part II, Boorde's Life, noting only, that of his Works I have here reprinted the two that seem to me the most likely to interest the general student of Tudor days—the *Introduction* and *Dyetary;* that I have added *Barnes in Defence of the Berde* on account of its connection with Boorde, its giving the substance of his lost Treatise on Beards, and its being unique, though it wants a leaf; and that I have extracted most of the chapters and bits of Boorde's *Breuyary* (and its second Part, the *Extrauagantes*) that contain his opinions on the England and Rome of his day, and things in general, besides showing his medical practice. That they'll amuse and interest the reader with a turn for such things, I can promise.

Of Boorde's *Introduction*, Dibdin rightly says, "This is probably the most curious and generally interesting volume ever put forth from the press of the Coplands." *Dibdin's Ames*, 1816, iii. 160. It is the original of Murray's and all other English Handbooks of Europe.

PART II. LIFE OF ANDREW BOORDE.

§ 18. For a sketch of Andrew Boorde's life and opinions we have little else than the materials he himself has left us in his Letters and Will, and in the pleasant little outbreaks he makes in unexpected places in his books. But as there has been a good deal of talk and gammon mixt up with the facts of his life, it may be as well at the outset to give a dry list of these facts, with the authority for each, and the page in which such authority will be found in the present volume. I must, however, warn the reader that I don't feel sure of my arrangement of Boorde's letters being the right one. It is only the best that I can make.

FACTS OF ANDREW BOORDE'S LIFE.

Born at Boord's Hill, in Holms dayle (Authority, *Peregrination*, p. 23, above).

Brought up at Oxford (Auth. p. 40, or *Introduction*, p. 210; *Pronosticacion* for 1545 A.D., p. 25).

Under age, admitted a Carthusian monk (Letter IV, p. 57).

1517 Accused of being conversant with women (Letter VII, p. 62).

1521 Dispensed from Religion by the Pope's Bull, that he might be Suffragan Bp. of Chichester, though he never acted as such (Letter V, p. 58).

1528? Letter I, to Prior of Hinton (p. 47).

1529 Is dispensed of Religion in Batmanson's days, by the *Grande Chartreux* (Letter V, p. 58).

Goes over sea to school (p. 58), that is, to study medicine (*Dyetary*, p. 226).

1530 Returns to England, and attends the Duke of Norfolk (*Dyetary*, p. 225).

1532? Goes abroad again to study (*Dyetary*, p. 226); getting a fresh license from Prior Howghton, after 16 Nov., 1531 (p. 47-8)

Returns to the London Charter-House.

? Lost book of Sermons written (*Breuyary*, p. 24).

1534 June 6. Takes the oaths to Henry VIII's supremacy (*Rymer*, xiv. 492; *Smythe's Hist. Charter-House*, p. 51-2).

Is in prison, in thraldom, ghostly and bodily, in the Charter-House (p. 52). Writes from there to Prior Howghton, who is confined in the Tower of London (Letter VI, p. 59).

Is set free by Cromwell (Letter VI, p. 59), whom he probably now visits at Bishop's Waltham in Hampshire (Letter VI, p. 59), and goes abroad a third time.

1535 In Catalonia, when Charles V took shipping to Barbary (Letter III, p. 56).

June 20. Letter II, from Bordeaux (p. 53).

July 2. In Toulouse (Letter III, p. 55).

After July 2. Boorde sick; can't get home (Letter III, p. 55).

Aug. 2. Letter IV, from the Grande Chartreux. Boorde, having renewed his License, declares himself clearly discharged from Religion or Monkery (p. 57).

Writes Cromwell a lost letter from London (p. 58).

1536 Letter V to Cromwell, before 1 April (p. 58).

" April 1, Letter VI, at Leith. Is practising and studying at Glasgow (p. 59).

Returns to London thro' Yorkshire (*Breuyary*, p. 61). Has 2 horses stolen. Sees Cromwell (p. 62).

1537 August 13, Letter VII, from Cambridge (p. 62).

Goes abroad the 4th time.

1542 In Montpelier. Gets drunk (*Barnes*, p. 309). Writes *Dyetary*, *Breuyary*, and *Introduction* (p. 14).

Returns to England, lives in London, denounces beards, and (?) writes a *Treatyse vpon Berdes* (Barnes, p. 307-8). Barnes answers him (p. 305-316).

1547 Lives in Winchester, ?acquires property there and elsewhere.
" Was late a tenant of a house in St Giles's, London (p. 64).
" *Breuyary, Dyetary* II, (?) *Astronamye* (written in 4 days), and *Introduction*, published (p. 13-24).
" Is accused of keeping 3 whores at Winchester (*Bp. Ponet*, p. 66).
Is imprisoned in the Fleet (p. 70).
1549 April 25, makes his Will in the Fleet, devising houses, &c., in Lynne, Pevensey, and in and about Winchester, besides chattels (p. 73).

§ 19. Expanding our List, we note first that Boorde, in his *Peregrination*,—printed by Hearne in the 2nd vol. of *Benedictus Abbas Petroburgensis de Vita et Gestis Hen. III et Ric. I*, &c. (1735, 8vo)—tells us in an entry under Sussex, at p. 777, where he was born: "Boords hill, the authours birth place, in Holms dayle."

Now Board Hill in Sussex is, and has long been, a well-known place as the residence of the Boordes. It is a small Elizabethan mansion, lately enlarged by its present owners, Major Macadam and his wife (formerly Miss Preston) and her mother, Mrs Preston. It is very pleasantly situated on one of those charming hills in the Wealden formation, with the ground falling away on three sides of it into a basin-like valley, and bounded by rising land in the distance. On my way back to town, the day after our most successful Volunteer Review last Easter Monday, I walked two miles north by west of Hayward's Heath Station, through lanes whose banks were all aglow with primroses, wood sorrel[1], and mallows (as I suppose), and was shown quickly over the house by Mrs Macadam. The earliest date in the wainscoted rooms of the house itself is 1601, and that is twice repeated, with the initials S. B., which must stand for Stephen Boorde, who was knighted, the son of the Stephen Boorde who heads Mr Lower's pedigree of the family in vol. vi of the *Sussex Archæological Collections*.[2] An earlier date, however,—namely, 1569,

[1] "Kiss me quicks" we call 'em, once said a man to me in Combe Hurst near Croydon.

[2] "Stephen Boord or Borde, whose name stands at the head of the pedigree as of 'the Hill' in Cuckfield, is described in his will, dated 10th February, 1566, as 'of Lindfield.' He directs his body to be buried in the church of

—is shown on an old black piece of oak taken off a barn pulled down by Major Macadam; and I have no doubt that in a house at this place, Andrew Boorde was born. For though the valley round it is not now called Holmsdale—so far as Mrs Macadam and the vicar of Cuckfield (pronounced Cookfield) know—yet it may have been so in former days, as two little streams run eastward, north and south of Board Hill, and the A.Sax. *holm* means 1. water, 2. a river island, a green plot of ground environed with water (Bosworth). It is clear too that the Hill, and not the Dale, is the feature on which Andrew Boorde dwells. He might have found some hundreds of hills in England with as much right to be included in his list as his "Boord's hill;" but he was born there, and so he brings it in. I therefore reject Mr Lower's suggestion,

"As Borde-Hill is certainly not in a dale, the probability is that the place indicated is a house not far distant, still called Holmesdale,

Lindfield, and gives to the repairs of that church and of Cokefelde, ten shillings each. He was interred in the south transept at Lindfield, where, on a marble slab, were formerly to be seen brasses representing himself, his wife, and their four sons and three daughters, with the following inscription:—

"'Stephen Boorde and Pernell his wyfe resteth here . . . . . after the troubles of this world, in assured hope of the resurrection: which Stephen decessed xxij day of August, in ye year of our Lord MCCCCC lxvij, and the said Pernell decessed xviij day of June in the yeare above engraven: whose souls we commende to Gods infinite mercy.'

"Of the children of the pair thus commemorated, George . . . . and Thomas became the progenitors of the two branches settled respectively at Board Hill and at Paxfield Park.

"At the time when the threatened Spanish invasion excited the patriotism and the liberality of our gentry, we find Thomas Boord of Paxhill and Stephen Boord of Boord Hill (afterwards knighted) contributing the sum of thirty pounds each towards the defences of the country."—M. A. Lower in *Suss. Arch. Collections*, vol. vi. p. 33, 37.

"From that period the two branches of the family seem to have pursued the steady and comparatively undiversified career of country gentlemen, forming respectable alliances, and continuing the name by a rather numerous progeny, as will be seen by the following pedigree. The Board Hill branch I have been unable to deduce below the year 1720; but the Lindfield branch I have traced down to its extinction in the male line on the death of William Board, Esq., in 1790. From that gentleman, through his youngest daughter and coheiress, the Lindfield estate passed to the Crawfurds. The late William-Board- Edw.- Gibbs Crawfurd, Esq., who died in 1840, left two daughters and coheiresses, the elder of whom is married to Arthur W. W. Smith, Esq., now of Paxhill, the old family seat of this branch. Both the lines produced several younger sons; and the name is by no means extinct in other counties, though it seems totally so in this."—*Sussex Archæological Collections*, pp. 200, 201, vol. vi. See a later note in Lower's *Worthies of Sussex*.

in later times a seat of the Michelbornes and Wilsons, and at present existing as a farm house."—*Worthies of Sussex*, p. 27,

and hold that, as Johnson defined *Dale* to be 'a low place between hills, a vale, a valley,' Boorde Hill may be fairly said to be in a dale, that is, to rise out of the low ground between it and the range of hills seen at a distance round it. It is on the south of Ashdown Forest, the remains of what was formerly called the Forest of Pevensel, which again was only part of the great forest of Anderida, that was 'coextensive, or nearly so, with the wealds in Sussex, Kent, and Surrey,' and in Bede's days 120 miles from east to west, and 30 miles from north to south.[1]

When Andrew Boorde was born at Boord's Hill (or Board Hill), we do not know; but it must have been before 1490 A.D., as by 1521 he was old enough to have been appointed Suffragan Bishop of Chichester, and to have got the Pope's Bull dispensing him from filling the office (p. 44, below). But I am anticipating.

§ 20. Where Boorde was brought up, he probably tells us in *The fyrst Boke of the Introduction of Knowledge*, cap. 35,

"What countrey man art thou?" *Cuius es.*
"I was borne in England, and brought up at Oxford."
Natus erum in Anglia, et educatus Oxoni[æ] . . .
"What is thy name?" *Cuius nominis es.*
"My name is Andrew Borde."
Andreas Parforatus[2] *est meum nomen.*

Now though this is part of an imaginary conversation, yet Boorde describes himself in his *Pronosticacion* for 1545 as 'of the Vniversity of Oxford' (p. 25, above), and his name is given in Wood's *Athenæ*, vol. i, p. 169, of Bliss's edition, as that of an Oxford man. Wood also—though he gives no authority for his statement, and I can find none in his *Fasti*[3]—states positively

[1] 'Ashdown Forest or Lancaster Great Park,' by the Rev. E. Turner, *Sussex Arch. Collections*, xiv. 35.

[2] *Borde* is also an early word for 'table,' and *Boorde* one for joke, play, jest.—See *Babees Book*, Index, &c.

[3] Alexander Hay, in his *History of Chichester*, 1804, p. 506, says that Boorde "completed his education at New-College, in Oxford; where for several years, he applied very closely and successfully to the study of physic. [No doubt, gammon.] Leaving Oxford he is said to have travelled into every kingdom in Europe, and to have visited several places in Africa. At

that Boorde took his M.D. degree at Oxford. We may therefore fairly conclude, that he was brought up at Oxford. though we cannot be certain of the fact.

§ 21. If we could trust Mr Lower's judgment, which I do not think we can,[1] the next notice of Andrew Boorde—or perhaps a prior one—shows him to have been in 1510 A.D. a *nativus*, or villein regardant[2]—attached to the soil, and sellable with it,—of Lord Abergavenny's manor of Ditchling, in Suffolk, holding goods and chattels, therefore of age (I assume), though childless, and being the son of John Borde. This villein Andrew Borde, Lord Abergavenny manumits or frees, and quits claim of his goods, by the following charter, the last in Madox:

O.A. An Enfranchisement of a Villain Regardant.

Omnibus Christi fidelibus ad quos præsens scriptum pervenerit, *Georgius Nevile* Dominus de Bergevenny,[3] salutem in Domino. Noveritis me præfatum *Georgium* manumisisse *Andream Borde* filium *Johannis* BORDE, nativum meum, Manerio sive Dominio meo de *Dychelyng*[4] in Comitatu *Sussex* spectantem; & eundem Andream liberum fecisse, & ab omni servitutis jugo, villinagio, & condicione servili liberum fecisse; Ita videlicet, quòd nec Ego præfatus Dominus de *Bergevenny* nec hæredes mei, nec aliquis alius pro nobis seu nomine nostro, aliquid Juris vel clamei in prædictum Andream, nec in bonis aut catallis suis, ad quascumque mundi partes divertent, exigere, clamare, vendicare, poterimus nec debemus in futuro; sed ab

Montpelier in France he took his degree of doctor of physic; and returning to England, was admitted at Oxford to the same honour in 1521." [No doubt, gammon too.]

[1] I speak with all respect for Mr Lower's great services to his county and to Literature; but in many points I cannot follow him.

[2] "The villein," says Coke, *on Littleton*, fol. 120 *b*, "is called *regardant* to the manour, because he had to do all base or villenous services within the same, and to gard and keepe the same from all filthie or loathsome things that might annoy it: and his service is not certaine, but he must have regard to that which is commanded unto him. And therefore he is called regardant, *a quo præstandum servitium incertum et indeterminatum, ubi scire non potuit vespere quale servitium fieri debet mane, viz. ubi quis facere tenetur quicquid ei præceptum fuerit* (Bract. li. 2, fo. 26, Mir. ca. 2, sect. 12) as before hath beene observed (vid sect. 84)." See my essay on "*Bondman*, the Name & the Class," in the Percy Folio Ballads and Romances, vol. ii. p. xxxiii—lxii.

[3] He was the 5th Baron by writ; succeeded to the title in 1492, on the death of his father; and died in 1535.—*Nicolas's Peerage*.

[4] The manor of Ditchling extends over a considerable portion of the parish of Cuckfield. M. A. Lower, in *Sussex Arch. Coll.* vi. 199.

omni actione juris & clamei inde simus exclusi imperpetuum, per præsentes. In cujus rei testimonium huic præsenti scripto sigillum meum apposui. Datum vicesimo septimo die Mensis Junii, Anno regni Regis *Henrici* octavi secundo.[1] G. Bergevenny."—Madox's *Formulare Anglicanum*, edit. 1702, page 420.

This, being englished, is,

"To all the faithful of Christ to whom this present writing shall come, *George Nevile*, Lord of Bergevenny, [wishes] salvation in the Lord. Know ye that I, the aforesaid *George*, have manumitted *Andrew Borde* (son of *John Borde*) my villein regardant to my Manor or Lordship of *Dychelyng* in the county of Sussex; and have made free the same Andrew; and have made him free from all yoke of serfdom, villenage, and servile condition; in such wise, to wit, that neither I the foresaid Lord of Bergevenny, nor my heirs, nor any other person for us, or in our name, may or shall hereafter require, claim, [or] challenge any right or claim to the foresaid Andrew nor to his goods or chattels, to whatsoever parts of the world they may turn; but that we shall be by these presents shut out for ever from all action of right and claim. In witness of which thing I have set my seal to this present writing. Dated on the 27th day of the month of June, in the 2nd year of the reign of King Henry the 8th. G. Bergevenny."

Now there is not an atom of evidence beyond the sameness of name and the nearness of place, to connect this manumitted villein Andrew Borde with our Andrew; and the reasons why I at first sight held, and still hold, that this villein is not our Andrew are, that our man himself tells us in his Letter II, p. 53 below, 'to Master Prior & the Couentt off the Charter-howse off London, & to all Priors & Couentes off the sayd Order in Ynglond' that he was 'receuyd amonges' them,—as a Carthusian monk,—under age, contrary to their Statutes. Lord Abergavenny's charter implies that his Andrew Borde was of age, and did hold, and could hold, property. Our Andrew, if an infant, couldn't have had such a charter made to him,—an infant couldn't (and can't) hold property;—our Andrew, if of age, was a monk; and, being so, couldn't have needed manumission, for his admission as a monk must have freed his person. The only supposition, says Professor Stubbs,—who has kindly helpt me here,—on which the Charter could apply to our Andrew is, that he was 21, that he was going to profess himself a monk, and that he

[1] The 2nd year of Henry VIII's reign was from 1510 to 1511.—*Nicolas.*

obtained the Charter for that purpose, as the Constitutions of Clarendon forbid any *nativus* or bondman being received as a monk[1] without his lord's leave.[2]

But our Andrew was not 21 before he became a monk; and he could not have taken in his lord about his age like he could the non-Sussex monks of the London Charter-house,—if indeed they wanted taking in.—Moreover, had he been a *nativus* in his youth, he would certainly have told the Prior and Convents this additional reason against his having been legally admitted into their order. We know that there were other Bordes in Sussex in our Andrew's time—as Dr Richard, and Stephen of the Hill, Cuckfield;[3]—and we may safely conclude that in 1510 there was another Andrew Borde than ours, namely, he whom Lord Bergevenny freed. Sir T. Duffus Hardy and Prof. Brewer both agree that that Lord's charter did not relate to any Carthusian monk, or any infant in law.

We may notice in passing, that the Monks' habit of enticing lads under age to join their orders, is known from Richard de Bury's reproof to them in 1344: "You draw boys into your religion with hooks of apples, as the people commonly report, whom, having professed, you do not instruct in doctrines by compulsion and fear as their age requires, but maintain them to go upon beggarly excursions, and suffer them to consume the time in which they might learn, in catching at the favours of their friends, to the offence of their parents, the danger of the boys, and the detriment of the Order."[4] (Translation of 1832, p. 40.)

[1] Compare the Friars, in Prof. Brewer's *Monumenta Franciscana*, p. 574, quoting the Cotton MS, Faustina D iv. 'No man shalbe resceived to the Order [of St Francis] but he have thes thingis . . that *he be not a bonde man borne* . . yf he be clerke, at the leste that he be goynge of xvi yere of age.'

[2] And sith, *bondemenne barnes* · han he made bisshopes,
And barnes bastardes · han ben archidekenes.
(ab. 1380. *Vision of Piers Plowman.* Whitaker's Text, Passus Sextus.)

[3] See pages 38-9 and 65.

[4] The Friars were as bad. In or about 1358 A.D. the University of Oxford also passed a Statute, reciting that the common voice and experience of the fact proved that 'the nobles and people generally were afraid to send their sons to Oxford lest they should be induced by the Mendicant friars to join their order,' and therefore enacting 'that, if any Mendicant friar shall induce or cause to be induced, any member of the University under 18 years of age to join the said friars, or shall in any way assist in his abduction, no

§ 22. The next notice that Boorde gives us of himself points to one of the evils of this taking lads into religious orders before they have passed through their hot youth, and known what sexual desire is. An old writer, the extract from whom I have unluckily mislaid, dwells very strongly on the mischief arising from this practice; and we must not therefore wonder to hear Boorde telling Lord Privy-Seal Cromwell, in a Letter to him (Letter VII, p. 62), dated 13 August, 1537 (as I judge),

> "ther be yn London certyn persons that owth me in mony & stuff liij[li] . . . . & doth slawnder me by-hynd my bak off thynges that I shold do *xx[ti] yers agone;* & trewly they can nott prove ytt, nor I neuer dyd ytt: the matter ys, *that I shold be conversant with women;* other matteres they lay nott to my charge."

Young blood was even younger blood in those days than now; but let us accept Andrew's denial of the truth of the slander.

§ 23. Our next notice is from Boorde's Fifth Letter, to Cromwell, —then a knight, and Master of the Rolls,—which must bear date before the 1st of April, 1536 (p. 59, below).

> "I was also, xv yeres passyd, dispensyd with the relygyon by the Byshopp of Romes bulles, to be Suffrygan off Chychester, the whych I never dyd execute the auctore."

Mr Durrant Cooper says that in 1521, Sherborne, Bishop of Chichester, was 80 years old, and it was for him that Boorde was appointed to act, but did not do so. His connection with Sussex no doubt led to his nomination for the office[1]; and we may suppose that his family was of some influence in the county. Professor Brewer tells me that no one could be made a Bishop—regular or suffragan—under 30 years of age; and we must therefore put back the year of Boorde's birth to before 1490. The phrase 'dispensyd with the relygyon' puzzles me. I don't know whether it means absolved wholly from the vows of the Carthusian Order, or only absolved for a time and a special purpose, like this acting as Suffragan, going abroad to study medicine, &c. (p. 47-8), the dis-

graduate belonging to the cloister or society of which such friar is a member, shall be permitted to give or attend lectures in Oxford or elsewhere, for the year ensuing.'—*Munimenta Academica*, ed. Anstey, i. 204-5.

[1] Prof. Stubbs does not believe that Boorde ever received episcopal orders.

pensed person continuing otherwise liable to the bidding of the head of his House and Order. The latter interpretation is favoured by Boorde's talk of renewing his license (Letter V, p. 58), and his returning to the Charter-house by 1534; the former, of absolute freedom, by his argument in the same Letter V, p. 58, that by the Pope's act, as well as the Carthusians', he was free of Religion.

§ 24. About this time—as likely before as after—I suppose that the Letter of Boorde's which Mr W. D. Cooper and I put first (p. 47, below), and Sir Hy. Ellis last, was written: that to Doctor Horde, Prior of the Charter-house at Hinton or Henton in Somersetshire. Why I put this Letter first (though it may be of 1535), is because of Boorde's saying in it, "yff I wyst the master Prior off London wold be good to me, I wold see yow more soner than yow wold be ware off." I take this to mean that Boorde was then in the London Charter-house, not yet 'dispensed of religion,' but subject to its strict rules, so that he could not go out of the gates of the monastery without the Prior's leave. Were this letter the last of Boorde's, as Sir Hy. Ellis makes it, and therefore written after 1537, Boorde wouldn't have cared twopence for the 'Master Prior off London.' Indeed, there wasn't one then, for on May 18, 1537, Prior Trafford and his brethren surrendered the London Charter-house into Henry's hands. (By the way, in connection with this first letter of Boorde's, I must mention Mr W. Durrant Cooper's unwitting practical joke with five of the set. Although they had been printed by no less a person than Sir Hy. Ellis, and in no less known a book than his *Original Letters*, no less than 15 years before 1861, yet Mr Cooper printed the Letters as "unpublished correspondence" in the collections of the Sussex Archæological Society for 1861 (vol. xiii, p. 262)—and I suppose read them as such to the Meeting at Pevensey, on Aug. 8, 1860—thus unconsciously taking in the 'young men from the country,' to say nothing of others for years, and for three weeks myself, who had read the letters in Ellis, made a note of their "trust yow no Skott," ii. 303, and then forgotten all about them. Having sinned myself in this way, I can't resist the temptation of giving a fellow-sinner a good-natured poke in the ribs.)

As in this First Letter, Boorde speaks of the 'rugorosite' of the Carthusian 'relygyon,' we may as well give an extract about that Order and its Rule.

The Carthusian Monks were a branch of the Benedictines, whose rule, with the addition of a great many austerities, they followed. . . Bruno, who was born at Cologne in Germany, first instituted the Order at Chartreux, in the diocese of Grenoble in France, about A.D. 1080; whence the Monasteries of the Order, instead of Chartreux houses, were in England corruptly called *Charter-houses.* The rule of the Carthusians, which is said to have been confirmed by Pope Alexander III as early as 1174, was *the most strict of any of the religious orders; the monks never eating flesh, and being obliged to fast on bread, water, and salt one day in every week: nor were they permitted to go out of the bounds of their Monasteries,* except their priors and procurators, or proctors, and they only upon the necessary affairs of the respective house.

The Carthusians were brought into England in 1180, or 1181, by King Henry II., almost as early as their establishment at Grenoble, and had their first house at Witham in Somersetshire. Their habit was all white, except an outward plaited cloak, which was black. Stevens, in his continuation of Dugdale's *Monasticon,* says there were but five nunneries of this austere order in the world, and but 167 houses of these monks. In England there was no nunnery, and but nine houses of this order. These nine houses were at Witham and Henton in Somersetshire, the Charter-house in London, Beauvale in Nottinghamshire, St Anne's near Coventry, Kingston-upon-Hull, and Mountgrace in Yorkshire, Eppworth in the Isle of Axholm, and Shene in Surrey.—*Penny Cyclopædia,* from *Tanner,* &c.

The Latin Statutes of the Order are given in Dugdale's *Monasticon,* ed. 1830, p. v-xii, from Cotton MS. Nero A iii, fol. 139, and are of such extreme strictness and minuteness as to behaviour, dress, meals, furniture of cells, &c.—telling the monks how to walk, eat, drink, look, and hardly to talk—that they must have nearly worried the life out of a man like Boorde. An English summary of the Carthusian Rules is given in Fosbroke's *British Monachism,* p. 71-2, ed. 1843, where also is the following extract:

"I know the Carthusians," says he (Guyot de Provins in the 13th century), "and their life does not tempt me. They have each [his own] habitation; every one is his own cook; every one eats and sleeps alone. I do not know whether God is much delighted with all this. But this I well know, that if I was myself in Paradise, and alone there, I should not wish to remain in it. A solitary man is always subject to bad temper. Thus I call those *fools* who wished me to immure myself in this way. But what I particularly dislike in the Carthusians is, that they are murderers of their sick. If these require any little extraordinary nourishment, it is peremptorily refused. I do not like religious persons who have no pity; the very quality, which, I think, they especially ought to have."—*Fosbroke's British Monachism*, p. 65, ed. 1843.

[[1] Letter I. ? Boorde in the Charter-house, London.]

"Vene*r*able fa*þ*er, p*re*cordyally I com*m*end me vnto yow w*ith* thanks, &c. I desyre yow to p*ra*y for me, & to p*ra*y all yo*ur* co*n*uentt to p*ra*y for me / for much co*n*fidence I haue in yo*ur* p*ra*yers; & yff I wyst[2] Mast*er* p*ri*or off london wold be good to me, I wold see yow more son*er* þen yow be ware off. I am nott able to byd þᵉ rugorosyte off yo*ur* relygyon. yff I myth be suffreyd to do w*h*at I myth, w*ith* outt int*er*rupcyon, I can tell w*h*at I had to do, for my hartt ys eu*er* to yo*ur* relygyon, & I loue ytt, & all þᵉ p*er*sons in them, as Iesus knowth me, and kepp yow. "Yours for eu*er*,

(*on back*) "To the ryght ven*er*able fa*þer* A. Bord.
p*ri*or off Hynton,[3] be *þ*is byll delyu*er*yd."

§ 25. Well, the 'rugorosyte' of the Carthusian rules—the no-meat, no-fun, and all-stay-at-home life—did not suit Andrew Boorde, the confinement injured his health, he wanted to be quit of the place, and let others see this. Accordingly Prior Batmanson—who was Prior, says Mr W. Durrant Cooper,[4] from 1529 to 16 Nov. 1531,—got Boorde a Dispensation from the Grande Chartreux, the General Chapter, as he calls it in another place (p. 48). Boorde says in his Fifth Letter, p. 58, below, written to Cromwell when Master of the Rolls, late in 1535 or early in 1536:—

"now I dyd come home by the grawnte Charterhouse, wher[5] y was dyspensyd of the relygyon in the prior Batmansons days."

In his Fourth Letter also (p. 57)—evidently written from the Grande Chartreux (Aug. 2, 1535?), and to the Prior of the London

[1] In the Record Office. [2] 'þᵉ' follows, but is scratcht out.

[3] "Master Doctor Horde." See the postscript to Letter III.

[4] *Sussex Arch. Collections*, xiii. But the last edition (1830) of Dugdale's Monasticon says, "William Tynbygh was made prior in 1499. He died in 1529. John Houghton succeeded in 1530," vol. vi, Pt. I, p. 9, col. 2. Charter-house, London. Yet Bale in his *Scriptores*, ed. 1548, gives 'Ioannes Batmanson, prior Carthusianoru*m* Londini, scripsit *Contra Erasmum*, *li.* I.' Fol. 254, back.

[5] This *wher* probably means *whence*, the dispensation having been sent, only, from the Grande Chartreux, and the place not visited by Andrew Boorde,

Charter-house and all other Priors of the Order in England,—Boorde dwells on the point of his dispensation from Religion, and the time of it, and says to his fellow-Carthusians:

"yow know þ*at* I had lycence before recorde to dep*ar*tt from yow / ȝett nott w*ith*stondy*n*g my co*n*scyence myȝth not be so satysfyd, but I thowth to vysett þe sayd reu*er*end faþ*er* [the Master of the Grande Chartreux], to know þe trewth whett*er* faþ*er* Ioh*a*n batma*n*son dyd impetratt for me of þe gen*er*all chapytt*er* þe lycence þ*at* dane george hath. þe trewth ys, þ*at* when dane george was dyspensyd w*ith* þe relygyon, I & anoþ*er* was dyspensyd w*ith* all / co*n*syderyng I can [not], nor neu*er* cowld, lyue solytary / & I among*es* yow intrusyd in a close ayre / myȝth neu*er* haue my helth."

This passage confirms the former one, and leaves no doubt that Boorde was abroad by 1529. There he studied medicine, "trauelled for to haue the notycyon & practes of Physycke in diuers regyons and countres,"[1] and

§ 26. Having, from the Continent, "returned into England, and [being] requyred to tary, and to remayne, and to contynue with syr Robert Drewry, knyght, for many vrgent causes,"[2] the Duke of Norfolk sent for Boorde, still "a young doctor"[3] (though full 40 years old), to attend him, A.D. 1530, "the yeare in the whiche lorde Thomas [Wolsey], Cardynal bishop of York, was commaunded to go to his see of York,"[4] to which he had been restored by Henry VIII after his first disgrace.

The head of all the Howards, the President of the Council, the uncle of Anne Boleyn, was an important patient, and Boorde hesitated at first to prescribe for the Duke without a consultation with his old physician, Dr Butte.[5] But as the old Doctor did not come,

[1] Preface to the *Dyetary*, ed. 1547 or -67, below, p. 225, col. 2.

[2] See note 3, p. 225, below.

[3] See the Preface to the *Dyetary*, p. 225, below. Boorde speaks again of when he was 'young,' in the *Breuyary*, Fol. lxxx, back: "In Englyshe, *Morbus Gallicus* is named the Frenche pockes: when that I was yonge, they were named the Spanyshe pockes." "This disease . . dyd come but lately into Spayne and Fraunce, and so to vs about the yere of our lord .1470." *ib.* Fol. lxxiv.

[4] A.D. 1530. Wolsey . . was now permitted to come nearer to the court; and he removed from Esher to Richmond. But Anne and her party took the alarm, and he was presently ordered to reside in the north of England, within his Archbishopric.—*Macfarlane's Hist.* vi. 182.

[5] This is our old acquaintance of the *Babees Book* Forewords, p. lxxviii, whose allowances for dinner and supper on every day of the week are given

Boorde, 'thankes be to God,' set his ducal patient straight, and was by his means allowed to wait on[1] Henry VIII.

§ 27. After this, urged by righteous zeal "to se & to know the trewth of many thynges,"[2] "to haue a trewe cognyscyon of the practis of Physycke,"[3] Boorde passed "ouer the sees agayne, and dyd go to all the vnyuersyties and scoles approbated, and beynge within the precinct of Chrystendome."[3] But, could he go abroad without a fresh license from the Prior of his House? Had his former dispensations by the Pope and the General Chapter of the Grande Chartreux rendered him free of his Order? Seemingly not; for, in his Fifth Letter to Cromwell, p. 58, below, written late in 1535, or early in 1536, Boorde says:—

> "I haue suffycyentt record that the prior off Charterhouse off London last beyng, off hys own meere mocyon, gaue me lycence to departe from the relygyon: whereuppon I wentt ouer see to skole, and now I dyd come home by the grawnte Charterhouse, wher y was dyspensyd of the relygyon in the prior Batmansons days.
>
> "att the sayd howse, in þe renewyng þat lycence, I browth a letter, yow [Cromwell] to do with me and ytt what you wyll."

This Prior "last beyng" must have been Howghton, who had been executed for denying the King's supremacy on April 27, 1535—according to Mr W. D. Cooper; on May 4, according to Stowe—and the first lines of the passage must refer to Boorde's 2nd journey abroad, and not his first, as they seem at first to do.

As to 'the vnyuersyties and scoles approbated' above, the only universities that Boorde mentions are, I think, Orleans, Poictiers,

at p. lxxix there, from *Household Ordinances*, p. 178-9. In Nicolas's Privy Purse Expenses of Henry VIII we find a payment of £10 to Dr Butts for Dr Thirlby (afterwards the first and only Bishop of Westminster), on Oct. 5, 1532. In his Index and Notes, p. 305, Nicolas notes that Henry 'sent Doctor Buttes, his graces physician,' to see Wolsey (Cavendish's Life of Wolsey, i. p. 220-2), and that 'Dr Butts is honourably commemorated by Fox as the friend of Bp Latimer. See also Gilpin's Life of Latimer, p. 42-5.'

[1] These words 'wait on can hardly mean 'attend professionally,' as there is no payment to Boorde in the *Privy Purse Expenses of Henry VIII* from Nov. 1529 to Dec. 1532, ed. Nicolas, 1827. Had Boorde attended Henry, we should no doubt have had an entry like that for Dr Nicholas, under Febr. 3, p. 192: "*Item* the same day paied to my lorde of Wilshire for a phisician called Doct*our* Nicholas, xx Angell*is*, vij li. x s."

[2] *Fyrst Boke*, chap. xxxii, Upcott's reprint, sign. Y 2, p. 204, below.

[3] Pref. p. 226, col. 1, below.

Toulouse, and Montpelier[1] in France; Wittenburg in Saxony.[2] The Italian ones he omits. At Orleans he dwelt for some time[3]; of his stay at Poictiers and Wittenburg (if any), he has left no record; in Toulouse he evidently stopt for a while,—"in Tolose regneth treue iustice & equite of al the places *that* euer I dyd com in;"[4]—and "at the last I dyd staye my selfe at Mountpyllyowre, which is the hed vniuersite in al Europe for the practes of physycke,"[5] or, as he says elsewhere, "Mu*n*tpilior is the most nobilist vniuersite of *the* world for phisicio*ns* & surgions. I can not geue to greate a prayse to Aquitane and Langwadock, to Tolose and Mountpiliour." And wherever he travelled, "in dyuers regyons & prouynces," he did "study & practyce physyk . . for the sustentacyon off [his] lyuyng."[6] Accordingly, we get, in such of his works as are left to us, little touches like the following: "For this matter [Scrofula . . in Englyshe . . named 'knottes or burres which be in chyldre*n*s neckes'[7]] in *Rome* and Mountpyller is vsed incisions" (instead of the pills and plaisters he has mentioned). "I, beinge long there [in Compostella in Navarre] . . was shreuen of an auncient doctor of diuinite, the which was blear [e]yed; and whether it was *to haue mi counsel in physicke or no,* I passe ouer, but I was shreuen of hym . ."[8] We shall see soon his practice in Scotland and Yorkshire, p. 61. Thus learning to do good, and doing it, the helper and friend of all he came across, Boorde, either in 1530-4, 1534-6, or 1538-42, went through almost the whole of Europe, and perhaps part of Africa, and pilgrimed it to Jerusalem, which he did not consider to be in Asia, as he tells us "as for Asia, I was neuer in [it]," *Fyrst Boke*, chap. vii. sign. I 2, back, p. 145, below.

The kindly nature of the man,—his willingness to help others at the cost of much hardship and danger to himself,—as well as his readiness to be off anywhere at any time, are well shown by his account of his sudden start from Orleans, and his journey to Compostella with 9 English and Scotch men whom he met:

[1] *Fyrst Boke*, chap. xxvii, sign. T .i. back, p. 191, below.
[2] *ib.* chap. xvi, p. 165. His disgust at the vices in Rome seems to have kept him from the Italian Universities. [3] *ib.* chap. xxxii, sign. Y 2, back, p. 205.
[4] *ib.* chap. xxvii. sign. U back, p. 194.
[5] Dedication to ed. 1547, Pref. p. 226, col. 2, below. [6] Letter VI, p. 59, below.
[7] *Breuiary*, Fol. C .iii. [8] *Fyrst Boke*, chap. xxxii, sign. Y 2, p. 204.

"whan I dyd dwell in the vniuersite of Orlya*n*ce, casually going ouer the bredge into the towne, I dyd mete with .ix. Englyshe and Skotyshe parsons goyng to saint Compostell, a pylgrymage to saynt Iames. I, knowyng theyr pretence, aduertysed the*m* to returne home to England, saying that 'I had rather to goe .v. tymes out of England to Rome,—and so I had in dede,—tha*n* ons to go from Orlyance to Compostel;' saying also that 'if I had byn worthy to be of the kyng of Englandes counsel, such parsons as wolde take such iornes on them wythout his lycences, I wold set them by the fete. And that I had rather they should dye in England thorowe my industry, than they to kyll them selfe by the way:' with other wordes I had to them of exasperacyon. They, not regardyng my wordes nor sayinges, sayd that they wolde go forth in theyr iourney, and wolde dye by the way rather than to returne home. I, hauynge pitie they should be cast a way, poynted them to my hostage, and went to dispache my busines in the vniuersyte of Orliaunce. And after that, I went wyth them in theyr iurney thorow fraunce, and so to burdious and byon; & than we entred into the baryn countrey of Byskay and Castyle, wher we coulde get no meate for money; yet wyth great honger we dyd come to Compostell, where we had plentye of meate and wyne; but in the retornyng thorow spayn, for all the crafte of Physycke that I coulde do, they dyed, all by eatynge of frutes and drynkynge of water, the whych I dyd euer refrayne my selfe. And I assure all the worlde, that I had rather goe .v. times to Rome oute of Englonde, than ons to Compostel: by water it is no pain, but by land it is the greatest iurney that an Englyshma*n* may go. and whan I returnyd, and did come into Aquitany, I dyd kis the ground for ioy, surrendring thankes to God that I was deliuered out of greate daungers, as well from many theues, as frome honger and colde, & *that* I was come into a plentiful country; for Aquitany hath no felow for good wyne & bred."—*Fyrst Boke,* chap. xxxii., p. 205, below.

That Boorde, though he hated water, and loved good ale and wine (p. 74), could live on little, we know from his description of Aquitaine (p. 194, below):

"a peny worth of whyte bread in Aquitany may serue an honest man a hoole Weke; for he shall haue, whan I was ther, ix. kakys for a peny; and a kake serued me a daye, & so it wyll any man, excepte he be a rauenner."

§ 28. The next notice that we have of Boorde is due to the Reformation. He must have returned to the Charter-house in London by the summer of 1534, for in Rymer's *Fœdera,* xiv. 491-2, we find that, on 29 May, 1534, Roland Lee, Bp of Coventry and Lichfield[1]

[1] See a good Memoir of him in Sir Henry Ellis's *Original Letters,* Third Series, 1846, vol. ii, p. 363-5.

(who married Henry VIII and Anne Boleyn), and Thos Bedyll, clerk, took the oaths of Johannes Howg[h]ton, the Prior of the Charter-house, and 13 other dwellers and servants there; and on the 6th of June following, at the Charter-house, Bp Lee and Thomas Kytson, knight, took the oaths of 19 Priests,—18th in the list of whom was *Andreas Boorde*—and 16 other persons. The names of all are given in Rymer, and reprinted in Smythe's *History of the Charter-house*, Appendix XVIII, p. 49, and the regular oath to Henry's supremacy that Boorde and all other conformers swore, is given in Latin in Smythe's *Appendix*, p. 49, and in English at p. 50-1.

§ 29. After thus conforming, Boorde seems to have remained at the Charter-house, and to have got into some trouble there, for which he was 'kept in thraldom bodyly and goostly,' 'kept in p*er*son [1] straytly.' His Prior, Howghton, was convicted of high treason in April 1535 for speaking against the king's supremacy, and on the 27th of April [2] was hanged, drawn, and quartered. While Howghton was in the Tower (? in 1534), before his execution, Boorde tells Cromwell that he wrote to Howghton, at his fellow-Carthusians' request (p. 60). Boorde's letter to Cromwell is dated Leith, 1 April [1536]:—

"when I was keppt in thrawldom in þe chart*er*howse, & knew noþ*er* þe kyng*es* noble act*es*, nor yow, then, stultycyusly thorow synystrall wordes, I dyd as many of þ*a*t ord*er* doth; butt aft*er* þ*a*t I was att lyberte, manyfestly I ap*er*seuyde þe yngnorance & blyndnes þ*a*t they & I wer yn: for I could neu*er* know no thyng of no man*er* off matt*er*, butt only by them, & they wolde cause me wrett full incypyently to þe p*ri*or of london, when he was in þe tower before he was putt to exicucyon; for þe which I trust yo*ur* mast*er*shepp hath p*ar*donyd me; for god knowth I was keppt in p*er*son [1] straytly, & glad I was to wrett att theyr request; but I wrott nothyng þ*a*t I thowght shold be agenst my p*r*ince, nor yow, nor no oþ*er* man."

§ 30. From this 'thraldom' of body and soul, Andrew Boorde was delivered by Cromwell, as the Vicegerent of Pope Henry VIII, —if I read aright another passage in this same Leith letter (p. 60), —and he then (I suppose) visited Cromwell at his seat at Bishops-Waltham in Hampshire, where Cromwell received him kindly:

"Yow haue my hartt, & shalbe sure of me to þe utt*er*must off my poer power, for I am neu*er* able to mak yow amend*es*; for wher

[1] ? *prison*. [2] p. 54.—Stowe says, convicted on April 29, and hanged on May 4.

I was in greatt thraldom, both bodyly and goostly, yow off y*our* gentylnes sett me att liberte & clernes off conscyence. Also I thank y*our* mast*er*shepp for y*our* grett kyndnes, þ*at* yow shewde me att bysheppe*s* waltam, & þ*at* yow gaue me lycence to come to yow ons in a qwartt*er*."

§ 31. After this, Boorde must have at once gone abroad on his third long tour, seemingly as an emissary of Cromwell's, to observe and report on the state of feeling about Henry VIII's doings, but no doubt studying and practising physic on his road. He also renewed his license at the Grande Chartreux, p. 58.

[[1] Letter II, from Bordeaux, 20 June, 1535.]

"After humly salutacyon, Acordyng to my dewte coactyd, I am (causeys consideryd) to geue to yow notycyon of certyn synystrall matters contrary to o*ur* realme of ynglond, specyally a-ȝenst o*ur* most armipotentt, perpondentt, circu*m*specte, dyscrete, & gracyose soue-reyng lord the Kyng*e*; for, sens my dep*ar*tyng from yow, I haue p*er*lustratyd normandy, frawnce, gascony, & Byon [2]; þe regyons also of castyle, byscay, spayne, paarte of portyngale, & returnyd thorow Arogon, Nauerne, & now am att burdyose. In the whych partyes, I hard of dyu*er*se credyble p*er*sons of þe sayd countryes, & also of rome, ytale, & almen, þ*at* the pope, þe emprowre, & all oþ*er* crystyn kynge*s*, w*ith* þer peple (þe french kyng except) be sett aȝenst o*ur* souereyne lord þe kyng*e*: apon the which, in all the nacyons þ*at* I haue trauellyd, a greatt army & navey ys preparyd: and few frendys ynglond hath in theys partes of Europe, as Iesus yo*ur* lou*er* knowth, who eu*er* haue y*our* master & yow, w*ith* þe holo realme, vnder hys vynge*s* of tuyssyon [3]! from burdyose, the xx day of Iune, by þe hond of yo*ur* sa[r]uantt & bedma*n*

"Andrew Boord.

"I humyly & precordyally desyre yo*ur* mastershepp to be good master (as yow eu*er* haue byn) to yo*ur* faythfull bedmen, mast*er* p*ri*or of the cherter howse of london, & to Mast*er* doct*er* Horde, p*ri*or of Hynton.

[*directed on back*] "To hys venerable master, Mast*er* Thomas Cromwell, secretory to o*ur* souereyngne lord the kyng, be þis byll d*y*rect*y*d.[4]"

[1] The originals of this and the following letters (except Letter IV) are preserved in the Record Office, vol. 4, 2nd Series, of Miscellaneous Letters, temp. Hen. VIII.

[2] It may be 'Lyon,' but is 'Byon,' I feel sure, for Bayonne. Cp. Boorde's *Introduction*, ch. xxxiii, p. 206.

[3] wings of defence.

[4] The word is 'dyrectyd' in the next two letters.

The postscript to the last letter raises a difficulty as to its date; for, says Mr Cooper,—using Smythe's *History of the Charter-House, &c.*:—

"In April, 1535, John Howghton the prior, with 2 other Carthusian priors, a monk of Sion, and the Vicar of Isleworth, were convicted of high treason.[1] On 27 April, Howghton, and on the 4th of May the others, were drawn, hanged, and quartered."

Perhaps Boorde supposed that a new Prior had been appointed, and askt Cromwell's favour for him on spec.

Prior Horde does not seem to have needed any intercession on his behalf, as he must have conformed willingly, and was used to bring other hesitaters round. Archbp Lee, writing to Cromwell on July 9, 1535 (III *Ellis*, ii. 344), about the Prior of the Charter-house of Mountgrace in Yorkshire, who was 'verie conformable,' reports of him:

"And forbicause ther bee in everie Howse, as he supposethe, some weake simple men, of small lernynge and litle discretion, he thinkethe it sholde doo mutche good if oure Doctor Hord, a Pryor of theyre religion, whom all the religion in this realme dothe esteme for lerning and vertue, were sent, not onlie to his Howse, but to all ordre Houses of the same religion; he saide (wiche I suppoase is true) they will give more credence, and woll rathre applie theire conscience to hym and his judgement, than to anie ordre, althowgh of greater lernynge, and the rathre if with hym be joyned also some ordre good fadre. This he desired me to move to you; and verelie I thinke it sholde doo mutche good. For manye of them bee verie simple men.'

And again in another letter of 8 Aug., 1535, after the Prior of Mountgrace has yielded and conformed, Archbp Lee repeats the Prior's request, 'that for the alureing of some his simple brodren, Doctor Hord, a priour of their religion, in whom they have greate confidence, maye come thidre. . . His commeng shall more worke in them than anye learneng or autoritie, as the Priour thinkethe, and I can well thinke the same.' III *Ellis*, ii. 345.

§ 32. During this tour in the summer of 1535, Boorde visited the Universities of Paris, Orleans, Poitou, Toulouse (where he was on July 2, 1535), and Montpelier, as well as Catalonia (he was there in

[1] His crime was 'delivering too free an opinion of the King and his proceedings, in regard to the supremacy, to speak against which was now made treason.'—Smythe's *Hist. Charter-House*, p. 73.

1535), noting the state of feeling towards Henry VIII. Then after his labour he fell sick, and wrote the next letter to Cromwell, late in 1535, or early in 1536. The phrase in the postscript "in thes partes" —cp. "in theys partes of Europe," p. 53—shows that the letter was written from abroad, from Spain, I suppose.

We get the approximate date for this letter from Boorde's mention of the Emperor Charles V's expedition against Barbarossa. Though Sir Hy. Ellis says that this was in 1534, it was in 1535:

"In 1535, Europe being at peace, Charles [the Fifth] sailed with a large armament for Tunis, where Khari Eddin Barbarossa, the dread of the Christians in the Mediterranean, had fortified himself. Charles, supported by his admiral, Andrea Doria, stormed La Goletta, and defeated Barbarossa: the Christian slaves in Tunis meantime having revolted, the gates of the city were opened, and the Imperial soldiers entering in disorder began to plunder and kill the inhabitants, without any possibility of their officers restraining them. About 30,000 Mussulmans of all ages and both sexes perished on that occasion. When order was restored, Charles entered Tunis, where he re-established on the throne Muley Hassan, who had been dispossessed by Barbarossa, on condition of acknowledging himself his vassal, and retaining a Spanish garrison at La Goletta. Charles returned to Italy in triumph, having liberated 20,000 Christian slaves, and given, for a time, an effectual blow to Barbarossa and his piracy. On his return to Europe, 1536, he found King Francis again prepared for war."—*Penny Cyclopædia*, vi. 500, col. 2, from Robertson's *History of Charles V*, &c.

"The emperor embarked at Barcelona for the general rendezvous of the rest of his forces. This was Cagliari, in Sardinia. The fleet sailed from this place on the 16th of July, 1535."—Robertson's *History of Charles V*, edit. 1857, vol. i. pp. 445, 446.

Letter III. [after 2 July, 1535.]

"Hone*r*able syr, aft*er* humily salutacyon, I c*er*tyffy yow þat sens I wrott to yo*ur* maste*r*shepp from burdyuse by þe se*r*uantt off s*ir* Iohan Arundell in cor[n]wall, I haue byn in dyue*r*ce regyons & vnyue*r*sytes for lerny*n*g, and I assewre yow þe vnyue*r*sytes off orlyance, pyctauens*is*,[1] Tolosa, mowntpyller, & þe reue*r*end faþ*er* off þe hed charte*r*howse, a famuse clark, & pa*r*tt[2] off þe vnyue*r*syte off pa*r*ys, doth hold *with* o*ur* soveryne lord þe kyng, in his act*es*, þat in so much att þe vysytacyon off o*ur* lady[3] last past in tolosa, in þe cheff skole, callyd petragorysensis, þe Kyng of Nauerr*e* & his qwene

[1] The MS mark of contraction is that for *ir*, as in S*ir*.

[2] MS ptt. Prof. Brewer and Mr W. D. Cooper read it 'Presidentt,' Sir H. Ellis rightly 'partt.'

[3] The *Visitation* is on July 2.

beyng presentt, þe gretyst articles þat any cowld lay a-genst *our* nobyll kyng wer disputyd & dyffynyd to þe hon*er* of *our* noble kyng, as I shall shew yow att my co*m*yng to yow. I was in cathalonya when þe emprowe tok sheppyng in-to barbary, the which emprow, *with* all oþ*er* kyng*es* in þe courtes of whom I haue byn, be *our* redoubtyd kyng*es* frend*es* & lou*ers*; incypyentt p*er*sons doth spek aft*er* þ*er* lerny*ng* & wytt. c*er*tyffyng yo*ur* mast*er*shepp aft*er* my laboure, I am syk, or els I wold haue come to yow & putt my selff fully in-to yo*ur* ordynance; as sone as I am any thyng recou*er*yd, I shall be att yo*ur* co*m*maundme*n*tt in all causis, god succ*u*ryng, who eu*er* kepp yow in helth & hon*er*,

"By yo*ur* bedma*n* Andrew bord, prest.

"I haue sentt to yo*ur* mast*er*shepp the seed*es* off reuberbe, the which come owtt off barbary. in thes p*ar*tes ytt ys had for a grett tresure. The seed*es* be sowne in March, thyn; & when they be rootyd, they must be takyn vpp, & sett eu*er*y one off them a foote or more from a noþ*er*, & well watred, &c.

[*directed on back*] "To the ryght hon*er*able Esquyre Mast*er* Thomas Cromeƚƚ, hygh secretory to *our* souereyne lord þe kyng & mast*er* of Rolls, be this lettres dyrectyd.

[*endorsed in a later hand.*] "Androwe bord, prest.
how king h. 8. is well esteemed
in ffraunce & other natyons."

On this Letter Sir Henry Ellis observes:

"The Postscript is perhaps the most curious part. Boorde not only sends to Cromwell the Seeds of Rhubarb from Barbary, where he says the plant was treasured, but with directions for transplanting the roots when grown, and rearing the Plant, two hundred years at least before the later cultivation of the Plant was known in England.

"Collinson, among the Memoranda in his 'Hortus Collinsonianus,' 8vo. Swansea, 1843, p. 45, says: 'True Rhubarb I raised from seed sent me by Professor Segisbeck of Petersburgh, in 1742:' by another memorandum it appears that the seeds really came from Tartary, and that four plants were transplanted next year."—*Original Letters*, Third Series, vol. ii, p. 300.

§ 33. Boorde refers in his last letter to the opinion of 'the reverend father of the head Charter-house, a famous clerk,' on Henry VIII's acts. I suppose that he ascertained it on his journey out from England. At any rate he tells us that he came home by the Grande Chartreux, "now I dyd come home by the grawnte charterhowse," Letter V, p. 58. While there, he wrote, as I judge, the following letter, dated August 2 [1535], to the Priors and Convents of his Order in England, telling them that the Father of the Head

Charter-house exhorted them to obey the King, and showing that he (Boorde) was free (as I suppose) of the Carthusian Order. He was evidently afraid that on his return to England, the London Charter-house would claim him again.

[Letter IV. 2 August, 1535.]

[1] *MS Cott. Cleop. E. iv. leaf* 54, *re-numbering* 70.

"After *p*recordyall reco*m*mendacyon. dere belouyd fath*er* in god, þe reuerend faþ*er* off þe hed cha[r]t*er*howse, doth salute yow in þe blessyng off Iesu chryst / adu*er*tysyng yow þ*at* yow loue god, & þ*at* in any vyse yow obay o*ur* souereyng lord þe kyng, he beyng very sory to here tell any wylfull or sturdy opynyons to be amonge*s* yow in tymes past to þe co*n*trary/. he desye[r]yth nothy*n*g off yow but only as I haue rehersyd, that yow be obedyent to o*ur* kyng, & þ*at* yow maak labore to yo*ur* frende*s* þ*at* yff any off yo*ur* frende*s* deye, or þ*at* any off *ther* frende*s* dey, þ*at* þe obytt off þ*em* may bytwyxt yow be sent / þ*at* þe ord*er* off charyte be not lost, p*ro* defu*n*cti*s* exorare. þe sayd reu*er*[en]d faþ*er* hath sentt to yow þe obytt off hys *pre*-dycessor / oþ*er* lett*er*s he wyll nott wrytt, nor he wold nott þ*at* yow to hym shold wrett / lest þe kynge*s* hyhnes shold be dysplesyd. as for me, yow know þ*at* I had lycence byfore recorde to dep*ar*tt from yow / ȝett nott w*ith*stondy*n*g my co*n*scyence myȝth not be so satysfyd, but I thowth to vysett þe sayd reu*er*end faþ*er*, to know þe trewth whett*er* faþ*er* Iohan batma*n*son dyd impetratt for me of þe gen*er*all chapytt*er* þe lycence þ*at* dane[2] george hath. þe trewth ys, þ*at* when dane george was dyspensyd w*ith* þe relygyon, I & anoþ*er* was dyspensyd w*ith* all / co*n*syderyng I can [not], nor neu*er* cowld, lyue solytary / & I amonge*s* yow intrusyd in a close ayre / myȝth neu*er* haue my helth. also I was receuyd amonge*s* yow vnd*er* age, contrary to yo*ur* statute*s* / wherfor now I am clerly dischargyd; not hauy*n*g þe byshopp of Romes dispensacyon; but yow þ*at* receuyd me to þe relygyon, for lefull & lawfull causes co*n*syderyd / haue dyspensyd w*ith* me. In wytnes þ*at* I do not fable w*ith* yow, specyally þ*at* yow be in all causis obedyentt to yo*ur* kyng. þe afforesayd reu*er*end father hath maad þe ryȝth hon*er*able esquyre mast*er* Cromeł, & my lord[3] of chest*er*, broþ*er* off all þe hole relygyon / *pr*aying yow þ*at* yow do no thyng w*ith* outt theyr counsell, as Iesus yo*ur* lou*er* knowth, who eu*er* keppe yow! wretyn in hast in þe cell of þe reu*er*end faþ*er* callyd Joh*a*n, & w*ith* hys counsyll, þe ij day of August, by þe hand off yo*ur* bedma*n*

"Andrew Bord[4], prest.

[1] Papers relating to the Reformation and Dissolution of the Monasteries.

[2] Dominus.

[3] ? A Prior. Henry VIII, when Prince of Wales, was Earl of Chester. The Bishopric of Chester was erected 4 Aug., 1542.

[4] Printed 'Bond' in the Cotton Catalogue.

[*on back*] "To master prior & the couentt off þe charterhowse off london, & to all priors & couentes off þe sayd order in ynglond."

On one corner of the back is written, "Androw Bord. to þe priour and Convent of Charterhouse in london &c' /"

§ 34. Boorde then returned to England, wrote from London to Cromwell a letter that is not now extant (so far as we yet know), and then the following excusatory missive, which shows that he did not feel satisfied himself that he *was* free from his Carthusian vows, but feared that Cromwell, notwithstanding his former release (p. 52), might hold him bound to them still.

[Letter V. ? before 1 April, 1536.]

"After humyle salutacyon with dew reuerence. Accordyng to my promyse, by my letters maade at burdyose, and also att london, þis presentt month dyrectyd to your mastershepp, I, Andrew Boorde, somtyme monk of the charterhowse of london, am come to your mastershepp, commynttyng me fully in to goddis handes & yours, to do with me whatt yow wyll. As I wrott to your mastershepp, I browth letters from by-ȝend see, but I haue nott, nor wyll nott, delyuer them, vnto the tyme yow haue seen them, & knowyng þe ouerplus of my mynd. I haue suffycyentt record þat þe prior off chartterhowse off london last beyng, of hys owne meere mocyon, gaue me lycence to departe frome þe relygyon: wheruppon I wentt ouer see to skole; & now I dyd come home by the grawnte charterhowse, wher y was dyspensyd of þe relygyon in the prior batmansons days.[1] att the sayd howse, in þe renewyng þat lycence, I browth a letter, yow to do with me and ytt what yow wyll, for I wyll hyd no thyng from yow, be ytt with me or agenst me. I was also xv. ȝeres passyd dyspensyd with þe relygyon by the byshopp of Romes bulles, to be suffrygan off chycester, the whych I neuer dyd execute þe auctore[2]; ȝett all þis nott-withstondyng, I submytt my-selff to yow; & yff yow wyll haue me to þat relygyon, I shall do as well as [I] can, god succuryng, who euer keppe your mastershepp in prosperuse helth and honer!

"By your be[d]man, þe sayd andrew prenomynatyd.

[*directed on back*] "Suo Honorifico Magistro Thomæ Cromell, Armigero, summo Secretario serenissimo nostro regi henrico octauo, Magistro que rotularum dignissimo, hæ litterulæ sint tradende."

[*endorsed* Andrew Boorde.]

§ 35. Cromwell's decision must have been in favour of Boorde's freedom from his monkish vows, for soon after his letter to Crom-

[1] Batmanson was Prior from 1529 to 16 Nov., 1531.—*Cooper.* [2] authority.

well, Boorde went to practise and study medicine in Scotland, where we find him on April 1, 1536. The authority for the year 1536 is Mr W. Durrant Cooper, who says (*Sussex Archæological Society's Collections*, vol. xiii, p. 266) of this next letter, that it "is not dated, but *the allusion to the vacancy in the office of prior of the Charter-house* enables me to fix 1st April, 1536, as the date of the letter."[1]

[Letter VI. Leith, 1 April, 1536.]

"After humly salutacyon, *with* dew reuerence, I certyffy your mastershepp þat I am now in skotlond, in a lytle vnyuersyte or study namyd Glasco, wher I study & practyce physyk, as I haue done in dyuerce regyons & prouynces, for þe sustentacyon off my lyuyng; assewryng yow þat in the partes þat I am yn, þe kynges grace hath many, ȝe, (& in maner) all maner of persons (exceppt some skolasty-call men) þat be hys aduersarys, & spekyth parlyus wordes. I resortt to þe skotysh kynges howse, & to þe erle of Aryn, namyd Hamylton,[2] & to þe lord evyndale, namyd stuerd, & to many lordes & lardes, as well spyrytuall as temporall, & truly I know þer myndes, for þei takyth me for a skotysh manes sone. for I name my selff Karre, & so þe Karres kallyth me cosyn, thorow þe which I am in the more fauer. shortly to conclude, trust yow no skott, for they wyll yowse flatteryng wordes, & all ys fal[s]holde.[3] I suppose, veryly,

[1] I can't find the date of Prior Trafford's appointment. Howghton was executed April 27, 1535 (or May 4, *Stowe*). Shortly after "And order for the charterhous of London" was made,—of which the first provision is

"that there be v or vj gouerners of temporll men, lernyd, wysse, & trusty, appoyntyd, wherof iij or ij of them shalbe contynually there to geder euery meale, and loge there euery nyght."—(Cott. MS Cleop. E. iv. leaf 27. Strype's *Memorials*, vol. i. pt. i. p. 303, &c.) See also Smythe's *Charter-house*. This Scheme does not seem to have been carried out.

[2] "James, son of the second Lord Hamilton, and of Mary, daughter of James II of Scotland, was created Earl of Arran in August, 1503, and died without issue."—*Cooper*.

[3] See a virtuous Scotchman's opinion to the contrary in chapter 13 of *The Complaynt of Scotland*, ab. 1548 A.D., p. 165, ed. 1801: "there is nocht tua nations vndir the firmament that ar mair contrar and different fra vthirs, nor is inglis men and scottis men, quhoubeit that thai be viht-in ane ile, and nythbours, and of ane langage. for inglis men ar subtil, and scottis men ar facile. inglis men ar ambitius in prosperite, and scottis men ar humain in prosperite. inglis men ar humil quhen thai ar subieckit be forse and violence, and scottis men ar furious quhen thai ar violently subiekit. inglis men ar cruel quhene thai get victorie, and scottis men ar merciful quhen thai get victorie. and, to conclude, it is onpossibil that scottis men and inglis men can remane in concord vndir ane monarche or ane prince, be-cause there naturis and conditions ar as indifferent as is the nature of scheip and voluis . . ." "i trou it is as onpossibil to gar inglis men and scottis men remane in gude accord vnder ane prince, as it is onpossibil that tua sonnis and tua sunnis can

þat yow haue in ynglond, by-ȝend x thowsand skottes, & *in*numer*a*ble oþer alyons, which doth (specyally þe skott*es*) much harme to þe kyng*es* leege me*n* thorowh þ*er* ewyll wordes[1]. for as I wentt thorow ynglond, I mett, & was in company off, many rurall felows, englich men, þat loue nott o*ur* gracyose kyng. wold to Iesu, þ*at* somc wer ponyshyd, to geue oþ*er* example! wolde to Iesu, also, þ*at* yow hade neu*er* an alyon in yo*ur* realme, specyally skott*es*, for I neu*er* knew alyon goode to ynglonde, exceppt þei knew p*ro*fytt & lucre shold co*m* to them, &c. In all þe p*ar*tes off crystyndom þ*at* I haue trawyllyd in, I know nott v. englysh me*n* inhabytou*rs*, exceppt only skolers for lerny*ng*.[2] I p*ray* to Iesu þ*at* alyons in ynglond do no more harme to ynglonde! yff I myght do ynglond any se*r*uyce, specyally to my soueryn lorde þe kyng, & to yow, I wold do ytt, to spend & putt my lyff in dang*er* & Iuberdy as far as any ma*n*, god be my Iuge. Yow haue my hartt, & shalbe sure of me to þe vtt*er*must off my poer power, for I am neu*er* able to mak yow amend*es*; for wher I was in greatt thraldom, both bodyly and goostly, yow of yo*ur* gentylnes sett me att liberte & clernes off conscyence. Also I thank yo*ur* maste*r*shepp for yo*ur* grett kyndnes, þ*at* yow sheude me att bysheppe*s* waltam,[3] & þ*at* yow gaue me lycence to come to yow ons in a qwartt*er*. as sone as I come home, I p*re*tende to come to yow, to submytt my selff to yow, to do w*ith* me w*h*at yow wyll. for, for lak of wytt, p*ar*adue*n*tt*er* I may in þis wrettyng say þ*at* shall nott contentt yow; but, gode be my Iudge, I mene trewly, both to my soue*r*rynge lord þe kyng & to yow. when I was keppt in thrawldom in þe chart*er*howse, & knew[4] noþ*er* þe kyng*es* noble act*es*, nor yow; then, stultycyusly thorow synystrall wordes, I dyd as many of þ*at* ord*er* doth; butt aft*er* þ*at* I was att lyberte, manyfestly I ap*er*seuyde þe yngnorance & blyndnes þ*at* they & I war yn: for I could neu*er* know no thyng of no man*er* off matt*er*, butt only by them, & they wolde cause me wrett full incypyently to þe p*ri*or of london, when he was in þe tower, before he was putt to exicucyon[5]; for þe which I trust yo*ur* maste*r*shepp hath p*ar*donyd me; for god knowth I was

be at one tyme to-giddir i*n* the lyft, be raison of the grit differens that is betuix there naturis & conditions."

[1] The dislike of Englishmen to aliens in Henry VIII's reign is testified by 'evil Mayday' in 1517, and numerous petitions and enactments. See my *Ballads from Manuscripts*, vol. i. p. 56-9, 104-7.

[2] In the 7th chapter of his *Boke of the Introduction of Knowledge* he says, "I have travelled round about Christendom, and out of Christendom, and I did never see nor know 7 Englishmen dwelling in any town or city in any region beyond the see, except merchants, students, and brokers, not there being permanent nor abiding, but resorting thither for a space."—*Cooper*. See also the extract from Torkington's Pilgrimage in the *Notes*.

[3] 'when I came to yow þer' follows, and is struck out.

[4] *orig.* know.

[5] Prior John Howghton was convicted of high treason on April 29, 1535, and executed on May 4 (*Stowe*).

keppt in person[1] straytly, & glad I was to wrett att theyr request; but I wrott nothyng þat I thowght shold be a-genst my prince, nor yow, nor no oþer man. I pray god þat yow may prouyde a goode prior for þat place of london; for truly þer be many wylfull & obstynatt yowng men þat stondyth to much in þer owne consaytt, & wyll not be reformyd, butt playth þe chyldryn; & a good prior wold so serue them lyk chyldryn. News I haue to wrett to yow, butt I pretende to be with yow shortly; for I am halff very[2] off þe baryn contry, as Iesu cryst knowth, who euer keppe yow in helth & honer. ffrom leth, a myle from Edynborowh, the fyrst day off Apryll, by the hand off your Poer skoler & seruantt

"Andrew Boorde, Preest.

[*directed on back*]
"To the right honerable esquire, Master Thomas Cromwell, hygh secretory to þe Kynges grace."

In his *Breuiary of Helth*, Boorde also tells us that he first prac tised Physic in Scotland, and stayed there a year:

"I dyd practyse phisicke fyrst in Scotlande; and after that I had taried there one yere, I returned then into England, and dyd come to a towne in Yorkeshire named Cuckold, where a bocher had a sonne that fel out of a hyghe haye ricke" [see below for the rest].—*The Seconde Boke of the Breuiary of Health, named the Extrauagantes*, Fol. xxiiii.;

that among his patients were two lords,

"Whan I dyd dwell in Scotlande, and dyd practice there Phisicke, I had two lordes in cure that had distyllacion like to nature; and so hath many men in al regyons."—*ib.* Fol. xxii., back;

and that though he was hated as an Englishman, yet his knowledge got him favour:

"Also, it is naturally geuen, or els it is of a deuellyshe dysposicion of a scottysh man, not to loue nor fauour an englishe man. And I, beyng there, and dwellyng among them, was hated; but my sciences & other polices did kepe me in favour that I did know theyr secretes."—*Fyrst Boke of the Introduction of Knowledge;* Taylor's reprint, sign. H.

§ 36. From Yorkshire, Boorde returned to London, and saw Cromwell, to whom he afterwards wrote the following letter from Cambridge, on Aug. 17, and in the year 1537, as I think certain, for

[1] Was '*prison*' meant? Or only that he was watcht, and kept in his cell?
[2] weary. The Scotch *w* and *v* of this time are used for one another.

he could hardly expect Cromwell to recollect such a trifle as meeting him, after the interval of more than a month or two; and Boorde would hardly allow more than that time to pass over before applying for help to recover his stolen horses.

[Letter VII. Cambridge, 13 August [1537].]

"Reuerently salutyd *with* loue and fere. I desyre yo*ur* lordshepp to co*n*tynew my good lorde, as eue*r* yow haue byn: for, god be my iudge, yff I know w*h*at I myght do þ*at* myght be acceptable to yow, I wold do ytt; for þ*er* ys no creature lyuy*n*g þ*at* y do loue *and* fere so much as yow, and I haue nott in þis world no refuge butt only to yow. when I cam to london owtt of skotlond, *and* þ*at* yt plesyd yow to call me to yow, as yow cam rydyng from westmestre, I had ij horsys stolyn frome me, & I can tell the p*er*sons þ*at* hath bowght them, butt I can nott recoue*r* my horse[s] althowh they þ*at* bowght þem dyd neue*r* toll for them, nor neue*r* bowth þem in no markett, butt p*r*iuetly. Also þ*er* be yn london c*er*tyn p*er*sons thatt owth me in mony *and* stuff .iiij[li]., þe which my frend*es* gaue me. I do aske my dewty off þem; & they callyth me 'appostata, & all to nowght,' & sayth they wyll troble me, & doth slawnd*er* me by-hynd my bak off thyng*es* þ*at* I shold do xx[ti] ȝers a-gone; & trewly they can nott p*r*oue ytt, nor I neue*r* dyd ytt; þe matt*er* ys, þ*at* I shold be co*n*ue*r*santt *with* wome*n*: oþ*er* matt*er*es they lay nott to my charge. I desyer yow to be good lord to me, for I wyll neue*r* co*m*playne forther then to yow. I thank Iesu cryst, I can lyue, althowh I neue*r* haue peny off ytt; but I wold be sory þ*at* they þ*at* hath my good, shold haue ytt: yff any off yo*ur* s*er*uantt*es* cowld gett ytt, I wold geue ytt to them. yo*ur* fayghtfull s*er*uantt, mast*er* watt*er* thomas, dwellyng in wrettyll,[1] knowth all þe hoole matt*er*, *and* so doth hys son, dwelly*n*g in þe temple. I co*m*mytt all to yow, to do *with* me & ytt what ytt shall plese yow; desyeryng yow to spare my rude wrettyng, for I do p*r*esume to wrett to yow upon yo*ur* gentylnes, as god knowth, who eue*r* kepp yow in helth and hon*er*! ffrome cambrydg, þe xiij day off August, by the hond off yo*ur* bedma*n*, & s*er*uantt to þe vtt*er*must off my poor power.

"Andrew Boorde, prest.

[*directed on the back*] "To the ryght hon*er*able lorde the lord of the pryue seale[2] be thys byll dyrectyd."

[*Endorsed* Andrew Boorde prste (*so*)]

Who were Walter Thomas of Writtle, and his son dwelling in the Temple?

[1] ? Writtle, Essex.

[2] Cromwell was created Keeper of the Privy Seal on July 2, 1536; Earl of Essex in 1539, and beheaded, 28 July, 1540.

§ 37. How soon after 1537 Boorde left England a fourth time for the Continent, and no doubt travelled about it, we cannot tell. The Dissolution of the Religious Houses in England in 1538 must have assured him of his freedom, and he probably used it to journey about, to see and know. The range of his travels at different times astonishes one. For though at first sight we may be inclined to think that there's a bit of brag in his talk about his travels 'round about Christendom, and out of Christendom' (*Fyrst Boke*, chap. vii.); yet I am convinced that he is quite honest in what he says, and that the words he sets down with his hand, tell the facts that he saw with his eyes. The very differences between his full treatment of certain places, &c., in a country, and his slurring over others of equal importance, prove it. Had we his full *Itinerary* left, instead of only the English part of it that Hearne printed in his Abbot of Peterborough's Lives of Henry III and Richard I (ii. 777, &c. A.D. 1735), I feel sure that Boorde's entries would contain all the countries he describes in his *Fyrst Boke*, except perhaps Turkey and Egypt. At any rate, there are touches in his descriptions of the following places which render it impossible to doubt that he had been there :—

England, p. 116.
  Wales, p. 125.
  Scotland, p. 135.
  Ireland, p. 131.

France, p. 190.
  Calais, p. 191.
  Boulogne, p. 209.
  Orleans, p. 191.
  Montpelier, p. 194.

Spain, p. 198.
  Castile, p. 199.
  Biscay, p. 199.
  Compostella, p. 205.
  Catalonia, p. 194.

Flanders, p. 146.
  Antwerp, p. 151.

Germany, p. 159.
  Tyrol, or Alps, p. 160.

Saxony, p. 164.
Denmark, p. 162.
Italy, p. 177.
  Lombardy, p. 186.
  Venice, p. 181.
  Rome, at least twice,[1] p. 177.
  Naples, p. 176.
Greece, p. 171.
Jerusalem, p. 218.

All these places, besides (as I believe) all the other countries mentioned in his *Fyrst Boke*, Boorde must have visited before he settled down in Montpelier,[2] and there by 1542 wrote his *Introduction*, *Dyetary*, *Breuyary*, and Treatise upon Beards (assuming that it existed). What he tells us about himself and these books has been already quoted on pages 15—26 above; and what Barnes says

[1] *Brev.* II. fol. iv. back, p. 76, below.

[2] I do saye as I do knowe, not onelye by my selfe, but by manye other *whan I did vse the seas.*—(*Brev.* ch. 381. Fol. C. xxii.)

about the books, and about Boorde's getting drunk at Montpelier,[1] earning a reputation by his books, and denouncing beards, will be found at p. 307, 309, below. The reader may as well turn on, and run his eye over the passages.

§ 38. I suppose that Boorde came back to England in 1542, when the first edition of his *Dyetary* was publisht (p. 12), and that he was also in England when he wrote his *Pronosticacion* for 1545 (p. 25). During this time he probably settled at Winchester; and if we suppose that then were left to him by his brother the houses and property in that town which he devises by his will, or the houses in Lynn (in Norfolk) which he also devises, or that he made money by practice as a physician, so that the 'lacke of money' which stopt the printing of his *Introduction* (p. 15) ceast, we can account for the publishing of that book in 1547 (or 1548), as well as of the second edition of the *Dyetary*, the *Breuyary*, and the *Astronamye*, which was evidently intended as a companion to the *Breuyary*, and was written in four days with one old pen without mending (p. 16, above). To superintend the passing of these books through the press—though I doubt whether he read his proofs—he ought to have been in London; and, most luckily, it is in 1547, or just before, that we find a "Doctor Borde" there, as the last tenant of the house appropriated to the Master of the Hospital of St Giles's, by Lord Lisle, to whom Henry VIII had in 1545 granted nearly all the possessions of the Hospital, part of the Reformation spoil. In 1547 Lord Lisle, by Henry's license, conveyed the Hospital property to Sir Wymonde Carew, and in the description of it, Dr Borde's name occurs.[2] The

[1] Compare the result as stated by Barnes with William Langley's Glutton in the *Vision of Piers Plowman*, Text B, Passus V, p. 76, l. 361-3, who

. . coughed up a caudel · in Clementis lappe;
Is non so hungri hounde · in Hertford schire
Durst lape of þe leuynges · so vnlouely þei smauȝte.

[2] Necnon unu*m* aliu*m* messuagiu*m*, parcellu*m* situs nup*er* d*icti* Hospitalis, unà cum pomeriis & gardinis eid*em* messuagio p*er*tinen*tibus* sive adjacen*tibus*, existen*tibus* in p*re*dicta parochia Sa*n*cti Egidii. nup*er* in tenura sive occupac*ione* Doctoris Borde.

The Licence to Lord Lisle is dated July 6, 1547. The original is, says Parton, "Among the records in the Lord Treasurer's Remembrancer's office, in the Exchequer, to wit, in the fifth part of the originals of the 38th year of the reign of King Henry the Eighth, Roll CV, and is printed in p. 35, note 32, of 'Some Account of the Hospital and Parish of St. Giles in the Fields, Middlesex, by the late Mr John Parton, Vestry-Clerk.' 1822."

unpleasant alternative that this Dr Borde may have been Dr Richard Borde of Pevensey, I am unable to negative.[1]

§ 39. Just at this time, at the culminating point of Boorde's life, the most serious charge of that life is brought against him, and this by no less a person than John Ponet, Bishop of Winchester,[2]—the

By this grant [of Henry VIII in 1545] all the possessions of the hospital of St Giles (not expressly mentioned in the exchange with the king) were vested in Lord Lisle. They consisted of the hospital, its site and gardens, the church and manor of St Giles.

After this grant Lord Lisle fitted up the principal part of the hospital for his own residence, leasing out other subordinate parts of the structure, and portions of the adjoining grounds, gardens, &c., and at the end of two years he conveyed the whole of the premises to John Wymonde Carewe, Esq., by licence from the king, in the last year of his reign.

The capital mansion or residence which Lord Lisle fitted up for his own accommodation, was situate where the soap manufactory of Messrs Dix and Co. now is, in a parallel direction with the church, but more westward. . . . . The house appropriated to the master of the hospital was situate where Dudley Cavet has been since built, and is mentioned as occupied by Dr Borde in the transfer from Lord Lisle to Sir Wymonde Carewe, which is said to have been afterwards the rectory house, being given by the Duchess for that purpose. 1834.—*R. Dobie, History of the United Parishes of St Giles-in-the-Field, and St George, Bloomsbury, 2nd ed.,* p. 23-5.

"The grant of the hospital by Henry VIII. to Lord Lisle simply describes it as 'All that the late dissolved hospital of St. Giles in the Fields, without the bars of London, with its appurtenances, &c., lately dissolved.' But his licence to that nobleman to convey the same to Wymond Carew, contains a description of part of these premises, sufficiently detailed to afford almost every information that can be desired. They are thus particularized :—

'All that mansion, place, or capital house, late the house of the dissolved hospital of St. Giles in the Fields;—and all those houses, gardens, stables, and orchards to the same belonging; and one other messuage (parcel of the site of the said late hospital), and the orchard and garden to the same belonging and adjoining, late in the tenure of Dr. Borde.'"—Parton's *Account of the Hospital and Parish of St Giles-in-the-Fields*, pp. 51, 52 (*printed in* 1822).

[1] "That Andrew was connected with Pevensey by residence [?] and property is well established. Contemporary with him, and probably a near kinsman, was another Doctor Borde, who held the vicarage of Pevensey, the vicarage of Westham, and the chantry of the chapel of Northye in the adjacent marsh. In the 'Valor Ecclesiasticus' of Henry VIII. [A.D. 1535] his valuable preferments are thus stated:

Pevensey.

Ric*ard*us Bord, doctor, vicarius ib*idem*, valet clare per annum &c. 18£ 6s. 8*d*.

Westham.

Ric*ard*us Bord, doctor, vicarius ib*idem*, valet &c. 21. 10. 10.

Cantaria de Northyde (sic).

Ric*ar*dus Bord, doctor, capellanus ibidem, valet &c. 2. 13. 4."

M. A. Lower, in *Sussex Arch. Coll.* vi. 200.

[2] He was appointed Bishop in May, 1551.—*Strype's Memorials*, vol. ii. Pt I. p. 483, ed. 1822.

very town that Boorde had lived in,—and who, therefore, must have known what Boorde's fellow-citizens said of the facts of the case. In his controversy with Stephen Gardiner, Ponet published a second book in 1555 (says Wood), whose title in the 'correctid and amendid' edition in the British Museum is—

"An Apologie fully avnsweringe by Scriptures and aunceant Doctors / a blasphemose Book gatherid by D. Steph. Gardiner / of late Lord Chauncelar[1], D. Smyth of Oxford / Pighius / and other Papists / as by ther books appeareth, and of late set furth vnder the name of Thomas Martin, Doctor of the Ciuile lawes (as of himself he saieth) against the godly mariadge of priests. Wherin dyuers other matters which the Papists defend be so confutid / that in Martyns ouerthrow they may see there own impudency and confusion.

By JOHN PONET Doctor of diuinitie, and Busshop of Winchester. Newly correctid and amendid.

The author desireth that the reader will content himself with this first book vntill he may haue leasure to set furth the next / wiche shalbe by Gods grace shortly. Yt is a hard thing for the to spurn against the prick. Act. 9."

At page 48 of this work Bp Ponet says:—

"And within this eight yere [that is, in or after 1547] / was there not a holy ma*n*, named maister Doctour boord, a Phisicion, that thryse in the week would drink nothinge but water / such a proctour for the Papists the*n* / as Martyñ the lawier is now? Who vnder the color of uirginitie / and of wearinge a shirte of heare / and hanginge his shroud and socking / or buriall sheete at his beds feet / and mortifyeng his body / and straytnes of lyfe / kept thre whores at o*n*ce in his chambre at Winchester / to serue / not o*n*ely him self / but also to help the virgin preests about in the contry, as it was prouid / That they might with more ease & lesse payn keepe theire blessed uirginitie. This thinge is so trew / and was so notoriously knowen / that the matter cam to examination of the iustices of peace / of whom dyuerse be yet lyuinge / as Sir Ihon Kingsmill / Sir He*n*ry Semar / etc. And was before them confessed / and his shrowd & sheart of hear openly shewed / and the harlots openly in the stretes / & great churche of Whinchester punished. These be knowen storyes, whiche Martin[2] and the Papists can not denye /"[3]—

[1] Sir Thomas More. [2] Stephen Gardiner.

[3] I add the continuation of the passage, which is somewhat violent and exaggerated, so that it may lessen, perchance, the effect of the charge against Boorde. "And they know well enoughe themselues / that there be of the lyke thousands / whiche I omitt for brefenes / that destroy this affection of Martin's prouinge him a false lyer in this point.—When the deuell by losenes

Ponet's *Apologie*, &c., pp. 48, 49; printed 1556.[1]

§ 40. Now we know, on the one hand, that "the way of a man with a maid" is one of the four things that Agur the son of Jakeh knew not (*Proverbs* xxx. 1, 18-19), and we all are in like case: we know that lechery is an old-man's sin,[2] and that Boorde had been charged with the same sin in early life, though he denied it; and we see that the bishop of Boorde's diocese and town brought the charge as one of public notoriety against Boorde's memory, appealed to witnesses then living, in confirmation of it, and (as I suppose, though I have not seen Ponet's first edition of 1555) re-affirmed the charge in the second edition of his book published in the year of his death (he died April 11, 1556). We know too that Boorde under-

of liuinge / appeareth in his owne forme / he can not so easyly deceaue the world as otherwise / wherfore who seeth not that he vseth to put on a vysor of holines / of the punishement of the body / and austeritie of lyfe as often as he myndeth thorowly to deceaue? Which thinge he hath most perfectly brought to passe in all the orders of Antichrist. Of Popes / Cardinals / Buschoppes / preests / monks / Chanons / fryers / etc. To the perfect establishment of buggery of whoredom, and of all vngodlynes / and to the vniuersall ruine of the true faith of Christs trew religion / & of all vertrew and godly lyfe. And for cumpassinge of this enterpryse / Doctor Martin the lawyer is become the deuils Secretary / who being taught by his master / taketh diligent heed throughout his book / that in no wyse he geue any kynde of praise or commendacion to matrimony in any kinde of peple. But termeth somtyme (carnall libertie) somtyme (the basest state of lyfe in the churche of God) somtyme (a color of bawdry) somtyme (that it is a let for a man to geue himself whollye to God). Somtyme that (it is a doubling / rather then a takinge away the desyer of flesh) making himself therin wyser then God, who gaue it for a remedye against the lasciuiousnes of the flesh, as God him selfe witnessed when he sayd *faciamus ei adiutorium* lette vs make Adam a helper. And in the leaues .121. & 122. he goethe aboute to proue by Saynte Paule that all menne should auoide mariadge. Wher-by he confirmeth the opinions of Montanus, Tatianus / and suche other abhominable heritiques."—Ponet's *Apology*, pp. 49, 50, 51.

[1] Strype's Memorials, vol. iii. Pt I. p. 529, reprints Ponet's attack on Boorde; "Ponet also expected these sanctimonious pretenders to a single life, by the horrible uncleannesses they were guilty of." Bp Ponet had previously written A Defence for Marriage of Priestes, 1549, but this (says our copier, Mr Wood) contains nothing about Andrew Boorde. Strype says that Ponet wrote this book in 1544, when an exile (*Memorials*, vol. iii. Pt I. p. 235). But see his *Cranmer*, i. 75, 475, 1058, and especially his *Life of Parker*, ii. 445, and foll. He or his editors confuse the layman's tract on which Parker's Defence of Priests' Marriages was founded, with Ponet's two tracts, though it has nothing to do with either of them, except being on the same subject.

[2] Boorde must have been at least 57 in 1547.

stood women,[1] witness his article on them in his *Breuyary*, Fol. lxxxii. back :—

"¶ The .242. Chapitre dothe shewe of a woman.

MVlier is the latin worde. In greke it is named *Gyuy*. In Englyshe it is named a woman; first, when a woman was made of God, she was named *Virago* because she dyd come of a man, as it doth appere in the seconde Chapitre of the Genesis. Furthermore now why a woman is named a woman, I wyll shewe my mynde. *Homo* is the latin worde, and in Englyshe it is as wel for a woman as for a man; for a woman, the silables co*n*uerted, is no more to say as a man in wo; and set wo before man, and then it is woman; and wel she may be named a woman, for as muche as she doth bere chyldren with wo and peyne, and also she is subiect to man, except it be there where the white mare is the better horse; therfore *Vt homo non cantet cum cuculo*, let euery man please his wyfe in all matters, and displease her not, but let her haue her owne wyl, for that she wyll haue, who so euer say nay.

☞ The cause of this matter.

This matter doth sprynge of an euyl educacion or bringynge vp, and of a sensuall and a peruerse mynde, not fearyng god nor worldely shame.

☞ A remedy.

☞ Phisike can nat helpe this matter, but onely God and greate sycknes maye subdue this matter, and no man els.

Vt mulier non coeat cum alio viro nisi cum proprio, &c.

☞ Beleue this matter if you wyll.

TAKE the gal of a Gote and the gal of a Wolfe, myxe them togyther, and put to it the oyle of Olyue, *ET VNG. virga.* Or els take of the fatnes of a Gote that is but of a yere of age. *ET VNG. virga.* Or els take the braynes of a Choffe, and myxe it with Hony. *ET VNG. virga.* But the best remedy that I do knowe for this matter, let euery man please his wyfe, and beate her nat, but let her haue her owne wyll, as I haue sayde."

We know, too, that medical students are apt to gain their knowledge of women's secrets—and Boorde knew plenty—by practical experiences inconsistent with a vow of chastity; and that in the 16th century, both at home and abroad, opportunities for indulgence must have been many, to a roving doctor. Still, the knowledge of women's external and internal arrangements shown by Boorde in his *Bre-*

[1] Compare the answer to the question what women most desire in *The Marriage of Sir Gawaine*, Percy Folio Ballads and Romances, i. 112. 'Item, I geue to all women, *soucreygntee*, which they most desyre; & that they neuer lacke excuse.'—*Wyll of the Deuyll.*

uyary may have been only professional, and got purely. He also knew all the Doctors' remedies for lechery,[1] and the penalty of indulgence by old men; though, as he says, "it is hard to get out of the flesh what is bred in the bone".[2] We know too that the Protestant parson, William Harrison, in his *Description of England*, printed in 1577, within 30 years of Boorde's death, called him "a *lewd* and vngratious priest," and in the 2nd edition of 1586-7 "a *lewd* popish hypocrite, and vngratious priest,"[3] using *lewd* in its modern sense. On the other hand, we know that Bp Ponet's charge was made at second hand, in a controversial book, and we have Anthony a Wood's suggested plea, above 140 years afterwards, in mitigation of the charge:

"He always professed celibacy, and did zealously write against such monks, priests, and friers, that violated their vow by marriage, as many did when their respective houses were dissolv'd by king Hen. 8. But that matter being irksome to many in those days, was the reason, I think, why a Calvinistical bishop (Joh. Ponet, B. of Winchester, who was then, as it seems, married), fell foul upon him, by reporting (In his *Apology fully answering*, &c. *Tho. Martin's Book*, &c., printed 1555, p. 32. See more in Tho. Martin) openly, that under colour of virginity and strictness of life, he kept three whores at once in his chamber at Winchester, to serve not only himself, but also to help the virgin priests, &c. about 1547. How true this is, I cannot say (though the matter, as the bishop reports, was examined before several justices of peace) because the book here quoted contains a great deal of passion, and but little better language, than that of foul-mouth'd Bale, not only against him (And. Borde), but also against Dr. Joh. Storie, Dr. Th. Martin, &c. The first of whom, he saith, kept a wench called Magd. Bowyer, living in Grandpoole in the suburbs of Oxon; and the other, another call'd Alice Lambe, living at the Christopher inn in the said city. But letting these matters pass (notwithstanding I have read elsewhere[4] that the said three whores, as the bishop calls them, were only

[1] See his chapter on *Priapismus*, p. 100, below.

[2] "And an olde man to fall to carnall copulacion to get a chylde, he doth kyll a man, for he doth kyl hym selfe, except reason with grace do rule hym. But oftymes in this matter old men doth dote, for it is harde to get out of the fleshe, that is bred in the bone. And furthermore I do saye *Qui multum coniunt diu viuere non possunt*, for it doth ingender dyuers infirmyties, specially if venerious persons vse carnall copulacion vpon a full stomake."—*Breuiary*, Fol. xxxi. back. See too p. 84, l. 4, below.

[3] See p. 106, below.

[4] Wood gives no reference, and I don't know what book or MS he alludes to.

patients that occasionally recurred to his hous), I cannot otherwise but say, that our author Borde was esteemed a noted poet, a witty and ingenious person, and an excellent physician of his time; and that he is reported by some to have been, not only physician to king Hen. 8, but also a member of the colledge of physicians at London, to whom he dedicated his *Breviary of Health*."—Athen. Oxon. I. 170, 171.[1]

but on the evidence before us I must confess myself unable, as judge, to ask, or hint to, the jury, to acquit the prisoner. Perhaps the publication or investigation of the Winchester records will throw further light on the matter. It is a painful business to wind up the record of a useful life with; but men are men. (See p. 85, No. VII.)

§ 41. We come now to the closing scene. Our lettered and widely-travelled healer of others' bodies, our preacher to others' souls, and reprover of others' vices, our hero sinned against and sinning, lies in the Fleet prison, sick in body, yet whole in mind. He is there, says Bp Bale in 1557-9, for his sin at Winchester, and has poisoned himself to save public shame:

"Quum sanctus hic pater, Vuintoniæ in sua domo, pro suis concœlibibus Papæ sacrificulis prostibulum nutriret, in eo charitatis officio deprehensus, uenenato pharmaco anno Domini 1548[2] sibijpsi

[1] The prior part of Wood's Memoir, with many mistakes, is as follows:

"Andrew Borde, who writes himself Andreas Perforatus, was born, as it seems, at Pevensey, commonly called Pensey, in Sussex, and not unlikely educated in Wykeham's school, near to Winchester, brought up at Oxford, (as he saith, in his *Introduction to Knowledge*, cap. 35), but in what house, unless in Hart-hall, I know not. Before he had taken a degree, he entred himself a brother of the Carthusian order, at or near to London? where continuing till he was wearied out with the severity of that order, he left it, and for a time applied his muse to the study of physic in this university. Soon after, having a rambling head, and an unconstant mind, he travelled through most parts of Europe (through and round about Christendom, and out of Christendom, as he saith, *Introduction to Knowledge*, cap. 7), and into some parts of Africa. At length upon his return, he settled at Winchester, where he practised his faculty, and was much celebrated for his good success therein. In 1541 and 1542, I find him living at Montpelier in France, at which time he took the degree of doctor of physic, and soon after being incorporated in the same degree at Oxon, he lived for a time at Pevensey, in Sussex, and afterwards at his beloved city of Winchester; where, as at other places [? invention or gammon, this 'other places'], it was his custom to drink water three days in a week, to wear constantly a shirt of hair, and every night to hang his shroud and socking or burial-sheet at his bed's-feet, according as he had done, as I conceive, while he was a Carthusian." [Why accept the hair-shirt, &c., and reject the whores, Mr Anthony?]

[2] Read 1549.

mortem accelerauit, ne in publicum spectandus ueniret."—Bale's *Scriptorum illustrium maioris Brytanniæ, Catalogus; Scriptores nostri Temporis* (after *Cent.* xii.) p. 105, edit. 1569.

Or, as Wood says:

"Joh. Bale, in the very ill language that he gives of Dr Borde, saith [1] that the brothelhouse which he kept for his brother-virgins being discovered, took physical poison to hasten his death, which was, as he saith, (but false [2]) in 1548. This is the language of one who had been a bishop in Ireland."—Wood's *Athen. Oxon.* I. 173, ed. Bliss, 1813.

He is there for his poverty,[3] says Mr Payne Collier, with that notorious daringness of invention that has made him read imaginary lines into MSS, and spelling into words, and has rendered him a wonder and warning to the editors of this age.[4]

[1] In lib. De Script. maj. Britan, p. 105, post cent. 12.

[2] Bale is wrong by less than a month; he wrote in old-style times.

[3] "poverty brought him to the Fleet prison, where, according to Wood (*Ath. Oxon.* I. 172, edit. Bliss) he died in 1549." (*Bibliographical Catalogue*, i. 327.) And yet Bliss gives Boorde's Will, showing all the houses and property that he left by it!

[4] To the Council of the Camden Society, who have lately put him among them, an object of honour, and (I suppose) a model for imitation.

As minor instances of this 'daring' of Mr Collier's, take the last four that I have hit on in following him over the first 61 pages of his print of the Stationers' Registers, and one song in a Royal MS. 1. The clerk has left out the subject of one ballad, and entered on leaf 22, back, 'a ballytt of made by nycholas baltroppe;' the *a* of *made* is not very decided, so that a hasty reader might take the word to be *mode.* Ritson (or the man he followed) so read it. Mr Collier prints the entry, leaves out the word *of*, and says, "We cannot suppose that Ritson saw the entry himself, and misread the words, 'A ballytt made,' 'A ballytt of mode.'" 2. On leaf 75 of the Register, the clerk has made a first entry of the printing a picture of a monstrous child born at Chichester, for which 4*d.* was paid; a second entry of one born in Suffolk, the sum paid for which is not put to it; and a third entry of the print of a monstrous pig, for which the usual 4*d.* was also paid. Mr Collier has run parts of the 1st and 2nd entries together, making one of the two, and put '[*no sum*]' at the end: he has then added the following note '[Perhaps the clerk of the Company did not know what ought to be the charge for a license for a publication of this kind' [though he had entered the iiij*d* just before]; 'but, when he made the subsequent entry, he had ascertained that it should be the same as for a ballad, play, or tract].' 3. On the back of leaf 84 of the MS, in an entry is '*our* salvation cōseeth [= consest[et]h] only in christe.' Mr Collier prints this 'cōseeth' as 'coseth,' and says we ought to read for 'coseth,' *consisteth.* 4. In MS No. 58 of the Appendix to the Royals in the British Museum is the song or ballad, 'By a bancke as I lay,' set to music. Mr Collier prints the words in his *Stat. Reg.* i. 193-4, makes two lines,

So fayre be seld on few
Hath floryshe ylke adew,

As we know the sad state of London prisoners in Elizabeth's time from Stubbes,[1]—and it was doubtless worse earlier—we may, if we like, conjecture that Boorde's illness may have been the "Sickenes of the prison" for which he prescribes in his *Breuyary*, Fol. xxvi. back.

"¶ The .59. Chapitre doth shewe of the
syckenes of the prisons.

CArcinoma is the greke worde. In englyshe it is named the sickenes of the prison. And some auctours doth say that it is a Canker, the whiche doth corode and eate the superial partes of the body, but I do take it for the sickenes of the prison.

☞ The cause of this infirmitie.

¶ This infirmitie doth come of corruption of the ayer, and the breth and fylth the which doth come from men, as many men to be together in a lytle rome, hauyng but lytle open ayer.

¶ A remedy.

☞ The chefe remedy is for man, so to lyue, and so to do, that he deserue nat to be brought into no prison. And if he be in prison, eyther to get frendes to helpe hym out, or els to vse some perfumes, or to smel to some odiferous sauours, and to kepe the prison cleane."

and observes on these "there is some corruption, for it seems quite clear that 'few' and 'adew' must be wrong, although we know not what words to substitute for those of the MS." Why not keep to the manuscript's own,—not misreading it, and foisting your own rubbish on to it?—

So fayre be feld on fen)
hath floryshe ylke a den).

These rashnesses arose, no doubt, from Mr Collier taking his careless copying as very careful work, not reading his proofs or revises with his MS, and yet finding fault with other people as if he had so read them.

A neat instance of Mr Collier's way of correcting a mistake of this kind occurs in his *Stat. Reg.* ii. xiv. Mr Halliwell, having in a note duly attributed the Ballad 'Faire wordes make fooles faine' to its writer, Richard Edwards, Mr Collier misses the note, and says (*Stat. Reg.* i. 87) that Mr Halliwell *was not aware* of Edwards's authorship. Having found afterwards that that gentleman's print showed his awareness of the fact, Mr Collier corrects his own mistake by saying (*Stat. Reg.* ii. 14) that Mr Halliwell did properly assign the ballad to Edwards, "a circumstance *to which we did not advert* when we penned our note."

Lastly, we have the beginning of the process that resulted in the imaginary words in the Dulwich MSS, in Mr Collier's printing the Stationers' clerk's "kynge of   " as "kynge of skottes" (*Stat. Reg.* i. 140, at foot). Here Mr Collier's insertion is the right one; but this importing his knowledge *without notice* into one MS, led to his importing his fancies into others, also without notice.

[1] *Anatomie of Abuses*, p. 141-2, ed. 1836, quoted in my *Ballads from MSS* (Ballad Soc., 1868), p. 33.

But whether Bale be right or wrong in the causes he assigns to Andrew Boorde's imprisonment and death, here is all that Boorde himself tells us :—

"IN THE NAME OF GOD, AMEN. The yere of *our* lorde God, a Thousande five hundreth ffortie and nyne, the xjth daye of Aprill, I, Andrewe Bord of Wynchester, in Hamshire, Doct*our* of Phisike,[1] being in the closse ward*es* of the Flete, *pri*soner in london, hole in mynde and sicke in body, make this my last will in maner and forme [following]. First, I bequeth my soule to Almyghtie God, and my bodie to be buried in erthe, where yt shall please my Executo*ur*. Also I bequeth vnto the poore prisoners now lying in the close ward*es* of the Flete, x s. Also I bequeth to Edwarde Hudson a fetherbed, a bolster, a paire of shett*es*, and my best coverlet. Also I bequeth and giue to Richard Mathew, to his heires and to his assignes, two tenement*es* or howses lying in the soocke in the towne of Lynne.[2] Also I giue and bequeth vnto the same Richard Mathew, to his heires and to his assignes, all those tenement*es* *with* thappurten*aunc*e*s* whiche I had by the deathe of my brother lying in Pemsey in Sussex. All whiche two tenement*es* in Lynne, whiche I hadd by the gifte of one Mr Conysby,[3] and those other tenement*es* in Pemsey whiche I had by my brother, *with* all and singuler ther appurten*aunc*e*s*, I will and giue, by this my last Wyll, vnto Richard Mathew, and to his heires and his assignes for ever (the deutye of the Lordes of the Fee always excepted). The residue of all my goodes vnbequethed, moveable and vnmoveable, I will and bequeth vnto Richarde Mathew, whom I make my Executour, and he to dispose as he shall thynke best for my soule and all *Christ*en soules. Also I giue and bequeth all my chattell*es* and houses lying abowte Wynchester or in Wynchester vnto Richard Mathew and his assignes. Witnesses vnto this wyll,

[1] He has dropt the "prest" of his letters.

[2] "The 'Soken' was used to distinguish the inhabited part of the parish of All Saints, South Lynn, which, though within the fortifications, was subject to the Leet of the Hundred of Freebridge-Lynn, from the Bishop's Borough of Lynn. *Ex inf.:* Alan H. Swatman, Esq., of Lynn. It was incorporated with the Borough, *temp.* Phil. & Mary."—*Cooper.*

[3] "Dr Borde's friend and benefactor at Lynn was William Conyngsby, Esq., some time Recorder of, and Burgess in Parliament for, that Borough,* who, in July, 1540, was made a justice of the King's Bench, and died in a few months. In addition to his house at Eston Hall, Wallington,† he resided in a mansion-house, in a street called the Wool-Market in Lynn. He was much trusted by the Crown and by Cromwell, to whom he addressed several letters preserved in the State-paper Office."—W. D. Cooper, in the *Sussex Archæological Society's Collections*, xiii. 268, 269.

* "Wm. Conysby was elected recorder of Lynn, pursuant to the new charter, on Monday the feast of St. Michael, 16th Hen. VIII., and was elected burgess to serve in parliament, for that borough, 31st March, 28th Hen. VIII. (*Ex inf.:* Alan H. Swatman, Esq.) He was afterwards a Judge (See Foss's *Judges*, v., 145.) I have not been able to identify Borde's houses."—*Cooper.*

† "He also owned West Linch Manor, in Norfolk."—*Cooper.*

WILLM. MANLEY, Gent. JOHN PANNELL. MARTEN LANE. HUMFREY BELL. EDWARD HUDSON. THOMAS WOSENAM. NICHOLAS BRUNE.

"Boorde's Will was proved in the Prerogative Court of Canterbury, by the oath of Richard Mathew, on the 25th of April, 1549: and the copy is in the register Poppulwell, 32."[1]

Boorde must thus have died very soon after the date of his Will, 11 April, 1549; but we have no record of where he was buried.

§ 42. *Portraits of Andrew Boorde.* No authentic portrait of Boorde exists besides that which he has left us in his works. Neither of the two old woodcuts of him in this volume (pages 143, 305) was ever drawn for him. The engraving of him in the 1796 edition of Scogin's Jests, after (?) Holbein's[2] picture, of a man carrying a bone (?) in one hand and a cylindrical jar in the other, is not authenticated. Readers who want to know Boorde must therefore go to his works, of which the two most characteristic and interesting are contained in the present volume. But his *Breuyary* has also many incidental passages containing statements of his opinions, notices of his travels, and touches of himself, which ought to be before the reader, and the chief of these I therefore extract here.

---

§ 43. *Characteristic Extracts from Boorde's Breuyary.*

a. Let us take first the passages in which Boorde speaks of himself or his tastes.

I. *Boorde hates water,*[3] *but likes good Ale and Wine.*

"This impediment [*Hidroforbia* or abhorynge of water] doth come, as many auctours doth say, of a melancoly humour, for the inpotent is named a melancoly passion; but I do saye as I do knowe, not onelye by my selfe, but by manye other, whan I dyd vse the seas, and of all ages, and of all complexions beynge in my company, that this matter dyd come more of coler than melancoly, considerynge that coler is mouable, and doth swimme in the stomake.

[1] Henry Poppulwell's will is the first in it.

[2] Mr R. N. Wornum says it is not Holbein's.

[3] He tells you also to wash your face only once a week if you want to clear it of spots. On the other days, wipe it with a Skarlet cloth. See Fol. xlix. and p. 95 here. See also p. 102, 'wype the face with browne paper that is softe.'

☞ A remedy.

☞ For this matter, purge Coler and melancoly humours; for I my selfe, whiche am a Phisicion, is combered muche lyke this passion, for I can not away with water, nor waters by nauigacion, wherfore I do leue al water[1], and to take my selfe to good Ale; and other whyle for Ale I do take good Gascon wyne, but I wyl not drynke stronge wynes, as Malmesey, Romney, Romaniske wyne, wyne Qoorse, wyne Greke, and Secke; but other whyle, a draught or two of Muscadell or Basterde, Osey, Caprycke, Aligant, Tyre, Raspyte[2], I wyll not refuse; but white wyne of Angeou, or wyne of Orleance, or Renyshe wyne, white or read, is good for al men; there is lytle read Renyshe wyne, except it growe about Bon, beyonde Colyn. There be many other wynes in diuers regions, prouinces, and countreys, that we haue not in Englande. But this I do say, that all the kyngdomes of the worlde haue not so many sondry kyndes of wynes, as be in Englande, and yet there is nothynge to make wyne of."—Fol. C.xxii.

Boorde does not love Whirlwinds. His opinion of Evil Spirits.

"¶ The .183. Chapitre dothe shewe of standynge vp of mannes heare.

HOrripilacio is the latin worde. In Englyshe it is named standyng vp of a mans heare.

☞ The Cause of this impediment.

¶ This impediment doth come of a colde reume myxte with a melancoly humour and fleume. It may come by a folyshe feare, when a man is by hym selfe alone, and is a frayde of his owne shadow, or of a spirite. O, what saye I? I shulde haue sayde, afrayd of the spirite of the buttry, whiche be perylous beastes. for suche spirites doth trouble a man so sore that he can not dyuers times stande vpon his legges. Al this notwithstandyng, with out any doute, in thunderynge and in lyghtenynge and tempestious wethers many euyl thynges hath ben sene and done; but of all these aforesayde thynges, a whorlewynde I do not loue: I in this matter myght bothe wryte & speake, the which I wyl passe ouer at this tyme.

☞ The seconde cause of this impediment.

¶ This impediment doth come of a faynte herte, and of a fearefull mynde, and of a mannes folyshe conceyte, and of a tymerous fantasy.

¶ A remedy.

¶ Fyrste, let euery man, woman, or chylde, animate them selfe vpon God, and trust in hym that neuer deceyued no man, that euer had, hath, or shal haue confidence in hym. what can any euyl spirite or deuell do any man harme without His wyll? And if it be my

[1] *Il n'a pas soif qui de l'eau ne boit:* Prov. Hee's not athirst that will not water drinke.—*Cotgrave,* A.D. 1611. See p. 255, below. [2] for 'Raspyce.'

Lorde Goddes wyl, I wolde all the deuyls of hell dyd teare my fleshe al to peces ! for Goddes wyll is my wyll in all thynges."—Fol. lxv, back.

Yet Boorde is afraid that Devils may enter into him. He is also shocked at the vicious state of Rome.

"The fyrst tyme that I did dwell in Rome, there was a gentylwoman of Germany the whiche was possessed of deuyls, & she was brought to Rome to be made whole. For within the precynct of S. Peters church, without S. Peters chapel, standeth a pyller of white marble grated rounde about with Yron, to the whiche our Lorde Iesus Chryste dyd lye in hym selfe vnto in [*so*] Pylates hall, as the Romaynes doth say, to the which pyller al those that be possessed of the deuyll, out of dyuers countres and nacions be brought thyther, and (as they saye of Rome) such persons be made there whole. Amonge al other, this woman of Germany, whiche is .CCCC. myles and odde frome Rome, was brought to the pyller ; I then there beyng present, with great strength and vyolently, with a .xx. or mo men, this woman was put into that pyller within the yron grate, and after her dyd go in a Preest, and dyd examyne the woman vnder this maner in the Italyan tonge :—'Thou deuyl or deuyls, I do abiure the by the potenciall power of the father, and of the sonne our Lorde Iesus Chryste, and by the vertue of the holy ghoste, that thou do shew to me, for what cause that thou doest possesse this woman !' what wordes was answered, I wyll not wryte, for men wyll not beleue it, but wolde say it were a foule and great lye, but I did heare that I was afrayd to tary any longer, lest that the deuyls shulde haue come out of her, and to haue entred into me, remembrynge what is specified in the .viii. Chapitre of S. Mathewe, when that Iesus Christ had made .ii. men whole, the whiche was possessed of a legio*n* of deuyls. A legio*n* is .ix. M. ix. C. nynety and nyne ; the sayd deuyls dyd desyre Iesus, that when they were expelled out of the aforesayd two men, that they myght enter into a herde of hogges ; and so they dyd, and the hogges dyd runne into the sea, and were drowned. I, consyderynge this, and weke of faith and afeard, crossed my selfe, and durst not to heare and se suche matters, for it was so stupendious and aboue all reason, yf I shulde wryte it. and in this matter I dyd maruel of an other thynge : yf the efficacitie of such makynge one whole, dyd rest in the vertue that was in the pyller, or els in the wordes that the preste dyd speake. I do iudge it shuld be in the holy wordes that the prest dyd speake, and not in the pyller, for and yf it were in the pyller the Byshops and the cardinalles that hathe ben many yeres past, and those that were in my tyme, and they that hath ben sence, wolde haue had it in more reuerence, and not to suffre rayne, hayle, snowe, and such wether to fal on it, for it hath no coueryng. but at last, when that I dyd consyder that the vernacle, the fysnomy of Christ, and skarse the sacrament of the aulter was in maner vncouered, & al .S. Peters churche downe in ruyne, & vtterly decayed, and nothyng set by; consydering, in olde chapels, beggers and baudes, hoores and theues, dyd ly within them; asses, and moyles dyd defyle within the precynct of the churche; and byenge and sellynge there was vsed within the precynt of the sayd churche, that it did pytie my hart and mynde to come and to se any tyme more the sayde place and churche. Then dyd I go amonges the fryers mendicantes, and dyuers tymes I dyd se *releuathes pro de-functis* hange vppon fryers backes in walettes; then I wente to other relygious houses, as to the Celestynes and to the Charter-house, and there I dyd se *nullus ordo.* And after that I dyd go amonges the monkes & chanons and cardynalles, and there I dyd se *horror inhabitans.* Then did I go rounde aboute Rome, and in euery place I did se Lechery and boggery [1], decceyt and vsery in euery corner and place. And if saint Peter and Paule do lye in Rome, they do lye in a hole vnder an Aulter, hauyng as much golde and syluer, or any other Iewell as I haue about myne eye; and yf it do rayne, hayle, or snowe, yf the wind stande Estwarde, it shal blowe the rayne, hayle, or snow to saynt Peters spelunke; wherfore it maketh manye men to thynke that the two holye Apostles shulde not lye in Rome, specially in the place as the Romaynes say they do lye. I do maruelye greatlye that suche an holye place and so great a Churche as is in all the worlde (except saynt Sophis churche in Constantinople), shulde be in such a vile case as it is in. Consyderynge that the bysshops of Romes palice, and his castel named Castel Angil standyng vpon the water or great ryuer of Tiber within Rome, and other of theyr places, and all that Car-

[1] "And lyghtlye there is none of theym [Cardinals and Prelates] withoute .iii. or .iiii. paiges trymmed like yonge prynces; for what purpos I wolde be loth to tell.—If I shoulde saye, that vnder theyr longe robes, they hyde the greattest pride of the worlde, it might happen some men wolde beleue it, but that they are the vainest men of all other, theyr owne actes doe wel declare. For theyr ordinarie pastime is to disguise them selfes, to go laugh at the Courtisanes houses, and in the shrouing time, to ride maskyng about with theim, which is the occasion that Rome wanteth no iolie dames, specially the strete called *Iulia*, whiche is no more than halfe a myle longe, fayre buylded on both sydes, in maner inhabited with none other but Courtisanes, some worthe .x. and some worthe .xx. thousand crownes, more or lesse, as theyr reputacyon is. And many tymes you shal see a Courtisane ride into the countrey, with .x. or .xii. horse waityng on hir.—Briefely by reporte, Rome is not without 40,000. harlottes, mainteigned for the most part by the clergye and theyr folowers. So that the Romaines them selfes suffer theyr wifes to goe seldome abrode, either to churche or other place, and some of theim scarcelye to looke out at a lattise window, wherof theyr prouerbe sayeth, *In Roma vale piu la putana, che la moglie Romana*, that is to say, 'in Rome the harlot hath a better lyfe, than she that is the Romaines wyfe.'—In theyr apparaile they are as gorgeouse as may be, and haue in theyr goyng such a solemne pace, as I neuer sawe. In conclusion, to liue in Rome is more costly than in any other place; but he that hathe money maye haue there what hym lyketh."—1549 A.D., Thomas's *History of Italye*, fol. 39 (edit. 1561).

dynalles palacis, be so sumptuously maynteyned, as well without as in maner within, and that they wyl se their Cathedral churche to lye lyke a Swynes stie. Our Peter pence was wel bestowed to *the* reedifieng of s. Peters Churche, the which dyd no good, but to noryshe syn & to maynteyne war. And shortly to co*n*clude, I dyd neuer se no vertue nor goodnes in Rome, but in Byshop Adrians days, which wold haue reformed dyuers enormities, & for his good wyl & prete*n*ce he was poysoned within .iii. quarters of a yere after he did come to Rome, as this mater, with many other matters mo, be expressed in a boke of my sermons. & now to co*n*clude, who so euer hath bene in Rome, & haue sene theyr vsage there (excepte grace do worke aboue nature, he shal neuer be good man after). be not these creatures possessed of the deuyl? This matter I do remit to the iudgement of the reders, for God knoweth that I do not wryte halfe as it is or was; but that I do write is but to true, the more pitie, as God knoweth."—*Extrauagantes*, Fol. iv, back.

On another page of his *Breuyary* he says:

"In Rome they will poyson a mannes sterope, or sadle, or any other thynge; and if any parte of ones body do take anye heate or warmenes of the poyson, the man is then poysoned." Fol. C.xvi. back.

Boorde is told of a Spirit by an Ancress at St Alban's.

"The .119. Chapitre dothe shewe of the Mare,
and of the spirites named *Incubus*
and *Succubus*.

E*Phialtes* is the greke worde. *Epialtes* is the barbarus worde. In latin it is named *Incubus* and *Succubus*. In Englyshe it is named the Mare. And some say that it is kynd of spirites, the which doth infect and trouble men when they be in theyr beddes slepynge, as Saynt Augustine saythe *De ciuitate dei*, *Capi*. 20. and Saynt Thomas of Alquine sayth, in his fyrst parte of his diuinitie, *Incubus* doth infeste and trouble women, and *Succubus* doth infest men. Some holdeth opynyon that Marlyn was begotten of his mother of the spirite named *Incubus*. Esdras doth speke of this spirite, and I haue red much of this spirite in *Speculum exemplorum*; and in my tyme at saynt Albons here in Englande, was infested an Ancresse of such a spirite, as she shewed me, & also to credyble persons.[1] but this is my opynyon, that this *Ephialtes*, otherwyse named the Mare, the

[1] Compare the curious set of depositions in a Lansdowne MS, 101, leaves 21-33, as to 'the Catt' which Agnes Bowker, aged 27, brought 'fforthe at Herboroghe, *within the* Iurisdiction of ye Archdeaconrie of Leicester, 22 Janu. 1568.' The vermilion drawing of 'the Catt,' its exact size, 'measured by a paire of compasses,' is given on the inside of the folio, leaf 32, back, and leaf 33. Agnes Bowker seems to have been delivered of a child, and to have substituted a flayed kitten in its place.

whiche doth come to man or woman when they be sleping, doth come of some euyll humour; consyderyng that they the which be thus troubled slepyng, shall thynke that they do se, here, & fele;—the thyng that is not true. And in such troublous slepyng a man shal scarse drawe his breth.

The cause of this impediment.

¶ This impediment doth come of a vaporous humour or fumosytie rysynge out and frome the stomake to the brayne; it may come also thorowe surfetynge and dronkennes, and lyenge in the bed vpryght; it may come also of a reumatyke humour supressyng the brayne; and the humour discendynge, doth perturbate the hert, bringyng a man slepynge into a dreame, to thynke that the which is nothynge, is somwhat; and to fele that thyng that he feleth not, and to se that thynge that he seeth not, with such lyke matters.

¶ A remedy.

☞ Fyrste, let suche persons beware of lyenge vpryght, lest they be suffocated, or dye sodenly, or els at length they wyll fall into a madnes, named Mania; therfore let suche persons kepe a good dyet in eatynge and drynkynge, let theym kepe honeste company, where there is honest myrth, and let them beware of musynge or studienge vpon any matter the whiche wyl trouble the brayne; and vse diuers tymes sternutacions with gargarices, and beware of wynes, and euery thyng the whiche doth engender fumositie.

☞ Yf it be a spirite, &c.

¶ I haue red, as many more hath done, that can tell yf I do wryte true or false, there is an herbe named *fuga Demonum*, or as the Grecians do name it *Ipericon*. In Englyshe it [is] named saynt Johns worte, the which herbe is of that vertue that it doth repell suche malyfycyousnes or spirites."—Fol. xlv.

Boorde has Cachexia, or a Bad Habit of Body.

" ☞ The .50. Chapitre dothe shewe of an infirmite the whiche is concurrant with an Hyedropsy.

C*Acecia*, or *Cacexia*, or *Cathesia*, be the greke wordes, In latin it is named *Mala habitudo*. In Englyshe it is named an euyl dweller, for it is an infirmitie concurrant with the hidropsies.

¶ The cause of this infirmytie.

¶ This infirmitie doth come thorowe euyll, slacke, or slowe digestion.

¶ A remedy.

☞ Vse the confection of Alkengi, and kepe a good dyet, & beware of drynkynge late, and drynke not before thou do eate somewhat, and vse temperate drynkes, and labour or exercise the body to swete. I was in this infirmite, and by greate trauayl I dyd make my selfe whole, more by labour than by phisicke in receyptes of medecines."—Fol. xxiii. back.

Boorde accidentally has the Stone, and cures himself of it.

"¶ The .207. Chapitre dothe shewe of the stone in the bladder

LIthiasis is the greke worde. In latin it is named *Calculus in vesica*, and *Lapis* is taken for all the kyndes of the stones. In Englysshe, *lithiasis* is the stone in the bladder. And some doth saye that *Nefresis* is the stone in the raynes of the backe, therfore loke in the Chapytre named *Nefresis*.

¶ The cause of this impediment.

This impedimente doth come eyther by nature, or els by eatynge of euyl and vyscus meates, and euyl drinkes, as thycke ale or beare, eatynge broyled and fryed meates, or meates that be dryed in the smoke, as bacon, martynmas biefe, reed hearynge, sprottes, and salt meates, and crustes of breade, or of pasties, and such lyke.

¶ A remedy.

☞ If it do come by nature, there is no remedy; a man maye mitigate the peyne, and breake the stone for a tyme, as shalbe rehersed. If it do come accidentally, by eatyng of meates that wyll ingender the stone, take of the bloud of an Hare, & put it in an erthen pot, and put therto .iii. vnces of Sa[xi]frage rotes, and bake this togyther in an Ouen, & than make pouder of it, and drynke of it mornynge and euenyng. For this mater, this is my practise: fyrste I do vse a dyet eatynge no newe bread, excepte it be .xxiiii. houres olde. I refuse Cake bread, Saffron bread, Rye bread, Leuyn bread, Cracknelles, Symnelles, and all maner of crustes; than I do drynke no newe ale, nor no maner of beere made with Hoppes, nor no hoote wynes. I do refrayne from Fleshe and fyshe, whiche be dryed in the smoke, and from salte meates and shell fyshes. I do eate no grosse meates, nor burned fleshe, nor fyshe. thus vsynge my selfe, I thanke God I dyd make my selfe whole, and many other. but at the begynnyng, whan I went about to make my self whole, I dyd take the pouder folowynge: I dyd take of Brome sedes, of Percilles sedes, of Saxfrage sedes, of Gromel sedes, of eyther of the*m* an vnce; of Gete stone a quarter of an vnce, of Date stone as much; of egges shelles that chekyn hath lyne in, the pyth pulled out, half an vnce; make pouder of al this, and drynke halfe a sponeful mornyng and euenynge with posset ale or whit wyne. Also the water of Hawes is good to drynke."—Fol. lxxii. (See p. 292, below.)

Boorde occasionally gets a Nit or a Fly down his Weasand, and commits the Cure to God.

"☞ The .356. Chapitre doth shewe of the Wesande or throte boll.

TRachea arteria be the latin wordes. In Englyshe it is named the wesande, or the throte bol, by the whiche the wynde and the ayer is conueyed to the longes; & if any crome of brede, or drop of drynke, go or enter into the sayde wesande, yf a man do not coughe he shulde be stranguled; and therfore, whether he wyl or wyll not, he must cough, and laye before hym that is in the throte and mouth; nor he can be in no quietnes vnto the tyme the matter be expelled or expulsed out of the throte, as it doth more largely appere in the Chapitre named *Strangulacio.*

¶ The cause of this impediment.

¶ This impedimente doth come of gredynes to eate or drynke sodeynly, not taking leysure; also it may come of some flye inhausted into a mans throte sodeynely, as I haue sene by other men as by my selfe; for a nytte or a flye comming vnto a mannes mouth, when he doth take in his breth and ayer, loke what smal thyng is before the mouth, is inhausted into the wesande, and so it perturbeth the pacient with coughynge.

☞ A remedy.

+ For the fyrst cause, be nat to gredy, eate and drynke with leyser, fearyng God; and as for the seconde cause, I do committe only to God: for this matter, coughynge is good."—Fol. C.xiiii. See too Fol. C.xxi. back.

Boorde can take-in other Phisicians by his Urine.

"There is not the wisest Phisicion liuynge, but that I (beynge an whole man) may deceyue him by my vryne; and they shall iudge a sicknes that I haue not nor neuer had, and all is thorowe distemperaunce of the bodye vsed the day before that the vryne is made in the mornynge; and this I do saye, as for the colours of vrynes, [vryne] is a strumpet or a harlot, and in it many phisicions maye be deceyued, but as touchynge the contentes of vrynes, experte phisicions maye knowe the infyrmyties of a pacient vnfallybly."—*Extrauagantes*, Fol. xxvi.

Boorde has seen Worms come out of Men.

"☞ The .364. Chapitre dothe shewe of diuers kyndes of wormes.

V*Ermes* is the latin worde. In greeke it is named *Scolices.* In Englishe it is wormes. And there be many kyndes of wormes. There be in the bodye thre sortes, named *Lumbrici, Ascarides,* and *Cucurbiti. Lumbrici* be longe white wormes in the body. *Ascarides* be smal lytle white wormes as bygge as an here, and halfe an ynche of length; and they be in a gutte named the longacion; and they wyl tycle in a mans foundement. *Cucurbiti* be square wormes in a mans body: and I haue sene wormes come out of a mans body lyke the fashion of a maggot, but they haue bene swart, or hauynge a darke colour. Also there be wormes in a mans handes named *Sirones,* & there be wormes in a mans fete named *degges;* then is there a rynge worme, named in latin *Impetigo*; And there may be wormes in a mans tethe & eares, of the which I do pretende to speke of nowe. As for all the other wormes, I haue declared theyr properties and remedies in theyr owne Chapitres.

☞ The cause of wormes in a mannes Eare.

¶ Two causes there be that a man haue wormes in his eares, the one is ingendred thorowe corruption of the brayne, the other is accidentall, by crepynge in of a worme into a mans eare or eares.

☞ A remedy.

☞ Instyll into the eare the oyle of bitter Almons, or els the oyle of wormewode, or els the iuyce of Rewe; warme euery thyng that must be put into the eare."

§ 43. β. Let us take, secondly, the notices of seven evils in England of which Boorde complains:—I. The neglect of fasting. II. The prevalence of swearing and heresies. III. The Laziness of young People. IV. The want of training for Midwives. V. Cobblers being Physicians. VI. The Mutability of Men's Minds. VII. The Lust and Avarice of Men:—adding his few allusions to the state of the poor (p. 86-7), and his one to early marriages (p. 87).

I. *The neglect of Fasting.*

α. "As for fastyng, that rule now a dayes nede not to be spoken of, for fastynge, prayer, and almes dedes, of charytie, be banyshed out of al regions and prouinces, and they be knockynge at paradyse gates to go in, wepynge and waylynge for the Temporaltye and spiritualtye, the which hath exyled them."—Fol. vii. back.

β. "Here it is to be noted that nowe a dayes few or els none doth set by prayer or fasting, regardyng not Gods wordes: in this mattere I do feare that such persons be possessed of the deuil, although they be not starke madde."—*The Extrauagantes*, Fol. iiii. back.

II. *The prevalence of Swearing and Heresies.*

"Do not you thynke that many in this contrie be possessed of the deuil, & be mad, although they be not starke mad? who is blynder then he that wil not se? who is madder then he that doth go about to kyl his owne soule? he that wil not labour to kepe the commaundementes of God, but dayly wil breke them, doth kil his soul. who is he that loueth God and his neyghbour, as he ought to do? but who is he that nowe a dayes do kepe their holydayes? & where be they that doth vse any wordes, but swearyng, lyeng, or slaunderynge is the one ende of theyr tale. In all the worlde there is no regyon nor countrie that doth vse more swearynge, then is vsed in Englande, for a chylde that scarse can speake, a boy, a gyrle, a wenche, now a dayes wyl swere as great othes as an olde knaue and an olde drabbe. it was vsed that when swearynge dyd come vp fyrst, that he that dyd swere shulde haue a phylyp, gyue that knaue or drabbe a phylyp with a club that they do stagger at it, and then they and chyldren wolde beware, after that, of swerynge, whiche is a damnable synne; the vengeance of God doth oft hange ouer them, and yf they do not amend and take repentance, they shalbe dampned to hell where they shalbe mad for euer more, worlde without ende. Wherfore I do counsayle al suche euyll disposed persons, of what degre so euer they be of, amende these faultes whyles they haue nowe leysure, tyme, and space, and do penance, for els there is no remedy but eternall punyshement.

A remedy.

Wolde to God that the Kynge our soueraygne lorde, with his most honorable counsell, wolde se a reformacion for this swerynge, and for Heresies, for the whiche synnes we haue had greate punyshment, as by dere price of corne and other vitayles; for no man can remedy these synnes, but God and our kynge; for there be a perilous nomber of them in Englande if they were diligently sought out; I do speke here of heretikes: as for swearers, a man nede not to seke for theym, for in the Kynges courte, and lordes courtes, in Cities, Borowes, and in townes, and in euery house, in maner, there is abhominable swerynge, and no man dothe go about to redresse it, but doth take swearyng as for no synne, whiche is a damnable synne; & they the which doth vse it, be possessed of the Deuill, and no man can helpe them, but God and the kyng. For *Demoniacus* loke in the Chapitre named Mania."—*The Extrauagantes*, Fol. vi.

III. *The Laziness*[1] *of young People.*[2]

"☞ The .151. Chapitre dothe shewe of an euyl Feuer
the whiche dothe cumber yonge persons,
named the Feuer lurden.

AMonge all the feuers I had almost forgotten the feuer lurden, with the which many yonge menne, yonge women, maydens, and other yonge persons, be sore infected nowe a dayes.

[1] 'the slowe worme and deadely Dormouse called Idlenes, the ruine of realmes, and confounder of nobilitie.' Louis, Duke of Orleans, to Henry IV, in the 5th year of his reign.—*Hall's Chronicle*, p. 33, ed. 1809.

[2] Compare Discipline's saying, in W. Wager's "The longer thou liuest, the more foole thou art," ab. 1568 A.D. (Hazlitt), sign. D iij back,

Two thinges destroye youth at this day,
Indulgentia parentum, the fondnes of parents,
Which will not correct there noughty way,
But rather embolden them in there entents.
Idlenesse, alas! Idlenesse is an other.
Who so passeth through England,
To se the youth he would wonder,
How Idle they be, and how they stand!

(★

¶ The cause of this Feuer.

¶ This feuer doth come naturally, or els by euyll and slouthfull bryngynge vppe. If it come by nature, then this feuer is vncurable, for it can neuer out of the fleshe that is bred in the bone; yf it come by slouthfull bryngynge vp, it may be holpen by dylygent labour.

¶ A remedy.

☞ There is nothyng so good for the Feuer lurden as is *Vnguentum baculinum*, that is to say, Take me a stycke or wan[d] of a yerde of length and more, and let it be as great as a mans fynger, and with it anoynt the bake and the shulders well, mornynge and euenynge,[1] and do this .xxi. dayes; and if this Feuer wyll net be holpen in that tyme, let them beware of waggynge in the Galowes; and whiles they do take theyr medecine, put no Lubberworte into theyr potage, and be[w]are of knauerynge aboute theyr hert; and if this wyl nat helpe, sende them than to Newgate, for if you wyll nat, they wyll brynge them selfe thither at length."—*Breu.* Fol. lv.

IV. *The want of training for Midwives.*

"If it do come of euyll orderynge of a woman whan that she is deliuered, it must come of an vnexpert Mydwyfe. In my tyme, as well here in Englande as in other regions, and of olde antiquitie, euery Midwyfe shulde be presented with honest women of great grauitie to the Byshop, and that they shulde testify, for her that they do present shulde be a sadde woman, wyse and discrete, hauynge experience, and worthy to haue the office of a Midwyfe. Than the Byshoppe, with the counsel of a doctor of Physick, ought to examine her, and to instructe her in that thynge that she is ignoraunt; and thus proued and a[d]mitted, is a laudable thynge; for and this were vsed in Englande, there shulde not halfe so many women myscary, nor so many chyldren perish[2] in euery place in Englande as there be. The Byshop ought to loke on this matter."—*The Extrauagantes*, Fol. xv. back.

V. *Cobblers being Physicians.*

"O lorde, what a great detriment is this to the noble science of phisicke, that ignoraunt persons wyl enterpryse to medle with the

A Christian mans hart it would pittie,
To behold the euill bringiug vp of youth!
God preserue London, that noble Citie,
Where they haue taken a godly ordre for a truth:
God geue them the mindes the same to maintaine!
For in the world is not a better ordre.
Yf it may be Gods fauour still to remaine,
Many good men will be in that bordre.

See the curious list of Fool's officers, 'A whole Alphabete' of them, 'a rable of roysterly ruffelers,' on the back of leaf F 4.

[1] See quaint W. Bulleyn on Boxyng, &c., *Babees Book*, p. 240-8.

[2] *orig.* perished.

ministracion of phisicke, that Galen, prince of phisicions, in his *Terapentike* doth reprehende and disproue, sayeng, 'If Phisicions had nothyng to do with Astronomy, Geomatry, Logycke, and other sciences, Coblers, Curryars of lether, Carpenters and Smythes, and al such maner of people wolde leaue theyr craftes, and be Phisicions,' as it appereth nowe a dayes that many Coblers be, fye on such ones! whervpon Galen reprehended Tessalus for his ignoraunce: for Tessalus smattered and medled with Phisicke, and yet he knewe not what he dyd, as many doth nowe a dayes, the whiche I maye accompte Tessalus foolyshe dyscyples."—*Breu.* Fol. ii. (Compare the First Chapter of the *Introduction of Knowledge.*)

VI. *The Mutability of Men's Minds.*

"¶ The .23. Chapitre doth shewe of a mannes mynde.

ANimus is the latin worde: In greke it is named *Thimos.* In englyshe it is a mannes mynde. The mynd of a man is very mutable and inconstant, more in one man then in another, but the moste parte myght be amended.

¶ The cause of this Mutabilitie.

☞ This mutabylytie doth come thorowe wauerynge and inconstant wyttes, lackynge loue and charytye to God, to a mannes owne selfe, and to his neyghbour, regardynge more, other[1] sensualytie or prodigalytie, couetys or lucre, then the welth and profyte of the soule. Yet the mynde of man is so occupied aboute worldly matters and businesses, that God and the soule of man is forgotten, by the whiche great daungers foloweth.

☞ A remedy.

¶ Fyrst, let euery man reconcyle hym selfe in and to God, and not to set by the worlde, but to take the worlde as it is, not beyng parmanente nor abydynge place, but to lyue as one shulde dye euery houre. And yf a man may haue this memory, he wyl not be mutable, nor set by the worlde, but be constant, hauynge euer a respect to God his creatour, and to his neyghbour, which is euery man where soeuer he dwell."—*Breu.* Fol. xv.

VII. *The Lust and Avarice of Men.*

"☞ The .340. Chapitre doth shewe of touchyng the whiche is one of the .v. wyttes.

TActus is the latin word. In greke it is named *Aphi.* In Englishe it is named touching or handlyng; and of handlyng or touching be ii. sortes, *the* one is venerious and the other is auaricious; the one is thorowe carnal concupiscence, & the other is thorowe cupiditie of worldly substance or goodes.

[1] other = or.

¶ The cause of these impedimentes.

¶ The fyrst impedimente doth come eyther that man wyll not call for grace to God not to displese hym, or els a man wyl folowe his luxurious sensualtie lyke a brute beaste. The seconde impediment, the which is auaryce or couetyse, wyll touch all thynges, and take as much as he can get, for al is fyshe that cometh to the nette with such persons.

☞ A remedy.

¶ For these matters I knowe no remedy, but onely God ; for there is fewe or none that doth feare God in none of these .ii. causes : if the feare of God were in vs we wolde not do so. Iesus helpe vs all ! AMEN."—*Breu.* Fol. C.x. [Does this mean 'guilty, and sorry for it'? p. 66.]

On the state of the poor there is hardly anything in Boorde's books. The chapters on Kybes, noticing the bad shoes of children, that on Croaking in the Belly, and that on Lowsiness—a point brought under our notice before by the *Babees Book* (p. 134, 209), and Caxton's *Book of Curtesye*—are the only ones I have noted.

Chilblains.

"☞ The .272. Chapitre dothe shewe of an impediment in the Heles.

P*Erniones* is the latin worde. *Pernoni* is the Barbarous worde. In Englyshe it is named the kybes in a mannes heales.

☞ The cause of this impedimente.

¶ This impediment most comonly doth infest or doth happen to yonge persons the which be hardly brought vp, goyng barefoted, or with euyll shoes ; and it dothe come of extreme colde and fleumatyke humours.

¶ A remedy.

¶ For the Kybes beware that the Snowe do nat come to the Heles, and beware of colde, nor prycke, nor pycke the Kybes : kepe them warme with wollen clothes, and to bedwarde washe the heles and the fete with a mans propre vrine, & with Netes fote oyle."—*Breu.* Fol. lxxxxi.

Croaking in the Belly.

"¶ The .309. Chapitre dothe shewe of crokyng in a mannes bely.

R*Vgitus ventris* be the latin wordes. In Englyshe it is named crokyng or clockyng in ones bely. In greke it is named *Brichithmos.*

☞ The cause of this impediment.

☞ This impediment doth come of coldenes in the guttes, or longe fastyng, or eatyng of fruites and wyndy meates, and it may come of euyl dyet in youth.

☞ A remedy.

☞ Fyrste, beware of colde and longe fastynge, and beware of eatynge of frutes, potages, and sewes, and beware that the bely be not constupated or costiue, and vse dragges to breake wynde."—*Breu.* Fol. C. back.

Lowsiness.

"☞ The .273. Chapitre dothe shewe of lyce in a mannes body or head or any other place.

P*Ediculacio* or *Morbus pediculorum* be the latin wordes. In greke it is named *Phthiriasis.* In Englyshe it is named lousines, and there be .iiii. kyndes, whiche be to say, head lyce, body lyce, crabbe lyce, and nits.

¶ The cause of this impediment.

This impediment doth come by the corruption of hote humours with sweat, or els of rancknes of the body, or els by vnclene kepynge, or lyenge with lousy persons, or els not chaungynge of a mannes sherte, or els lyenge in a lousy bedde.

☞ A remedy.

☞ Take of the oyle of Baye, an vnce and a halfe; of Stauysacre made in fyne pouder, halfe an vnce; of Mercury mortified with fastynge spetyll, an vnce; incorporate al this togyther in a vessel vpon a chafynge dyshe of coles, and anoynt the body. I do take onely the oyle of Bayes with Mercury mortified, and it doth helpe euery man and woman, excepte they be not to rancke of complexion."—Fol. lxxxxi.

The custom of mere boys marrying, which Stubbes reproves so strongly in his *Anatomie of Abuses*, p. 100, ed. 1836 (quoted in my *Ballads from MSS*, p. 32), Boorde only notices incidentally:

"And let boyes, folysh men, and hasty men, the whych be maryed, beware howe that they do vse theyr wyues when they be with child."—*Breu.* Fol. viii.

§ 43. γ. Thirdly, we may take some of Boorde's opinions.

Boorde on the Tongue and its greatest Disease.

"¶ The .208. Chapitre doth shewe of a mannes tonge.

L*Ingua* is the latin worde. In greke it is named *Glossa*, or *Glotta.* In Englyshe it is named a tonge. The tonge of man is an instrument or a member, by the whiche not onely tastyng, but also the knowledge of mans mynde by the spekyng of the tonge, is brought to vnder-standynge, that reason may knowe the truth frome the falshod. and soe conuerse. The tonge is the best and the worste offycyall member in man: why, and wherfore, I do remit the matter to the iudgement of the reders. But this I do say, that the tonge may haue dyuers impedimentes besyde sclaunderynge and lyenge, the which is the greatest impediment or syckenes of all other diseases, for it doth kyll the soule without repentaunce. I passe ouer this matter, and wyll speake of the sickenesses whiche may be in mannes tonge, the which maye swell, or elles haue fyssures, or wheales, or carnelles, or the palsey."—*Breu.* Fol. lxxi. back.

Boorde on Mirth and Men's Spirits.

"The .163. Chapitre dothe shewe of
Ioye or myrthe.

G*Audium* is the latin worde. In Englyshe it is named ioye or myrth. In Greke it is named *Hidonæ.*

The cause of myrthe.

Myrth commeth many wayes: the princypal myrth is when a man doth lyue out of deadly syn, and not in grudg of conscience in this worlde, and that euerye man doth reioyce in God, and in charitie to his neyghbour. there be many other myrthes and consolacions, some beynge good and laudable, and some vytuperable. laudable myrth is, one man or one neyghboure to be mery with an other, with honesty and vertue, without sweryng and sclaunderyng, and rybaldry speaking. Myrth is in musycall instrumentes, and gostly and godly syngyng; myrth is when a man lyueth out of det, and may haue meate and drinke and cloth, although he haue neuer a peny in his purse; but nowe a dayes, he is merye that hath golde and syluer, and ryches with lechery; and all is not worth a blewe poynte.

¶ A remedy.

¶ I do aduertise euery man to remember that he must dye, how, whan, and what tyme he can nat tel; wherfore let euery man amende his lyfe, and commyt hym selfe to the mercy of God."—*Breu.* Fol. lviii. back.

"☞ The .329. Chapitre doth shewe of a mannes Spirites.

S*Piritus* is the latin word. In Greke it is named *Pnoæ* or *Pneuma.* In Englyshe it is named a spirite. I do not pretende here to speake of any spirite in heauen or in hell, nor no other spirite, but onely of the spirites in man, in the which doth consyst the lyfe of man, & there be thre, naturall, anymal, and vytall: the naturall spyrite resteth in the head, the animall spirite doth rest in the lyuer, and the vital spirite resteth in the hert of man.

To conforte and to reioyce these spirites.

Fyrste lyue out of syn, and folowe Christes doctrine, and than vse honest myrth and honest company, and vse to eate good meate; and drynke moderatly."—Fol. C.vii.

" To comforte the stomake, vse Gynger and Galyngale, vse myrth and well to fare; vse Peper in meates, & beware of anger, for it is a shrode hert that maketh al the body fare the worse."—Fol. C.viii. back.

Boorde on the Heart of Man, and on Mirth.

"¶ The .86. Chapitre doth shewe of
the herte of man.

COr is the latin worde. In Greke it is named *Cardia.* In Englyshe it is an herte. the herte is the principal member in man; And it is the member that hath the fyrste lyfe in man, and it is the laste thynge that dothe dye in manne. The herte dothe viuifycate all other members, and is the grounde and foundacion of al the vitall spirites in man, and doth lye in the mydle of the bodye, and is hote and drye. And there is nothyng so euyl to the herte as is thought and care, and feare: as for other impedimentes that be longynge to the herte, [they] dothe appere in theyr Chapitres, as *Cardiaca.*

To comfort the herte.

There is nothynge that doth comforte the herte so much, besyde God, as honeste myrth and good company. And wyne moderately taken doth letyfycate and dothe comforte the herte; and good breade doth confyrme and doth stablyshe a mannes herte. And all good and temperate drynkes the which doth ingender good bloud doth comforte the herte. All maner of cordyalles and restoratiues, & al swete or dulcet thinges doth comfort the hert, and so doth maces and gynger; rere egges, and poched egges not harde, theyr yolkes be a cordiall. Also the electuary of citrons, *Rob de pitis, Rob de ribes, Diambra Aromaticum mustatum, Aromaticum rosatum,* and so is *Electuarium de gemmis,* and the confection of *Xiloaloes,* and such lyke be good for the hert."—*Breu.* Fol. xxxv.

Boorde on Pain and Adversity.

"¶ The .99. Chapitre dothe shewe of peyne or dolour.

DOlor is the latin word. In Greke it is named *Lype.* In Englyshe it is named peyne or dolour, the whiche may be many wayes, as by syckenes of the body, or disquietnes of a mannes mynde.

The cause of this peyne.

Dyuers tymes of greate pleasure doth come greate peyne, as we se dayly that thorowe ryot and surfetyng and sensualytie doth come dyuers sickenesses. Also with sport and playe, takyng great heate, or takynge of extreme colde doth ingender diseases and peyne.

Also for lacke of pacyence many mens and womens myndes be vexed and troubled.

¶ A remedy.

☞ If a man wyll exchewe many peynes and dolours, lette hym lyue a sober lyfe, and [not] distemper nor disquyed the body by any excesse or sensualite. And let hym arme hym selfe with pacience, and euermore thanke God what soeuer is sente to man; for if aduersitie do come, it is either sent to punysse man for synne, or els probacion: and with sorowe vse honest myrth and good company." —*Breu.* Fol. xxxviii. back.

Boorde on Intemperance.

" ☞ The .214 Chapitre doth shewe of intemperance.

LVxus is the latin word. In Greke it is named *Asotia.* In Englishe it is named intemperance. . Temperance is a morall vertue, and worthely to be praysed, consideryng that it doth set all vertues in a due order. Intemperance is a greate vyce, for it doth set euery thyng out of order; and where there is no order there is horror. And therfore this worde *Luxus* may be taken for all the kyndes of sensualitie, the whiche can neuer be subdued without the recognition and knowledge of a mannes selfe, what he is of him selfe, and what God is. And for asmuch as God hath geuen to euery man liuing fre wil, therefore euery man ought to stand in the feare of God, and euer to loke to his conscience, callynge to God for grace, and dayly to desyre and to praye for his mercye; and this is the best medecyne that I do knowe for intemperance."—Fol. lxxiii. back.

Boorde on Drunkenness.

"The .110. Chapitre dothe shewe of dronkennes.

E*Brietas* is the latin worde. In Greke it is named *Mæthæ.* In Englyshe it is named dronkennes.

¶ The cause of this impediment.

☞ This impedimente doth come eyther by wekenes of the brayne, or els by some greate hurte in the head, or of to much ryotte.

¶ A remedy.

☞ If it do come by an hurt in the head, there is no remedy but pacience of all partes. If it do come by debilite of the brayne & head, drynke in the mornynge a dyshe of mylke, vse a Sirupe named *Sirupus acetosus de prunis*, and vse laxatiue meates, and purgacions, if nede do requyre, and beware of superflous drynkynge, specially of wyne and stronge ale and beere, and if anye man do perceuye that he is dronke, let hym take a vomite with water and oyle, or with a fether, or a Rosemary braunche, or els with his finger, or els let hym go to his bed to slepe."—Fol. xlii.

Boorde on Man and Woman, which be reasonable Beastes.

"¶ The .182. Chapitre doth shewe of a man.

HOmo is the latin worde. In Greke it is named *Anthropos* or *Anir.* In Englyshe it is named a man or a woman, which be resonable beastes; and man is made to the similitudenes of God, and is compacke and made of .xv. substances. Of bones, of grystles, of synewes, of veynes, of artures, of strynges, of cordes, of skyn, of pannycles, pellycles, or calles, of heare, of nayles, of grece, of fleshe, of bloud, and of mary within the bones. a man hath reason with Angelles, felynge with beastes, lyuynge with trees, hauyng a beyng with stones."—Fol. lxiiii. back.

Boorde on Marriage.

"And here is to be noted for maried men, that Aristotle sayth, *Secundo de Anima,* that euery parfyte thynge is, whan one may generate a thynge lyke to hymselfe; for by it he is assimiled to the immortall. God. *Auicene De naturalibus* glorified natural procreacion. And for this cause God made man and woman, to encrease & multiply to the worlds ende. For this matter loke further in the *Extrauagantes* in the ende of this boke."—Fol. xxxii.

Boorde on the Words of late-speaking Children.

"Chyldren that can not speake vnto the tyme that they do come to a certein age, doth speke these .iii. wordes: *Aua, Acca, Agon. Aua* doth signifye father; *Acca* doth signifye ioye or myrth; *Agon* doth signifye dolour or sorow. All infantes doth speke these wordes, if a man do marke them; and what *wa* doth signifye when they crye, I coulde neuer rede of it; if it do signifye any thynge, it is displeasure, or not contented."—*Extrauagantes,* Fol. xxvi. back.

Boorde on the Kings Evil.[1]

"¶ The .236. Chapitre doth shewe of the Kynges euyll.

MOrbus regius be the latin wordes. In Englyshe it is named the kynges euyll, which is an euyl sickenes or impediment.

[1] See Brand's Antiquities, ed. Ellis, iii. 140—150. Boorde also believed in kings hallowing Cramp-rings as a remedy for Cramp: see his *Introduction,* p. 121, below; and Fol. C.vi. back, of his *Breuyary:*

"☞ The kynges maiestie hath a great helpe in this matter in halowynge Crampe rynges, and so gyuen without mony or peticion. Also for the Crampe, take of the oyle of Lyllyes and Castory, yf it do come of a colde cause. If it do come of a hote cause, anoynte the synewes with the oyle of waters Lyllyes, and wyllowes, and Roses. If it do come of any other cause, take of the oyle of Euforbium, and Castory, and of Pyretory, and confecte or compounde al togyther, and anoynt the place or places, with the partes adiacent."

☞ The cause of this impediment.

☞ This impediment doth come of the corruption of humours reflectynge more to a pertyculer place then to vnyuersall places, and it is muche lyke to a fystle; for and yf it be made whole in one place, it wyl breke out in an other place.

Bp Percy in his *Northumberland Household Book*, p. 436, ed. 1827, has the following note on Creeping to the Cross, and hallowing Cramp-Rings:— "This old Popish ceremony is particularly described in an ancient Book of the Ceremonial of the Kings of England, bought by the present Dutchess of Northumberland, at the sale of manuscripts of the late Mr Ansti's, Garter King of Arms. I shall give the whole passage at length, only premising that in 1536, when the Convocation under Henry VIII. abolished some of the old superstitious practices, this of Creeping to the Cross on Good-Friday, &c., was ordered to be retained as a laudable and edifying custom.—See Herb. Life of Henry VIII.

'The Order of the Kinge, on Good Friday, touchinge the cominge to Service, *Hallowinge of the Crampe Rings*, and Offeringe and Creepinge to the Crosse.

'Firste, the Kinge to come to the Chappell or Closset, withe the Lords, and Noblemen, waytinge upon him, without any Sword borne before him, as that day. And ther to tarrie in his Travers until the Byshope and the Deane have brought in the Crucifixe out of the Vestrie, and layd it upon the Cushion before the highe Alter. And then the Usher to lay a Carpett for the Kinge to Creepe to the Crosse upon. And that done ther shal be a Forme sett upon the Carpett, before the Crucifix, and a Cushion laid upon it for the Kinge to kneale upon. *And the Master of the Jewell Howse ther to be ready with the Crampe Rings in a Bason of Silver*, and the Kinge to kneele upon the Cushion before the Forme, *And then the Clerke of the Closett be redie with the Booke concerninge the Hallowinge of the Crampe Rings, and the Amner* [i. e. Almoner] *moste kneele on the right hand of the Kinge holdinge the sayd booke.* When that is done, the King shall rise and goe to the Alter, wheare a Gent. Usher shall be redie with a Cushion for the Kinge to kneele upon: *And then the greatest Lords that shall be ther to take the Bason with the Rings, and beare them after the Kinge to offer.* And thus done, the Queene shall come downe out of her Closset or Traverse, into the Chappell, with La[dies] and Gentlewomen waytinge upon her, and Creepe to the Crosse: And then goe agayne to her Closett or Traverse. And then the La[dies] to Creepe to the Crosse likewise; And the Lords and Noblemen likewise.'

"On the subject of these Cramp-Rings, I cannot help observing, that our ancient kings, even in those dark times of superstition, do not seem to have affected to cure the King's Evil; at least in the MS. above quoted there is no mention or hint of any power of that sort. This miraculous gift was left to be claimed by the Stuarts: our ancient Plantagenets were humbly content to cure the Cramp."—Boorde's words abolish this inference of the Bishop's. Brand, *Antiquities*, ed. Ellis, iii. 150, col. 2, quotes Boorde's *Introd.* and *Brev.* on this subject, and has other good references, iii. 160, i. 87 (quoting Percy), i. 89, the last of which quotes a letter of "Lord Berners the accomplished Translator of Froissart . . to my Lorde Cardinall's grace," 21 June, 1518: "If your grace remember me with some Crampe Ryngs, ye shall doo a thing much looked for."

¶ A remedy.

• For this matter let euery man make frendes to the Kynges maiestie, for it doth pertayne to a Kynge to helpe this infirmitie by the grâce the whiche is geuen to a Kynge anoynted. But for as muche as some men dothe iudge diuers tyme a Fystle or a French pocke to be the kynges Euyll, in suche matters it behoueth nat a Kynge to medle withall, except it be thorowe and of his bountifull goodnes to geue his pytyfull & gracious counsel. For kynges, and kynges sones, and other noble men, hath ben eximious Phisicions, as it apperth more largely in the *Introduction of Knowlege*, a boke of my makynge, beynge a pryntyng with Ro. Coplande."—*Breu.* lxxx. back.

Boorde on the Five Wits, and Men being Reasonable Beasts.

"¶ The .321. Chapitre doth shewe of the .v. wittes in man.

S*ensus hominis* be the latin wordes. In Greeke it is named *Esthisis anthropon*. In Englyshe it is named the sences or the wyttes of man. And there be .v. which be to saye, heryng, felynge, seynge, smellynge, and tastynge; and these sences may be thus deuyded, in naturall, anymall, and ractionall. The naturall sences be in all the members of man the which hath any felyng. The animall sences be the eyes, the tonge, the eares, the smellynge, and all thynges perteynyng vnto an vnreasonable beast. The racionall sences consisteth in reason, the which doth make a man or woman a reasonable beaste, which by reason may reuyle vnresonable beastes, and al other thynges beyng vnder his dominion. And this is the soule of man, for by reason euery man created doth knowe his creatour, which is onely God, that created al thynges of nothyng. Man thus created of God doth not differ from a beaste, but that the one is reasonable, which is man, and the other is vnresonable, the whiche is euery beast, foule, fyshe, and worme. And for as much as dayly we do se and haue in experience that the moste part of reasonable beastes, which is man, doth decay in theyr memory, and be obliuious, necessary it is to know the cause, and so consequently to haue a remedy.

¶ The cause of this impedimente.

☞ This impediment doth come eyther naturally or accydentally.

☞ A remedy.

If naturally a mans memory is tarde of wyt and knowlege or vnderstandyng, I know no remedy; yf it come by great study or solicitudnes, breakyng a mans mynde about many matters the which he can nat comprehende by his capacite, and although he can comprehend it with his capacite, and the memory fracted from the pregnance of it, let hym vse odiferous sauours and no contagiouse ayers, and vse otherwhyle to drynke wyne, and smel to Amber de grece: euery thyng whiche is odiferous doth comfort the wittes, the memory, and the sences; and all euyll sauours doth hurt the sences and the memory, as it appereth in the Chapitre named *Obliuio*."—Fol. C.iiii.

Boorde on Wounds.

"☞ The .377. Chapitre doth shewe of woundes.

V*Vlnus* or *Vulnera* be the latin wordes. In Greke it is named *Trauma* or *Traumata*. In Englyshe it is named a wounde or woundes: and there be dyuers sortes of woundes, some be newe and freshe woundes, and some be olde woundes, some be depe woundes, and some be playne woundes, and some fystuled, and some be festered, some be vlcerated and some hath fyssures, and some hath none.

☞ The cause of woundes.

¶ Most comonly woundes doth come thorowe an harlot, or for an hounde; it doth come also thorowe quarelynge, that some hote knauyshe bloude wolde be out; & dyuers tymes woundes doth come thorowe dronkennes, for when the drynke is in, the wytte is out, and then haue at the, and thou at me: fooles be they that wold them part, that wyl make such a dronken marte.

¶ A remedy.

☞ If it be a grene wounde, fyrste stanche the bloude; and yf the wounde be large and wyde, styche it, and after that lay a playster, and let it lye .xx. houres or more, tha*n* open it, and mundify it with white wyne. And if the wounde be depe, vse siccatiue playsters made with Olibanum, Frankensence, Literge, Yreos, the bran of Bones, and *Aristologia rotunda* and suche lyke. If the wounde be playne, take of the rotes of Lyllies, of pome Garnade rynes, of Galles, of Aloes or suche lyke. If the woundes be indifferent, the wounde mundified, vse the pouder of Myrtylles and Rose leues, and suche lyke; and let the pacient beware of venerious actes & of contagious meates and drynkes."—Fol. C.xxi.

Boorde on Obliviousness.

"☞ The .253. Chapitre doth shewe of an impediment named Obliuiousnes.

O*Bliuio* is the latin worde. In Greke it is named *Lithi*. In Englyshe it is named obliuiousnes or forgetfulnes.

¶ The cause of this impediment.

This impedimente doth come of reume or some ventosytie, or of some colde humour lyenge about the brayne; it may come of solicitudenes, or great study occupyenge the memory so much that it is fracted; and the memory fracted, there muste nedes then be obliuiousnes; & it may come to yonge men and women when theyr mynde is bryched.

☞ A remedy.

Fyrst beware and eschewe all suche thynges as do make or ingender obliuiousnes, and than vse the confection of Anacardine, & smel to odiferous and redolent sauours, and vse the thynges or medecines the whiche is specified in the Chapitre named *Anima* and *Memoria.* ✱ A medecine for Bryched persones, I do nat knowe, except it be *Vnguentum baculinum,* as it dothe appere in the Chapitre named the feuer Lurden."—Fol. lxxxv. back (p. 83, above).

Boorde on Dreams.

" SOmnia is the latin worde. In Greke it is named *Enipnia.* In Englyshe it is named dreames.

¶ The cause of this impediment.

☞ This impedement doth come most comonly of wekenes or emptynes of the head, or els of superfluous humours, or els of fantasticalnes, or collucion, or illusyons of the deuyll; it maye come also by God thorowe the good aungell, or such lyke matters: but specially, of fraction of the mynde and extreme sickenes doth happen to many men.

A remedy.

¶ For this matter vse dormitary, and refraine from such matters as shulde be the occasion of such matters, and be not costiue. &c."—*Extrauagantes,* Fol. xxvii.

Boorde on the Face.

" The .133. Chapitre dothe shewe of a mannes face.

FAcies is the Latin worde. In Greke it is named *Prosopon.* In Englyshe it is named a face, the which is the fayrest thing that euer God made in the compasse of a fote; and it is a wonderfull thynge to beholde, consyderynge that one face is not lyke another. The face may haue many impedimentes. The fyrst impedyment is to se a man hauyng no berde, and a woman to haue a berde. In the face maye be moles, wertes, the morphewe, ale pockes, sauccfleme, dandruffe, skurfe, scabbes, pockes, mesele, fystles, cankers, swellynges. For all suche matters loke in the Chapitres of the infyrmyties.

¶ A remedy to mundifie the face.

☞ To clere, to clense, and to mundifie the face, vse stufes and bathes, and euery mornyng after keymyng of the head, wype the face with a Skarlet cloth, and washe not the face ofte, but ones a weke anoynt the face a lytle ouer with the oyle of Costine, and vse to eat *Electuary de aromatibus,* or the confection of Anacardine, or the syrupe of Fumitery, or confection of Manna, and do as is wrytten in the Chapitre named *Pulchritudo.*"—*Breu.* Fol. xlix.

§ 43. δ. Fourthly, let us see Boorde as a physician: some of the cases in which he specially notes his own treatment of diseases.[1] But we should observe, first, that he does not, like a very popular modern medical work for mothers, insist that for every little ailment the right treatment is "Send for a duly qualified medical man." For blisters (or boils) "the whiche doth ryse in the nyght vnkyndely," Boorde says (fol. lxxxv.),

"☞ Fyrst, for this matter, beware of surfetyng, and late eating and drynkyng. And for this impediment, I do neither minister medecines nor yet no salues, but I do wrap a lytle clout ouer or aboute it; and as it dothe come, so I do let it go; for and a man shulde, for euery tryfle sycknes and impediment, runne[2] to the Phisicion or to the Chirurgion, so a man shuld neuer be at no point with hymselfe, as longe as he doth lyue. In great matters aske substancial counsell; and as for small matters, let them passe ouer."

And he repeats the advice again, under "A White Flawe," Fol. lxxxx. back.

"I wolde not councel a man for euery tryfle sycknes to go to Phisike or Chierurgy: let nature operate in suche matters in expulsynge suche humours, and medle no further."

So also under "A Blast in the Eye," Fol. C.xxi. back, he says:

"I myghte here shewe of many salubriouse medecines, but the best medecine that I do knowe is, to lette the matter alone, and medle nat with it, but were before the eyes a pece of blacke sarcenet, and eate neyther garlycke nor onyons, nor drynke no wynes nor stronge ale, and it wyll were awaye."

Boorde's treatment of Itch:—A good Pair of Nails.

"¶ The .292. Chapitre doth shewe of Itchynge.

P*Rurigo* is the latin word. In Englyshe it is named itching of a mans body, skyn, or fleshe.

¶ The cause of this impedimente.

¶ This impediment doth come of corrupcion of euyll bloud, the which wolde be out of the fleshe; it may also come of fleume myxt with corrupt bloud, the which doth putrifie the fleshe, and so consequently the skyn.

[1] See that of Stone, p. 80.

[2] shulde runne, *orig.*

☞ A remedy.

This I do aduertise euery man, for this matter to ordeyne or prepare a good payre of nayles, to crache and clawe, and to rent & teare the skynne and the fleshe, that the corrupt bloud maye runne out of the fleshe; and vse than purgacions and stuphes & sweates; and beware, reuerberate not the cause inwarde with no oyntment, nor clawe nat the skyn with fyshye fyngers, but washe the handes to bedwarde."—*Breu.* Fol. lxxxxvi. back.

So under *Pruritus* he says:

"For this mater ordeyne a good payre of nayles and rent the skyn and teare the fleshe and let out water and bloude."—Fol. lxxxxvi.

Boorde's treatment of Tertian Fever.

"The medecines the whiche dothe helpe the Feuer causon, wyl helpe a Feuer terciane. Fyrste purge coler, and .iii. or .iiii. houres before the fytte dothe come, I do thus. I cause a man to lye in his doublet, and a woman in her waste cote, then do I cause them to put on a payre of gloues, & with .ii. garters I do bynde the wrestes of the armes, and do lay theyr armes and handes into the bedde, & do cast on clothes to brynge theym to a sweate before the fyt do come .iii. or .iiii. houres; and out of Gose quylles, one put into an other, they do take theyr drynke, because they shall take no ayer into the bed; then I do geue them fyrst an ale brue, and suffer them to drynke as muche Posset ale as they wyl; & when the burnyng do begyn, I do withdrawe the clothes; and thus I do .iii. courses, & haue made many hundredes whole; but theyr good dayes I do nat suffre them to go in the open ayer."—Fol. li.

Boorde's treatment of Scurf.

"¶ For this matter I do take .iii. vnces of Bores grece, the skynnes pulled out; than I do put to it an vnce of the pouder of Oyster shelles burnt, and of the pouder of Brymstone, and .iii. vnces of Mercury mortified with fastyng spetyl; compounde al this togyther, & anoynt the body .iii. or .iiii. tymes, & take an easy purgacion." —Fol. lxxiii.

Curding of Milk in Women's Breasts.

"If the mylke be curded in the brestes, some olde auctours wyll gyue repercussiues; I wolde not do so, I do thus: I do take Dragagant[1], and gumme Arabycke, and do compounde them with the whyte of rawe egges, and the oyle of violettes, and do make a playster. Or els I do take pytch, and do lyquifye it in the oyle of Roses, puttynge a lytle doues dunge to it, and dregges of wyne or ale, and make playsters."—Fol. lxxv.

[1] Tragacanth, a gum.

Pregnant Women's unnatural Appetite.

"An vnnaturall appetyde is to eate and drynke at all tymes without dewe order, or to desyre to eate rawe and vnlefull thynges, as women with chylde doth and such lyke. . . . .

¶ A remedy for women that haue vnlefull lustes.

¶ I have knowen that such lustes hath ben put awaye by smellynge to the sauer of theyr owne shoes, when they be put of. In such lustes, it is best that women haue theyr desyre, if it may be gotten, for they shal neuer take surfet by such lustes."—Fol. xvi. back.

Ulcer in the Nose; and how then to blow your Nose.

"☞ The .264. Chapitre doth shewe of an
vlcer in the Nose.

OZenai is the Greke worde. In latin it is named *Vlcera narium.* In Englyshe it is named an Vlcer or sore[1] in the nose.

¶ The cause of this impediment.

¶ This impedimente doth come of a fylthy and euyll humour, the which doth come from the brayne and heade, ingendred of reume and corrupte bloud.

¶ A remedy.

+ In this matter, reume must be purged, as it dothe appere in the Chapitre named Reuma; than, pycke not the nose, nor tuche it not, excepte vrgent causes causeth the contrary, & vse gargarices and sternutacions. I wyll councell no man to vse vehement or extreme sternutacions for[2] perturbatyng the brayne. Gentyl sternutacions is vsed after this sorte. Fyrst, a man rysynge from slepe, or comynge sodenly out of a house, and lokynge into the element or Sonne, shal nese twyse or thrise, or els put a strawe or a ryshe into the nose, and tyckle the ryshe or the strawe in the nose, and it wyl make sternutacions: the pouder of Peper, the pouder of *Eliborus albus*, snuft or blowen into the nose, dothe make quycke sternutacions. But in this matter I do aduertise euery man not to take to muche of these pouders at a tyme, for[2] troblynge the seconde principal member[3] whiche is the brayne. and they the whiche wyll not nese, stope the nosethrylles with the fore fynger and the thome vpon the nose, and nat within the nosethrylles; and if they wold, they can not nese, al maner of medecines natwithstandyng; howe be it, I wolde councell all men takyng a thynge to prouoke suche matters to make no restrictions."—Fol. lxxxviii. back.

[1] sere, *orig.* [2] for fear of, to prevent. [3] menber, *orig.*

Boorde's cure for Asthma.

"☞ A confection of muske is good. Also *loch*[1] *de pino, loch de squilla, loch alfescera* be good, and so is the sirupe of Isope, and the sirup of Calamint. For I haue practised these thynges, and haue sped wel. Fyrst I haue made a ptysane vnder this maner. Take of Enula campane rootes, pycked and made clene, and cut in slyces, vii. vnces; of the rootes of Fenell washed, and the pyth pulled out .vi. or .vii. vnces; of Anes sedes halfe a pounde, of fygges halfe a pounde; of greate reasons, the stones pulled out, a quartron of a pounde; of Isope thre good handfulles, of barly clensed .v. handefulles; seth al this togither in two galons of runnyng water, to halfe a galon. And .xv. dayes I haue gyuen to my pacyent, mornynge, noone, and nyght, ix. sponefulles at a tyme; and at the .xv. dayes ende I haue geuen pylles of Cochee, and after that I haue ministred Dyasulfur, and haue made many whole. Also the confection of Philonii of the fyrst inuencion is good: And so is to anoynt the stomake with the oyle of Philosophers, named in latin *Oleum philosophorum*. And beware of Nuttes, Almons, Chese and mylke, and colde. And the pylles of Agarycke is good for this sycknes."—Fol. xx.

Boorde's treatment of Palsy.

"☞ Fyrst, vse a good dyet, and eate no contagious meates; and yf nede be, vse clysters, and anoynt the body with the oyles of Laury and Camomyll; but whether the Palsy be vniuersal or perticuler, I do anoynte the body with the oyle of Turpentine compounde with Aqua vite, and vse fricacio*ns* or rubbynges with the handes, as one wolde rub with grece an olde payre of Botes, not hurtynge the skyn nor the pacient. And I do gyue the pacient Treacle with the pouder of Peper, or els Mitridatum with Peper; or els take of Diatriapiperion. And if one wyll, he may rub the pacient with the rotes of Lylyes brayed or stamped; after that vse drye stuphes, as the pacient is able to abyde. Or els, take a Foxe, and with the skynne and all the body quartered, and with the herte, lyuer and lunges, and the fatnes of the intrayles, stones and kydnes, sethe it longe in runnynge water with Calamynt and Balme and Carawayes, and bath the pacient in the water of it; and the smell of a Foxe is good for the Palsy."—Fol. lxxxxi.

Wood-powder, Boorde's remedy for Excoriation.

"☞ Anoynt the place with *Vnguentum cerisinum*, or washe the place ofte with the water of Roch alome, and then caste vpon the place the pouder of a Poste; and if one wyll not washe the place with the water of Roche alome, washe the place then with white wyne, and vse the fyne pouder of a Poste, and there is nothynge wyll skyn so sone as it wyll do. Parauenture some persons readyng this

[1] lozenge.

boke, specially this mater, wyl laughe me to scorne; but for all that, for skynnynge of a place there is nothyng shal skyn so sone as it wyl do if it be vsed, excepte the place be to muche vlcerated, but for a mans yerd and other secrete places, I haue proued this pouder to be the most best."—Fol. xlix.

Boorde's remedy for Fatness, Fogeyness, or such lyke.

"☞ The best remedy that I do knowe is to vse purgacions, and with mete and potages of sewes is to eate muche Peper, and vse electuary of Lachar, and vse gargarices and sternutacions, as it is specified in the Chapitre named Ozinei."—Fol. lxxxxiii.

Boorde on Priapismus. a.

"☞ The .282. Chapitre dothe shewe of inuoluntary standynge of a mannes Yerd.

P*Riapismus* is the Greke worde.. In latin it is named *Erectio inuoluntaria virge.* In Englyshe it is named an inuoluntary standyng of a mans yerd.

The cause of this impediment.

This impediment doth come thorow calidite and inflacions from the raynes of the backe, or els it dothe come of inflacions of the vaynes in the yerde and stones; it may come by the vsage of venerious actes.

¶ A remedy.

☞ Fyrst, anoynt the yerde and coddes with the oyle of Iuneper; and the oyle Camphoric is good. And so is *Agnus castus* brayed, and made in a playster, and layde vpon the stones. and let prestes vse fastyng, watchynge, euyll fare, harde lodgynge, and greate study, and fle from al maner of occasions of Lechery, and let them smel to Rue, Vineger and Camphire."—Fol. lxxxxiii. back.

β. Erection of the yerde to synne. A remedy for that is to leape into a greate vessel of colde water, or to put Nettles in the codpeece about the yerde and stones. Fol. C.ii.[1]

Web in the Eye.

"+ In this matter there is .ii. wayes to make one whole. The first is by wyndynge or cuttyng awaye the webbe with an instrument. And the other is by a water to corrode & to eate away the webbe. it maye be remedied by the iuyce of Horehounde, Oculus Christi, and Diaserys, iniected into the eye, but I take only the iuyce of Horehonde; & the iuyce of Lycoryce iniected in the eye is very good."—Fol. lxxxxvii. back.

[1] See also the end of Chapter 77 on *Coitus*, Fol. xxxii.

Impediment in the Eye.

"I myghte here shewe of many salubriouse medecines, but the best medecine that I do knowe is to lette the matter alone, and medle nat with it, but were before the eyes a pece of blacke sarcenet, and eate neyther garlycke nor onyons, nor drynke no wynes nor stronge ale, and it wyll were awaye."—Fol. C.xxi. back.

Boorde on the Gut-caul.

"☞ The .384. Chapitre doth shewe of a Pannicle
the whiche shalbe rehersed.

Z*Irbus* is the latin worde. In Englyshe it is a pannycle or a caule compounde of ii. thyn tunicles of dyuers artoures, and vaynes and fatnesse; it doth couer the stomake and the guttes, and it doth kepe the heet of them, and doth defende the cold: this pellicle or pannycle or caule may be relaxed or broken.

The cause of this impedimente.

¶ This impediment doth come of some great strayne, brose, or fall, or some greate lyft, or suche lyke thynges.

¶ A remedy.

☞ Fyrst make incision, and after that cauterise the abstraction; and I haue sene the cut cauterised, that the fluxe of bloud shuld nat folowe. The ouerplus of my mynde in this matter, and all other matters, I do commyt it to the industry of wyse and expert Phisicions and Chierurgions."—Fol. C.xxiii.

For the sake of Chaucer's Somonour,

That hadde a fyr reed Cherubynnes face,
ffor *sawcefleem* he was, with eyen narwe.
(*Canterbury Tales*, Group A, § 1, ll. 624-5, Ellesmere MS, Chaucer Soc., p. 18)

I add Boorde's two chapters on the disease.

A Saucefleume Face.

"¶ The .170. Chapitre dothe shewe of a
saucefleume face.

G*Vtta rosacea* be the latin wordes. In Englyshe it is named a sauce fleume face, which is a rednes about the nose and the chekes, with small pymples: it is a preuye signe of leprousnes.

☞ The cause of this impediment.

¶ This impedyment doth come of euyl dyet, and a hote lyuer, or disorderynge a mans complexion in his youth, watchynge and syttynge vp late.

¶ A remedy.

Fyrst, kepe a good dyet in meates & drynkes, drynke no wyne, feade nat of freshe befe, eate no shell fyshes, beware of Samon & Eles, and egges, and qualyfie the heate of the Lyuer and the stomake with the confection of Acetose, and than take this oyntement: take of Bores grece .ii. vnces, of Sage pouned smal, an vnce and a halfe, of Quycke syluer mortified with fastynge spetyll, an vnce; compounde all this togyther, and mornynge and euenynge anoynte the face, & kepe the chamber .vii. dayes: or els, take of Burre rotes and of Affodyl rotes, of eyther .ii. vnces, of white vinegar .ii. vnces, of Auripigment .ii. drames, of Brymstone a drame; make pouder of al that, that shulde be made pouder of; than put al togyther, and let it stande .xxiiii. houres, and after that anoynte the nose and the face."—Fol. lx. back.

"☞ The .311. Chapitre dothe shewe of a Saucefleume face.

S*Alsum flegma* be the latin wordes. In Englyshe it is named a sausefleume face, whiche is a token or a preuy sygne of leprousnes.

☞ The cause of this infirmite.

This infirmytie doth come eyther of the calydytie or heate of the lyuer, or els of the malice of the stomake: it doth most comonly come of euyll dyet, and late drynkynge, and great surfetynge.

☞ A remedy.

¶ Take of Bores grece—the skyn and straynes clene pycked out—an vnce, of Sage fynely stamped an handfull, of Mercury mortified with fastyng spetyl, an vnce; incorporate al this togyther, and anoynt the face to bedward. In the mornyng wype the face with browne paper that is softe, and washe nat the face in .vi. or .vii. dayes, and kepe the pacient close out of the wynde."—Fol. C.i. back.

§ 43. ε. Fifthly, and lastly, let us see our author in his serious aspect.

" ¶ The .22. Chapitre doth shewe of the soule of man.

A*Nima* is the latin worde. In Greeke it is named *Psichae*. In Englyshe it is named the soule of manne. The soule of man is the lyfe of the bodye, for when the soule is departed from the body, the body is but a deade thynge that can not se, heare, nor feele. The soule can not be felte nor sene, for it is lyke the nature of an Angell, hauynge wyll, wyt, wysdome, reason, knowledge and vnderstandynge, And is partaker of good or euyll, as the bodye and it doth or hath deserued or operated. The soule also is a creature made with man and connexed to man, for man is of .ii. natures, which is to say, the nature of the soule, and the nature of the body, whiche is fleshe and bloud, the fleshe or body is palpyble and may be sene and felte. The soule is not palbyble nor can not be sene nor felt, but both beyng together nowe and shalbe after the generall resurrection in tyme to come, doth, and shal do, fele ioy or payne, &c.

It is not the soule onely doth make a man, nor the body of a man is a man, but soule and bodye connexed or ioyned together maketh a man. And the one decepered from the other be of .ii. natures as I haue sayd, vnto the tyme that they do mete againe at the day of dome. Ther fore let euery man in this lyfe so prouide by the meryte of Chrystes passion that soule and body beynge perfite man may enter into euerlastynge ioy and glory to be in heauen with God. The electuary of *Gemmis:* and the confection named *Alchermes* be good to comforte the soule or the spirites of man, soule and body beynge together here in earth."—Fol. xiiii. back.

"☞ The Apendex to all the premisses
that foloweth.

LOrdes, Ladies, and Gentylmen, learned and vnlerned, of what estate or degree so euer you be of, thynke not that no man can be holpen by no maner of medecynes, yf so be God do sende the sicknes; for he hath put a tyme to euery man, ouer the which tyme no man by no art nor science can not prolonge the time: for the nomber of the monthes and dayes of mans lyfe, God knoweth. But this aforesayde tyme, these monthes and dayes, a man may shorten or abreuiate many wayes, concerning that God hath geuen man in this lyfe free wyl, the whiche of his ryghteousnes, as longe as we do lyue, he can not take it awaye from vs. Nowe, we hauyng this free wyll, dyuers tymes we do not occupy it to the wyll of God, as it appereth, both for soule and body; we do kyll our soules as much as doth lye in vs, when that we do breake any of his commaundementes, or do synne deadly; for that matter he hath prouided a spirituall medecine, whiche is, repentaunce with penaunce. Also we do kyll our bodyes as much as lyeth in vs (excepte that a man do kyl hym selfe wylfully, as many dayly doth, contrary to Goddes wyll) as wel the one as the other, when a manne doth abreuiate his lyfe by surfettynge, by dronkennes, by pencyfulnes, by thought and care, by takynge the pockes with women, and leprousnes, and many other infectious sickenesses, beside robbyng, fyghtyng, kyllyng, and many other myschaunces, whiche is not Goddes wyl that such thinges shuld be done; but God, knowynge at the begynnyng of the creacion of the worlde, that man wolde be prone many wayes to abreuiate his lyfe, made then prouision that man might be holpen, by his grace, and then, the vertue the whiche he dyd gyue to herbes, wedes, trees, rootes, frutes, and stones. The propertie and vertue of the whiche, fewe men or none doth knowe them, except doctours of phisicke, and such as doth Labour to haue the knowledge of theyr operacions. And this knowledge notwithstandynge, let no man thynke that there is no Phisicion nor

Chierurgion can make a man sodenly whole of his infyrmytie, as Chryst and his disciples and manye other sayntes dyd; for they must haue leysure tyme and space as theyr lerning and practise is; for sycke men and women be lyke a pece of rustye harnys, the whiche can not be made bryght at the fyrst scourynge; but lette a man continewe in rubbynge and scourynge, and than the harnys wyll be bryghte; so in lyke maner a sycke man can not be made whole of his malady or syckenes the fyrst day, but he must continewe with his medecines. But here let euery man that is sycke, beware of blynd Phisicions and Chierurgions the which be ignoraunt, and can not tel what thynges doth parteyne to their science; and therfore let al men be ware of vagabundes and ronagates that wyl smatter with Physicke, for by such persons many sycke men haue ben deceyued, the more pytie, God knoweth! who helpe vs al nowe and euer! Amen!"

"¶ A Preamble to sicke men and to those that be wounded.

I Do aduertise euery sicke man, and al other men the which hath any infirmitie, sickenes, or impediment, aboue all thynges to pacyfye hym selfe, or to arme hym selfe with pacyence, and to fyxe his harte and mynde in Christes death and passion, and to call to his remembrance, what peynes, what aduersyte, and what penury, and pouerty Chryst dyd suffer for vs. And he that can thus pacyfy him selfe, and fele his owne peyne in Chrystes passyon, shall mittigate his peynes and anguyshe, be it neuer so greate. And therefore let euery sycke person stycke as fast to Christe in his peynes and sickenes, as Christ dyd stycke fast to the Crosse for our sinnes and redempcion. And then if the pacient wyl haue any councell in Phisicke: fyrste let hym call to him his spirituall Phisicion, which is his goostly father, and let him make his conscience cleane, and that he be in perfyte loue and charitie; and yf he haue done any wronge, let him make restitucion yf he can; and yf he be in dette, let him loke to it, and make a formal wyl or testament, settyng euery thynge in a dewe order for the welth of his soule,—wyse men be sure of theyr testamentes makynge many yeres before they dye, and dothe renewe it once a yere as they increase or decrease in gooddes or substance.—All these aforesayd thynges goostly and godly prouided for the soule, Then let the pacient prouyde for his body, and take councel of some expert phisicion, howe & in what wyse the body may be recouered of his infirmitie, and than to commyt his body to the industry of his Phisicion, and at al tymes redy to folow the wil, mynde, and councell of his Phisicion, for who so euer wyll do the contrary, saynt Augustine sayth, *Seipsum interimit qui precepta medici obseruare non vult*, that is to saye, He doth kyll hym selfe that doth not obserue the commaundement of his Phisition."

(The reader should now turn to the *Hindwords*, p. 317.)

§ 44. If any one groans over the length of these extracts, he can relieve himself by skipping them, and losing the chance of knowing Boorde well. But if he reads them all through, as well as the books following, I think he'll find Andrew Boorde worth knowing, a man at times of great seriousness and earnestness, yet withal of a pleasant humour; reproving his countrymen's vices, and ridiculing their follies; exhorting them to prepare for their latter end, and yet to enliven their present days by honest mirth. A man eager to search out and know the truth of things, restless in that search, wandering far and often to see for himself. Yet a man bound by many of the superstitions of his time, though also free from many; not "a lewd Popish hypocrite and ungratious priest," as Harrison calls him, but a man genuine in his piety as well as his love of good ale and wine, and mirth; clever, able to take-in a Scotchman; at times weak and versatile, showing off occasionally, readily helping strangers, chancing to get drunk, falling into sexual excess—having, like his sex, "bursts of great heart and slips in sensual mire,"—yet sound at the core, a pleasant companion in many of England's most memorable days, worthy, with all his faults, of respect and regard from our Victorian time. Any one who would make him a mere Merry-Andrew, or more of that than anything else, is a bigger fool than he would make Boorde. (See the *Hindwords*, p. 317.)

§ 45. That Boorde and his writings were esteemed by his contemporaries, we have seen, by his appointment as Suffragan Bishop of Chichester, his attendance on Sir Robert Drury and the Duke of Norfolk, his waiting on Henry VIII, his connection with Cromwell, Barnes's account of great people resorting to him, the evident references to his books in Wilson's *Rhetorique* (p. 116, below), "doctor Boords breuiary of health" being in Captain Cox's Library,[1] and Harrison's mention of the *Introduction of Knowledge*, and of the *Dyetary* (if 'parks' mean 'pleasure for harte & hynde, &c.'):—

"An Englishman, indeuoring sometime to write of our attire, made sundrie platformes for his purpose, supposing by some of them to find out one stedfast ground whereon to build the summe of his

[1] It's the last in the list of the Captain's books. See p. 30 of my edition of *Captain Cox*, or *Laneham's Letter*, for the Ballad Society, 1870.

discourse. But in the end (like an oratour long without exercise) when he saw what a difficult peece of work he had taken in hand, he gaue ouer his trauell, and onelie drue the picture of a naked man, vnto whome he gaue a paire of sheares in the one hand, and a peece of cloth in the other, to the end he shuld shape his apparell after such fashion as himselfe liked, sith he could find no kind of garment that could please him anie while togither, and this he called an Eng-
Andrew Boord lishman. Certes this writer (otherwise being a lewd [[1] popish hypocrite] and vngratious priest) shewed himself herein not to be [altogether] void of iudgement, sith the phantasticall follie of our nation, [euen from the courtier to the carter] is such, that no forme of apparell liketh vs longer than the first garment is in the wearing, if it continue so long and be not laid aside, to receiue some other trinket newlie deuised by the fickle-headed tailors, who couet to haue seuerall trickes in cutting, thereby to draw fond customers to more expense of monie . . . the Morisco gownes, the Barbarian sleeues, [the mandilion worne to Collie weston ward, and the short French breches] make such a comelie vesture, that *except it were a dog in a doublet*, you shall not see anie so disguised, as are my countrie-men of England."—*Harrison's Description of England*, ed. 1586, p. 171-2.

"these daies, wherein Andrew Boorde saith there are more parks in England than in all Europe (ouer which he trauelled in his owne person)," *ib.* p. 205, col. 2. See below, p. 274.

Traditions of Boorde linger in Sussex,[2] whose anti-nightingale forest of St Leonards, its keepers and nigh-dwellers he knew,[3] and the Sussex Archæological Society has revived the memory of him in our day. Though Warton thought that his *Dyetary* was the only work that would interest posterity, yet Upcott's reprint of his *Introduction* showed that that book too had plenty of amusement and information in it (see p. 36, above), while the present volume testifies to the value of both works, as well as that of the *Breuyary*, which contains some of his most characteristic passages, and will, I hope, soon find an antiquarian doctor as an editor.

§ 46. The present reprint of the *Fyrst Boke of the Introduction of Knowledge* is made, as I have said at p. 19, from Mr Christie-Miller's unique copy of William Coplande's first edition printed at the Rose-Garland in Fleet Street in 1547 or -8, collated with his second of 1562 or -3, printed in Lothbury. My thanks are due 1. to

[1] The square brackets [ ] show the new matter inserted in the 2nd edition of 1586. [2] M. A. Lower, in *Sussex Archæol. Collections*, vi.

[3] *Introduction*, p. 121.

Mr Christie-Miller for his kindness and hospitality to Mr Hooper and myself; and 2. to the Committee of the Chetham Library, and their Librarian, Mr Jones, for lending me their very rare Lothbury volume, and enabling Mr W. H. Hooper to copy all the cuts in it, of which Upcott had only a few copied. The reader will see that the same cut often serves for men of different countries. Mr Hooper says:

"A Man with a hawk, and a Peasant with long-handled bill over his shoulder, are used, Chap. 6, p. 143, in the Lothbury edition (B) for 'Norway and Islonde,' Ch. 8, p. 146; both in A (the Rose-Garland edition) and B, for 'Flaunders,' changing places right and left; and the hawker appears again at Ch. 14, 'high Almayne,' in both A and B.

A dinner party illustrates Ch. 9, p. 148, 'Selande and Holand,' and Ch. 13, p. 155, 'base Almayne,' in both A and B.

A man with a cloak very jauntily thrown over his shoulder represents in B, Ch. 16, p. 165, 'Saxony;' Ch. 30, p. 198, 'Spaine;' Ch. 33, p. 206, 'Bion;' and Ch. 38, p. 217, 'Egypt.'

† A bearded man in a skull-cap and long coat, Ch. 19, p. 170, is 'Hungary,' and Ch. 26, p. 188, a Genoese; at Ch. 19, p. 170, he is in company with a bird in a tree that appears at Ch. 15 as a production of 'Denmarke.'

A turbaned figure, half-length, is in both A and B, as, Ch. 20, p. 171, 'Greece;' Ch. 23, p. 175, 'Italy;' and Ch. 24, p. 181, 'Venis;' with two little groups in this last instance.

A crowned head, half-length, stands in B for (Ch. 21) 'Sicell;' Ch. 28, p. 194, 'Catalony;' Ch. 31, p. 199, 'Castile & biscay;' Ch. 32, p. 202, 'Nauer;' while in A, two cuts do duty for the four countries.

A grave and learned individual in a long robe stands alone, Ch. 25, for 'Lombardye,' p. 186; and at Ch. 35, p. 209, he enacts 'The latyn man' so well that the 'englyshma*n*' takes off his hat to him.

† The foresaid long-coated man in Ch. 19 and 26 is very like the man labelled Dr Boorde in *Barnes in the defence of the Berde;* so like that I think it is hardly worth while to cut another.

The cuts for this book seem to have been got together from all quarters. The Englishman in the first chapter may have been cut for the work: there is a bluff King-Hal sort of a look about him that suggests the period.[1] But the Irishman is so knocked about that it is certain he is 'written up to,'[2] as the publishers have it now-a-days. They look to me an odd lot in every sense of the word; for some seem printed from the wood, while others are from *casts*, e.g. the Scot is bruised at the edges, and the ends of the ground-lines are thickened, just as old 'stereos' wear. Some of the blocks seem

[1] The cut of the Frenchman, p. 190, seems to me of the period too.—F.

[2] No! The Irishman's parasites were well known.—F.

to be much older than the date of the book, as they are wormed, and damaged by use."

On turning to Wynkyn de Worde's print of *Hyckescorner*, for my edition of Laneham, I found, on the back of the title, two of our *Introduction* cuts. The man who in the Lothbury edition does duty for Saxony, Spain, Bayonne, and Egypt, p. 165, 206, &c., figures in *Hyckescorner* as "Imagyna[cyon]," while the long-coated man used by Copland for the Hungarian (p. 170), and the Genoese (p. 188), and by Wyer for Boorde (p. 305), is Wynkyn de Worde's "Pyte." In *The Enterlude of Youth*, printed by William Coplande at Lothbury (after the Rose-Garland *Introduction*), Boorde's Dane, p. 162, is used for "Humility" (though he has no name over his head); and Boorde's Bohemian, p. 166, is used for "Youth."

In like manner the cut used for Andrew Boorde himself[1], *Introduction*, Ch. VII, p. 143, below, is merely an old cut of some one else, with a corner cut out, and Boorde's name let down into it; a fact obscured by Upcott's woodcutter, who evidently thought the break in the top line ugly, and so filled it up. This "portrait (as is well observed by Herbert, in his MS memoranda) is introduced for one of Skelton in the frontispiece to 'Certaine bookes compiled by maister Skelton, Poet Laureat, printed by Kynge and Marshe.'"—*Ames* (ed. Dibdin, 1816), iii. 160. Many of the Boorde cuts are used in the titleless copy of the Shepherd's Kalendar in the British Museum, which I claim as Copland's (p. 25, above); and most have, no doubt, an earlier continental history. That on p. 208 is part of Wynkyn de Worde's 'Robert the Deuyll.'

Again, the 2-men cut of Galen and another man in Boorde's *Dyetary*, p. 232, below, is used on the title-page of a little tract in 4 leaves in the British Museum, "Imprynted by me Rycharde Banckes," and called "The practyse of Cyrurgyons of Mountpyller: and of other that neuer came there." It is chiefly on the treatment of skull-wounds.

[1] The cut on the title-page of the *Introduction*, which Mr W. C. Hazlitt calls one 'of two serving-men conversing,' is stated by him to have been copied on the title-page of '*The doctrynall of good seruauntes.* Imprynted at London in Flete strete, at the sygne of Saynt Johan Euangelyste, by me Johan Butler [*circa* 1550] 4to. 4 leaves. In verse.' Dr Rimbault re-edited this tract for the Percy Society. The cut is also in *Frederyke of Jennen.*

To our member, Mr Henry Hucks Gibbs,—an old friend and helper of Herbert Coleridge and myself in our Dictionary work since 1858,—I am indebted for the ready loan of his copy—unique, so far as I know—of the 1542 edition of Boorde's *Dyetary* from which the reprint in the present volume is taken. It has been collated with the undated edition by Robert Wyer in the British Museum, and also with the edition of 1547 (colophon 1567) by Wyllyam Powell. Mr W. H. Hooper has copied the cuts for this tract too, and wishes to call attention to the two of St John at the end of it and on the title-page. That on the title is evidently from a cast of the block of that in the colophon, which cast has been cut down, and had another ornament put at the side of it, with a line atop, just as Mr Hooper has made the facsimile now. Mr Hooper has further evidence which proves clearly to him as a woodcutter, that our old printers in the 16th century could cast, and used casts, as we do, though of course to a less extent.

Of the big initial letters used in the *Dyetary*, Mr Hooper has cut all but five, of which he thought the designs much less good than those he has cut, and one extra-big A of the same pattern as the smaller one used on page 234, &c., below, which latter he has copied. The only other alterations in the text are, that the contractions have been expanded in italics according to our rule,—ā as a*n*, y[t] as *that*, &c.,—and that the first letters of proper names, and the stops, have been conformed to modern usage.

§ 47. For all the materials of these Forewords I am indebted to Boorde's own books, and to the workers who have preceded me in the field, Wood, Bliss, Ellis, Lower, Cooper, Rimbault, Hazlitt, &c. To the latter I feel grateful, though I have expressed freely some of my differences from them. My task has been only to get to their authorities, keep to these without straggling into guesses, and work into them Boorde's own statements in his different books. The number of supposes and probables is still lamentably great; I hope they will be lessened by the future volumes of Professor Brewer's admirable Calendar, or some other antiquarian publication of this age, which is setting itself, with more or less vigour, to get at all the facts it can about the men and speech of Early and Middle England.

The notes I have added would have been longer and better, had I been at home among my books, but this, and divers other bits of work, have dawdled on during our four-months' stay here, from the time when I began to write in the garden, with the lovely lilacs round me, and the hum of bees, till all the roses have gone, and the fresh green of the grass is brown. Games with my boy, long walks with my wife under "the glad light green" of Windsor-Park beeches lit by the golden sun, strolls down the long Rhododendron-Walk with its glorious masses of mauve towering high on either hand, over Runnymede, starred with wild flowers, canopied with sunsets of wondrous hue; rows on the Thames, dotted with snowy swans sailing over the ever-varying green of water-plants; gaily-coloured races at Ascot, picnic at the truly-named Belvedere; drives, visits, dances—oh fair-haired Alice, how well you waltz!—chats, pleasant outdoor country-life: who can work in the midst of it all? I can't.

And now comes the angry roar of war to trouble one's sweet content, to make one feel it wrong almost to think of private pleasure or Society's work. What interest can one take in printers' dates, or Boorde's allusions, when the furious waves of French vain-glory, driven by the guilty ambition of a conscienceless adventurer, are dashing against the barriers of German patriotism, striving to deluge thousands of innocent homes in blood?—May this Napoleon and his followers be humbled to the dust!—Still, the Forewords, &c., take up one-third of this book, and that is a fair share for an editor to fill. A great number of most troublesome little points have started up in the course of the work, and my ignorance of monastic rule, Continental countries, coins, languages, medicine, and botany, has made me leave many of these points to future students of the book to settle. I hope, however, that Andrew Boorde will be henceforth better known to English readers than heretofore, and only regret that some of the mirth he loved so well, has not crept into these foregoing pages, through all the bright sights and sweet sounds that have been before and around me while this work has been going on. But one does not get lighter-hearted as one gets older, alas!

Walnut-Tree Cottage, Egham,
July 30, 1870.

¶ The fyrst boke of the

Introduction of knowledge. The whych

dothe teache a man to speake parte of all maner of
languages, and to know the vsage and fashion of
all maner of countreys. And for to know the
moste parte of all maner of coynes of money, the whych is currant in euery region.
Made by Andrew Borde, of Phy-
sycke Doctor. Dedycated to
the right honorable & gracious lady Mary dough-
ter of our souerayne
Lorde kyng Henry
the eyght.

✠

¶ To the ryght honorable and gracyous lady Mary doughter of our souerayne Lorde kyng Henry the .viii. Andrew borde of phisyk doctor, doth surrender humble commendacion wyth honour and helth.

AFter that I had dwelt (moste gracyous Lady) in Scotlande, and had trauayled thorow and round about all the regions of Christynte, & dwelling in Mou*n*tpyler,[1] remembryng your bountyful goodnes, pretended to make thys first booke, named "the Introduction of knowledge" to your grace, the whyche boke dothe teache a man to speake parte of al maner of languages; and by it one maye knowe the vsage and fashyon of all maner of countres or regions, and also to know the moste part of all maner of coynes of mony, that whych is currant in euery prouince or region; trustyng that your grace will accept my good wyll and dylygent labour in Chryste, who kepe your grace in health and honour. Fro Mountpyler the .iii. daye of Maye, the yere of our Lorde .M.CCCCC.xlii.

¶ The Table of thys booke foloweth.

THe fyrst chapter treateth of the naturall disposicyon of an Englyshman, and of the noble realm of England, and of the mony that there is vsyd. [And of Cornwall, p. 122] (p. 116)

The seconde chapter treateth of the naturall dysposycion of Walshmen, and of the countre of Wales, teching an Englyshe man to speake some Walshe. (p. 125)

The thyrd chapter treateth of the naturall dysposicio*n* of an Irysh man, and of the kyngdomeshyp of Irland, and also teachyng an Englyshe man to speake some Irysh, and of theyr mony. (p. 131)

[1] Contractions in the original are expanded here in italics, as '*that*' for 'y^t;' capitals are put to some proper names; foreign words are printed in italics; modern stops are put, and hyphens.

[1] The fourthe chapter treateth of the naturall disposycyon of a Scotyshe man, and of the Kingdom of Scotland, and the speche of Scotland, and of their mony. (p. 135)

The .v. chapter treateth of Shotlande[2] and of Fryselond, and of the naturall dysposycion of the people of the countreys, and of[3] theyr money. (p. 139)

The .vi. chapter treateth of Norway & of Islond, and of the[4] naturall disposycion of the people of the countreys, and of theyr speche, and of theyr money. (p. 140)

The .vii. chapter treateth of the Auctor, *the*[5] which went thorow and rounde about Christendome; and what payne he dyd take to do other men pleasure. (p. 143)

The .viii. chapter treateth of Flaunders, and of the naturall disposicion[6] of Fleminges, and of their money, and of[7] theyr speche. (p. 146)

The .ix. chapter treateth of Seland & Holand, & of the natural disposicion of the people, & of theyr spech, and of their money. (p. 148)

The .x. chapter treateth of Braban, & of the naturall disposicion of Brabanders, & of their money & speche. (p. 150)

The .xi. chapter treateth of Gelderla*n*d and of Cleuela*n*d, and of the natural disposicion of the people of that[8] cou*n*treys, and of[9] their money and speche. (p. 152)

The .xii. chapter treateth of Gulik & Lewke,[10] & of the naturall disposycion[6] of the people of the[8] countreys, and of their money, and of their speche. (p. 155)

The .xiii. chapter treateth of base Almayn, and of the natural disposicion of the people of that countrey, and of[7] theyr money, and of[7] theyr speche. (p. 155)

The .xiiii. Chapter treateth of high Almayn, & of the naturall disposicion of the people of that cou*n*trey, and of[7] theyr mony, and of their spech.[11] (p. 159)

[1] sign. A .ii. [2] Scotlande A; Soctlande B.
[3] A has only "of;" B only "and." [4] theyr AB. [5] of Auctor yͤ AB.
[6] dispocion A; a mistake made 4 or 5 times more. [7] B leaves out "of."
[8] for "those." [9] B leaves out "and of."
[10] Julich or Juliers (the town is between Aix and Cologne) and Liège.
[11] and speche B.

The .xv. chapter treateth of Denmarke, and of the[1] na[2]turall disposicion of the people of the countrey, and of the money and speche. (p. 162)

The .xvi. chap. treateth of Saxsony, & of the natural disposicion of *the* Saxons, & of their money, & of their spech. (p. 164)

The .xvii. chapter treateth of the kingdom of Boem, and of the disposicion of the people of the countrey, and of theyr money, and of their speche. (p. 166)

The .xviii. chapter treateth of the kingdom of Poll, & of the naturall disposicion of the people of the countre, & of theyr mony, and of theyr speche. (p. 168)

The .xix. chapter treateth of the kingdome of Hungry, and of the natural disposicion of the people of theyr cou*n*trey, and of theyr money, and of their speche. (p. 170)

The .xx. chapter treateth of the land of Grece, & of Co*n*stantinnople, and of the natural disposicion of the people of the countrey, and of theyr mony and speche. (p. 171)

The .xxi. chapter treateth of the kyngdom of Sycel & of Calabry, and of the disposicion of the people of the countrey, and of theyr mony and speche. (p. 175)

The .xxii. chapter treateth of the kingdom of Naples, and of the disposicion of the people of the countrey, and of theyr money and speche. (p. 176)

The .xxiii. chapter treateth of Italy and of Rome, and of the disposicion of the people of the countrey, and of theyr money, and of theyr speche. (p. 177)

The .xxiiii. chapter treateth of Venys, & of the disposicion of the people of *the* cou*n*trey, & of[3] their money & spech. (p. 181)

The .xxv. chapter treateth of Lombardy, & of *the* natural disposicions of the people of the countrey, & of theyr money, and of theyr speche. (p. 186)

The .xxvi. chapter treateth of Ieene and of the Ieneueys,[4] and of theyr spech, and of theyr money. (p. 188)

The .xxvii. chapter treateth of Fraunce, and of other [5]prouinces

[1] that AB. [2] sign. A .ii. back. [3] B leaves out "& of."
[4] Genoa and the Genoese. [5] A .iii. not signed.

the which be vnder Fraunce, and of the disposicion of the people, and of their mony and speche. (p. 190)

The .xxviii. chapter treateth of[1] Catalony, and of the kyngdom of Aragon, and of the disposicion of the people, and of theyr money, and of theyr speche. (p. 194)

The .xxix. chapter treateth of Andalosye, and of the kingdome of Portingale, and of the dysposicion of the people, and of theyr speche, and of theyr money. (p. 196)

The .xxx. chapter treateth of Spayne, & of the disposycion of a Spayneard, and of the[2] money and of the[2] speche. (p. 198)

The .xxxi. chapter treateth of the kyngdome of Castel[3] and of Byscaye[4], and of the dysposycion of the [5]people of that countrey, and of[5] theyr money and spech. (p. 199)

The .xxxii. chapter treateth of the kyngdome of Nauer, and of the disposicion of the people, and of[6] theyr money and theyr speche. (p. 202)

The .xxxiii. chapter treateth of Bayon, and Gascoyn, and of lytle Britayn, and of the disposici*on* of the people of those countreys, and of theyr mony and of[6] their spech. (p. 206)

The .xxxiiii. chapter treateth of Norma*n*dy & Picardy; of the disposicion of the people, & of their money & spech. (p. 208)

The .xxxv. chapter treateth of the Latyn man and of the Englysh man, and where Latine is most vsed. (p. 209)

The .xxxvi. chapter treateth of Barbari, and of the blake Mores, and of[6] Moryske speche. (p. 212)

The .xxxvii. chapter treateth of Turkey, & of the Turkes, and of their money and of[6] their speche. (p. 214)

The .xxxviii. chapter treateth of Egypt, and of the Egypciens, & of[6] their speche. (p. 217)

The .xxxix. chapter treateth of Iury and of the Iues, and of[6] their speche. (p. 218)

¶ Thus endeth the table.

[1] B leaves out "of." [2] and their B. [3] Castle B (Castille).
[4] Bascaye H. [5—5] people and B. [6] B leaves out "and of."

[1]

¶ The fyrst chapter treateth of the naturall dysposicion of an Englyshman, and of the noble realme of England, & of the money that there is vsed.

I'm naked, as I can't settle what to wear.

¶ I am an English man, and naked I stand here,
Musyng in my mynde what rayment I shal were;
For now I wyll were thys, and now I wyl were that;
Now I wyl were I cannot tel what.

I like new fashions.

All new fashyons be plesaunt to me;
I wyl haue them, whether I thryue or thee.[2]

[1] A .iii. back.

[2] See chapter xxii. below, p. 177. The Neapolitan says: "Al new fashyons to Englond I do bequeue." Wilson, speaking of books, says: "And not onely are matters set out by description, but men are painted out in their colours, yea, buildynges are set forthe, Kingdomes and Realmes are portreed, places & times

Now I am a frysker, all men doth on me looke;
What should I do, but set cocke on the hoope?
What do I care, yf all the worlde me fayle?
I wyll get a garment, shal reche to my tayle;
Than I am a minion, for I were the new gyse.
[1]The next[2] yere after this I trust to be wyse,
Not only in wering my gorgious aray,
For I wyl go to learnyng a hoole somers day;[3]
I wyll learne Latyne, Hebrew, Greeke and Frenche,
And I wyl learne Douche, sittyng on my benche.
I do feare no man; all men feryth me;
I ouercome my aduersaries by land and by see;
I had no peere, yf to my selfe I were trew;
Bycause I am not so, dyuers times I do rew.
Yet I lake nothyng, I haue all thynge at wyll;
Yf I were wyse, and wolde holde my self styl,
And medel wyth no matters not[4] to me partayning,
But euer to be trew to God and [to] my kynge.[5]
But I haue suche matters rolling in my pate,
That I wyl speake and do, I cannot tell what;
No man shall let me, but I wyl haue my mynde,
And to father, mother, and freende, I wyl be vnkynde;
I wyll folow myne owne mynd and myn old trade;
Who shal let me, the deuyls nayles vnpared?
Yet aboue al thinges, new fashions I loue well,
And to were them, my thryft I wyl sell.
In all this worlde, I shall haue but a time;
Holde the cuppe, good felow, here is thyne and myne!

I'll get a garment to reach to my tail.

Next year I'll take to learning.

All men fear me.

I lack nothing.

I *will* do as I like.

Who'll stop me?

I *do* love new fashions.

are described. The Englishman for feeding and chaunging for (*sic*) apparell: The Dutchman for drinking: The Frenchman for pride & inconstance: The Spanyard for nimblenes of body, and much disdaine: the Italian for great wit and pollicie: the Scottes for boldnesse, and the Boeme for stubbornnesse."—1553. Wilson's *Art of Rhetorique*, edit. 1584, fol. 181-2.—W. C. Hazlitt.

[1] A .iiii. not signed. [2] B leaves out "next."

[3] See note [1], next page. [4] A leaves out B's "not."

[5] B leaves out this line: because of the "kynge," I suppose, as Queen Elizabeth was reigning in 1562 and 1563.

¶ The Auctor respondith.

Englishmen! strive for learning, and stop swearing;

¶ O good Englyshe-man, here what I shall say:
Study to haue learnyng,[1] with vertue, night and day;
Leue thy swearyng, and set pryde a syde,
And cal thou for grace, that with thee it may byde;
Than shall al nacions, example of the[2] take,
That thou hast subdued syn, for Iesus Christes sake.
And werkes of mercy, and charyte, do thou vse;
And al vyces and syn, vtterly refuse;

then all countries will come to you to learn the truth.

Than al countreys a confluence wyl haue to thee,
[3] To haue knowledge of trueth and of the veryte,
Of lernyng of Englyshe, of maners also.
Iesus I beseche, to kepe thee from all wo,
And send thee euer fortune, and also much grace,
That in heauen thou mayst haue a restyng place.

Is our land good, our people bad? No.

Englishmen are as good as any men;

and English lands, there's none like.

But no corn should be exported.

¶ The Italyen and the Lombarde say, *Anglia terra—bona terra, mala gent*. That is to say, "the land of England is a good land, but the people be yl." But I say, as I doo know, the people of England be as good as any people in any other lande and nacion *that* euer I haue trauayled in, yea, and much more better in many thynges, specially in maners & manhod. as for the noble fartyle countrey of England, hath no regyon lyke it; for there is plentye of Gold & Siluer. For Gold, Siluer, Tin, Lead & Yron, doth grow there. Also there is plenty of fisshe, flesshe and wylde foule, and copiousnes of woll & cloth. And if they wold kepe their corne within their realme, they had ynough to finde themself without scarcite, & of a low price. Though they haue no wines growing within the realme—*the* which they might haue yf they would,—yet there is no

[1] On the contempt for learning in England in Henry VIII's time, see the Forewords to the *Babees Book*, p. xii-xiv, the Additions to it of 1869, the Preface to *Quene Elizabethes Achademy*, &c. p. ix, x, and Starkey's *Dialogue on England in Henry VIII's Time*, E. E. T. Soc. 1870, p. 182-6, &c. On the Swearing in England, see p. 82-3 above.   [2] thee B.   [3] A .iiii. back.

realme *that* hath so many sortes of wines as they. The region is of such fertilite *that* they of the countrey nede not of other regions to helpe the*m*. Englishme*n* be bolde, stro*n*g, & mighty; the wome*n* be ful of bewty, & they be decked gayly. They fare su*m*ptiously: God is serued in their churches deuoutli; but treason & deceyt among the*m* is vsed craftyly, *th*e more pitie; for yf they were true wythin the*m*selfs, thei nede not to feare although al nacio*n*s were set agai*n*st the*m*; specialli now, co*n*syderi*n*g our noble prynce hath, & dayly dothe[1] make noble defe*n*ces, as castels, bulwarkes, & blokhouses, so *that*, almost, his grace hath munited, & in maner walled England rounde aboute, for *th*e sauegard of the realme, so that the poore subiectes may slepe and wake in saufegard, doing theyr busines without parturbaunce.

Englishmen are mighty; English women full of beauty.

But treason is in the land.

Were we true to ourselves, we need fear none.

Our King builds castles too.

[2] ¶ In England there be manye noble Cities and townes, Amonges *th*e whyche the noble citie of London precelleth al other, not onely of that region, but of all other regyons; for there is not Constantynople, Venis, Rome, Florence, Paris, nor Colyn, can not be compared to Londo*n*, the qualities and the quantite consydred in al thynges. And as for the ordre of the citie in maners, and good fashyons, & curtasy, it excelleth al other cities and townes. And there is suche a brydge of pulcritudnes, that in all the worlde there is none lyke.[3] In Englande is a metropolytane, the whych is a patriarke; and ther be now but few; for there was a patriarke of Ierusalem, ther is a patryarke at Constantinople, & there is a patryarke at[4] Venis; but al these aforesayde patriarkes hath not, one for one, so many bysshops vnder them as the patriarke or metrapolytan

The noble city of London excels all others; and

its bridge is the fairest in the world.

The Metropolitan of England is a Patriarch,

with more bishops than any other.

[1] ? this applies rather to 1542 than 1547. See *Notes*. Boorde notices that 7 castles were built, and 5 renewed by Henry.—*Forewords*, p. 23, near the foot.

[2] sign. B .i.

[3] This bridge was the first stone London Bridge, begun by Peter of Colechurch, A.D. 1176, finished in 1209, and which lasted till the New Bridge was built in 1825. For many centuries it was the wonder of Europe.—*Chronicles of London Bridge*, 2nd ed. 1839.

[4] A leaves out B's "at."

Universities. Oxford and Cambridge. Ports and Havens. The speeches spoken in England: French, Welsh, Cornish, Irish, Northern or Scottish; and all kinds by aliens. The wonders of England: hot baths at Bath; salt wells; Stonehenge;

of England. In England is the thyrd au*n*tyke[1] vniuersite of the worlde, named Oxford. And there is another noble vniuersitie called Cambrige. There is also in Englande more nobiler[2] portes and hauens tha*n* in any other region; there is Sandwiche, Douer, Rye, Wynchelse, Hastynges, Pemsey, Bryght-Hemston,[3] Arndel, Chychester, Porche mouthe, Southhampton, Dartmouth, Exmouth, and Plommouth. I do not recone no hauens nor portes betwixt Cornewall, Deynshire, and Wales, but beyond Cornewal and Wales, as saynt Dauys, Carnaruan, Umarys,[4] Abarde,[5] Cornewal, Weschester, Cokersend, and Cokermouth, Carlel, Barwyke, Newcastell, Bryllyngtone, Hull, Bostowe, Lyn, Yermouthe, and Harwyche, and dyuers other portes and hauyns, long to reherse. ¶ In Englande, and vnder the dominion of Engla*n*d, be many sondry speches beside Englyshe: there is Frenche vsed in Engla*n*d, specyally at Calys, Gersey, and Jersey: In Englande, the [6]Walshe tongue is in Wales, The Cornyshe tongue in Cornewall, and Iryshe in Irlande, and Frenche in the Englysshe pale. There is also the Northen tongue, the whyche is trew Scotysshe; and the Scottes tongue is the Northen tongue. Furthermore, in England is vsed all maner of languages and speches of alyens in diuers Cities and Townes, specyally in London by the Sea syde. ¶ Also in England be manye wonderfull thynges: Fyrst, there is at Baath certayne waters, the whyche be euer hote or warme, and neuer colde; wynter & Somer, they be euer at a temperat heate. In wynter the poore people doth go into the water to kepe themself warme, and to get them a heate. ¶ In England be salt wel waters; of the whych waters, Salte is made. ¶ Vpon the playn of Salysbury is the stonege, whyche is certayne

[1] ancientest. [2] noble B. [3] Bryght, He*m*ston A; Brighthelmstone or Brighton.
[4] ? Beaumaris, on the east coast of Anglesey.
[5] ? Aberystwith, on the west coast of Cardiganshire, or Aberffraw, west coast of Anglesey, &c. [6] sign. B .i. back.

great stones, some standyng, and some lyenge ouer-thawart, lyeng and hangyng, that no Gemetricion can set them as they do hange. And although they stande many a hondred yeares, hauyng no reparacion nor no solidacio*n* of morter, yet there is no wynde nor wether that doth hurte or peryshe them. Men say that Marlyn brought to that place the sayd stones by the deuels helpe & crafte. (Merlin built Stonehenge.)

¶ In the Forest of saynt Leonardes in Southsex there dothe neuer synge Nightyngale; althoughe the Forest rounde aboute in tyme of the yeare is replenysshed wyth Nightyngales, they wyl syng rounde aboute the Forest, and neuer within the precyncte of the Forest, as dyuers kepers of the Forest, and other credible parsons dwellyng there, dyd shew me. A forest, St Leonard's, that no nightingale will sing in.

¶ In dyuers places in England there is wood the which doth turne into stone. ¶ The kynges of England, by *the* power that God hath gyuen to th*em*, dothe make sicke me*n* whole of a sycknes called the kynges euyll.[1] ¶ The [2]Kynges of Englande doth halowe euery yere Crampe rynges,[3] *the* whyche rynges, worne on ones fynger, dothe helpe them the whyche hath the Crampe. Wood that turns into stone. Cramp-Rings hallowed by our Kings.

¶ There is no regyon nor countrey in al the world that theyr money is onely gold & syluer, but only Englande; for in England all theyr money is golde & syluer. There Golde is fyne and good, specyally the souerayns, the Ryals, and the halfe Ryals; the olde noble, the Aungels and the halfe aungels, is fyne golde. But the nobles of twenty grotes, and the crownes and the halfe crownes of Englande, be not so fyne Golde as the other is. Also Golde of other regyons, and some Syluer, yf it be good, doth go in England. The syluer of England is Grotes, halfe grotes, Pens, halfe pens, and there be some Fardynges. ¶ In England doth grow golde, and England's the only country with only gold and silver money. Our gold coins. Our silver coins. Our mines.

[1] See *The Breuyary of Health*, fol. lxx, and *Forewords*, p. 91-93 above.
[2] sign. B .ii. [3] See the *Forewords*, p. 91-2.

Mines in England. English speech is base.

Syluer, Tyn, Leade, and Irone. ¶ The speche of Englande is a base speche to other noble speches, as Italion, Castylion, and Frenche; howbeit the speche of Englande of late dayes is amended.[1]

¶ The apendex to the fyrst Chapter, treatinge of Cornewall, and Cornyshe men.

I can brew beastly beer

¶ Iche cham a Cornyshe man, al[e] che can brew;
It wyll make one to kacke, also to spew;
It is dycke and smoky, and also it is dyn;
It is lyke wash, as pygges had wrestled dryn.[2] 4

like hogwash.

Iche cannot brew, nor dresse Fleshe, nor vyshe;
Many volke do segge, I mar many a good dyshe.
Dup the dore, gos[3]! iche hab some dyng to seg, 7
'Whan olde knaues be dead, yonge knaues be fleg."

I'm very hungry!

Iche chaym yll afyngred,[4] iche swere by my fay
Iche nys not eate no soole[5] sens yester daye;
[6] Iche wolde fayne taale ons myd the cup;
Nym me a quart of ale, that iche may it of sup. 12

give me a quart of ale. I've fish and tin,

A, good gosse, iche hab a toome,[7] vyshe, and also tyn;
Drynke, gosse, to me, or els iche chyl begyn.

but suffer cold and hunger

God! watysh great colde, and fynger iche do abyd!
Wyl your bedauer, gosse, come home at the next tyde.
Iche pray God to coun him wel to vare, 17
That, whan he comit home, myd me he do not stare
For putting a straw dorow his great net.
Another pot of ale, good gosse, now me fet; 20

I'll go to law for a straw.

For my bedauer wyl to London, to try the law,
To sew Tre poll pen, for waggyng of a straw.
Now, gosse, farewell! yche can no lenger abyde;
Iche must ouer to the ale howse at the yender syde;

[1] Boorde evidently didn't appreciate the Anglo-Saxon words of our speech as he did his own long Latin and Greek coinages.
[2] therein: as *dyn* above is "thin," *dycke*, "thick." [3] gossip, mate.
[4] a-hungered. [5] soul, flavouring, meat; p. 138, l. 21.
[6] sign. B .ii. back. [7] at home.

And now come myd me, gosse, I thee pray,
And let vs make mery, as longe as we may.

¶ Cornwal is a pore and very barren countrey of al maner thing, except Tyn and Fysshe. There meate, and theyr bread, and dryncke, is marde and spylt for lacke of good ordring and dressynge. Fyrres and turues is theyr chief fewel; there ale is starke nought, lokinge whyte & thycke, as pygges had wrasteled in it,

Cornwall has only tin and fish. (See *Notes.*)

Their food is spoilt by bad cooking.

Their ale is awful stuff;

[1] smoky and ropye,
and neuer a good sope,
in moste places it is worse and worse,
pitie it is them to curse;
for waggınge of a straw
they wyl go to law,
and al not worth a hawe,
playinge so the dawe.

they'll go to law for wagging of a straw.

¶ In Cornwall is two speches; the one is naughty Englyshe, and the other is Cornyshe speche.

And there be many men and women the whiche cannot speake one worde of Englyshe, but all Cornyshe. Who so wyll speake any Cornyshe, Englyshe and Cornyshe doth folow.

Many Cornish people can't speak a word of English.

One. two. thre. foure. fyue. six. seuen. eyght. nyne.
Ouyn. dow. tray. peswar. pimp. whe. syth. eth. naw.
[2] Ten. aleuyn. twelue. thertene. fourtene. fyftene.
Dec. vnec. dowec. tredeec. peswardeec. pympdeec.
Syxtene. seuentine. eyghtyne. nyntene. twenty.
Whedeec. sythdeec. ethdeec. nawdeec. Igous.
One and twenty. two and twenty. three and twenty.
Ouyn war igous. dow war Igous. tray war ygous.
Fouer and twenty, &c.
peswar ygous: and so forthe tyl you come to thyrty.

The Cornish numerals.

¶ No Cornysheman dothe nomber aboue .xxx. and is named. *Deec warnegous.* And whan they haue tolde thyrty, they do begyn agayn, "one, two, and

30 is their highest number.

[1] Printed as prose. [2] B .iii. not signed.

thre," And so forth. and whan they haue recounted to a hondred, they saye *kans*. And if they nomber to a thousand, than they saye *Myle*.

A talk in Cornish and English.

God morow to you, syr! *Dar day dew a why, serra!*
God spede you, mayde! *Dar zona de why math-tath.*[1]
You be welcome, good wyfe!
Welcom a whe gwra da
I do thanke you, syr. *Dar dala de why, syra.*
How do you fare? *Vata lew genar why?*
Well, God thanke you, good master!
Da dar dala de why, master da!
Hostes, haue you any good meate?
Hostes, eus bones[2] *de why?*
Yes, syr, I haue enowghe. *Eus, sarra, grace a dew.*
Giue me some meate, good hostes!
Rewh bones[2] *de vy, hostes da!*
Mayde, giue me bread and drinke!
Math-tath,[1] *eus me barow ha dewas!*
Wife, bringe me a quarte of wine!
Gwrac, drewh quart gwin de vy!
Woman, bringe me some fishe!
Benen,[3] *drewh pyscos de vi!*
[4] Mayde, brynge me egges and butter
Math-tath,[1] *drewgh me eyo*[5] *hag a manyn de vi*
Syr, much good do it you!
Syrra, betha why lowe weny cke!
Hostes, what shal I paye?
Hostes, prendra we pay?
Syr, your rekenyng is .v. pens.
Syrra, iges rechen eu pymp in ar.
How many myles is it to london?
Pes myll der eus a lemma de Londres?
Syr, it is thre houndred myle.
Syrra, tray kans myle dere.

[1] *Mahthoid* P. (John W. Peard). [2] *Boos* P [3] *Beuen* AB. (*Bennen* P.)
[4] B .iii. back. [5] *oye*, an egg; pl. *oyow* P.

God be with you, good hostes! — A talk in Cornish and English.
Bena tewgena a[1] *why hostes da!*
God gyue you a good nyght!
Dew rebera vos da de why!
God send you wel to fare!
Dew reth euenna thee why fare eta!
God be wyth you! *Dew gena why!*
I pray you, commend me to all good felowes.
Meesdesyer,[2] *why commende me the olde matas*[3] *da.*
Syr, I wyl do your commaundement.
Syrra, me euyden gewel ages commaundement why.
God be with you! *Dew gena why!*

[4]

¶ The second chapytre treateth of Wales. And of the natural disposicion of Welshmen. Teaching an Englyshman to speake some Welsh.

I Am a Welshman, and do dwel in Wales,
I haue loued to serche boudgets, & looke in males; — I like thieving.

[1] *Dew genew*, P. [2] ? *Maz den syra*, good man Sir, good Sir, P.
[3] ? *maynys*, pl. of *mayn*, an intimate, P. [4] B .iiii. not signed.

I don't like work, and I do like prigging.

I loue not to labour, nor to delue nor to dyg;
My fyngers be lymed lyke a lyme twyg;
And wherby ryches I do not greatly set,
Syth all hys fysshe that commeth to the net.

I'm a gentleman and love the Virgin Mary.

I am a gentylman, and come of brutes blood;
My name is, ap Ryce, ap Dauy, ap Flood.
I loue our Lady, for I am of hyr kynne;
He that doth not loue hyr, I be-shrew his chynne.
My kyndred is ap hoby, ap Ienkin, ap goffe.

I go bare-legged.

Bycause I do go barlegged, I do cach the coffe;
And if I do go barlegged, it is for no pryde;
I haue a gray cote, my body for to hyde.

I love Roasted Cheese. (p. 129.)

[1] I do loue cawse boby,[2] good rosted[3] chese;
And swyshe swashe metheglyn I take[4] for my fees;

My Harp is my treasure;

And yf I haue my harpe, I care for no more;
It is my treasure, I do kepe[5] it in store;

It's made of mare-skin and horse-hair.

For my harpe is made of a good mares skyn,
The stringes be of horse heare, it maketh a good din;
My songe, and my voyce, and my harpe doth agree,

I sing like a bumble-bee.

Muche lyke the hussyng of a homble be;
Yet in my countrey I do make good pastyme,
In tellyng of prophyces whyche be not in ryme.

South Wales is better than North, for food.

Wales is deuided into two partes, whyche be to saye, North Wales, and South Wales. South Wales is better than North Wales in many thinges, specially for wyne, Ale, Breade, and wylde foule; yet bothe the countreys be very barayne, for there is muche waste, & wast ground, consydering there is maryses, & wylde and high moun-

Mountains: Snowdon and Manath Deny.

taynes. The mountayne of Snowdon is the hyghest mountayne of Wales. There is another hyghe mountain [in] Walles, called Manath deny, vpo*n* the toppe

[1] B .iiii. back.

[2] See the anecdote in 'The Hundred Merry Tales' (*Notes*) of St Peter getting the bothering Welsh out of heaven by shouting "*Cause bobe*" outside the gate, and then locking the gate on them when they'd rusht out.

[3] roted A; rosted B.

[4] toke B.

[5] I kepe B.

of the which is a fayre fountayne. And yf the winde be any thyng vp, yf a man do stande at the top of the hyl in any place, and do cast his hat or cap downe the hyll, the cap or hat shall flye bacwarde, and not forwarde, although a man stande in neuer so came[1] a place, as they of *the* countrey doth tel me. A wonder of Manath Deny.

There is a wel in Wales called "Saynte Wenefrydes Well." Walshe men sayth that if a man doth cast a cupe, a staffe, or a napkyn, in the well, it wyll be full of droppes or frakils, and redyshe like bloude; the whyche is false, for I haue proued the contrary in sondry tymes. St Winifred's Well: (See *Notes*.) Welshmen lie about it.

¶ In Wales there hath ben many goodly & stronge Castels, and some of them stande yet. The Castels and the Countre of Wales, and the people of Wales, be muche lyke to the Castels and the countrey and the people of Castyle and Byscaye; [2]for there is muche pouerty, and many reude and beastlye people, for they do drynke mylke and whay; they do fare ful euel, and theyr lodgynge is poore and bare, excepte in market townes, In the whych is vsed good fashion and good vytales, good meate, wine, and competent Ale, and lodgynge. Wales is like Castille and Biscay. The people are very poor and beastly.

North Wales and Sowth Wales do vary in there speche, and in there fare, and maners. Sowth Wales is best; but for all the variaunce of the premisses, they can not speke .x. wordes to-gyther of Welshe, but "deauol," that is to say, "the deuyl," is at the ende of one of the wordes, As "the foule euyll," whyche is the fallyng syckenes,[3] is at the ende of euery skottysh mans tale. South Wales is better than North. Welshmen always swear by the Devil, and Scotchmen by the Foul Evil.

In Wales in diuers places is vsed these two stulticious[4] matters. the fyrste is, that they wyl[5] sell there lams, and theyr calues, and theyr corne the whyche is not sowen, and all other newynges, a yere before that they be sure of any newynge; and men wyl bye it, trustynge vppon hope of suche thynges that wyl come. The seconde The Welsh do stupid things: 1. Sell all produce a year in advance.

[1] ? calm. [2] sign. C .i. [3] See p. 136, line 4.
[4] stulticious in, B. [5] well A; wel B.

9

2. When a friend dies, they cry out, "Darling, why did you do it? Come back, or we'll die with you!"

stulticious matter is, that yf any of theyr frendes do dye, & whan they shall be buried and put in to the graue, in certayne places they wyl cry out, making an exclamacion, and sayeng, "O venit[1]!" that is to saye, "O swetynge! why dost thou dye? thou shalt not go from vs!" and wil pul away the corse, sayeng, "venit! we wyl die with the, or els thou shalt tary with vs!" wyth many other folyshe wordes, as the Castilions and the Spaniardes do say & do at the burieng of theyr frendes[2]: thys dyd I se & here in Rithen and Oswoldestre, and other places.

I saw this at Ruthin and Oswestry.

The Welsh think too much of their kin; some are thieves; there are many bastards and priests' sons; but that's stopt now

¶ The Walsh men be hardy men, stronge men, & goodly men; they woulde be exalted, & they do set muche by theyr kynred & prophecyes; and many of them be louynge and kyndharted, faythful, & vertuous. And there be many [3]of them the whyche be lyght fyngered, & loueth a purse; but this matter latly is reformed. but lechery in manye places is to much vsed, Wherfore ther be many bastards openly knowen; and many prestes sonnes aboundeth in the countre, specially in North Wales; but that is nowe reformed, considring the restriction of the kynges actes, that prestes shal haue no concubynes.[4] who so wyll lerne to speake some Welshe, Englyshe and Welshe foloweth. And where that I do not wryte true Welshe, I do write it that euery man may rede it and vnderstand it without any teachynge.

The Welsh numerals.

One.	two.	thre.	four.	fyue.	syx.	seuyn.	eyght.
Eun.	*daw.*	*try.*	*pedwar.*	*pimp.*	*wheeth*[5].	*saygth.*	*oweyth.*

Nyne.	ten.	aleuen.	twelue.	thyrtene.	fourtene.
nau.	*deek.*	*vnardeek.*	*deuardeek.*	*tryardeek.*	*pedwardeek.*

Fyftene.	syxtene.	seuyntene.	eyghtene.
pympdeek.	*vnarbundeek.*	*dauarbundeek.*	*tryarbundeek.*

Nyntene.	twentye.	one and twenty.	two and twenty.
pedwarbuntheek.	*igain.*	*vnar igayn.*	*deuar igayn.*

[1] Lat. *benedictus*, D. (B. Davies.) [2] See p. 200. [3] sign. C .i. back.
[4] Statute 31 Hen. VIII, chap. 14, A.D. 1539. See 'Notes.' [5] *wheech* D.

Welsh Numerals.

Therty. forty. fyfty. syxty. seuenty.
thegarhigen. deugen. degadugen. trygen. degatrygen.
Eyghty. nynety. a.C. two .C. M.
pedwarugen. degapedwarugen. kant. dekant. Myl.

A talk in Welsh and English.

¶ God spede, fayre woman!
Deu ven-dicko[1], *gwen wraac!*
Good morow, fayr mayd! *Deyth dawh theet·morwyn!*
¶ God nyght, masters all! *Nos daw, masters igeet.*
Syr, can you speke any Welshe?
Sere, auedorowgh weh Gamraac?
Ye, syr, I can speke some Welshe.
Ede, oh sere, medora heth[2] *dyck.*
Mayden, come hether, and gyue me some roste chese!
Morwyn, therdomma moes imi gawse boby!
Tarry a lytle, man, and you shall haue enowgh.
[3] *Arow heth*[4] *dycke, gower wheh gooh dygan.*
Wyfe! hath preestes wyues in Wales?
Wraac, oes gwrath[5] *yn Kymery?*
Hold thy peace! they haue no Wyues now.
Tau son! neth os mor[6] *gwragath irrowan.*
Syr, wyll you lend me a horse to ryde to London?
Sere, a rowhe imi margh euer hogeth klynden?
You shall haue a horse. *Wheh agewh ar margh.*
Syr, how far is it to London? *Sere, pabelthter*[7] *klinden?*
Syr, it is .ix. myle. *Sere, now*[8] *mylter.*
Is this the ryght way to the towne?
Ay hon yoo yr forth yr dre?
Wher is the best In & best lodging?
Ple may I cletty gore yne?
At Iohn ap Dauy ap Ryse house.
In hy Iohan ap Dauyth ap Rys.
Hostes, god saue you!
Vey cleto wraac, Duw ah erosso[9] *why!*

[1] Lat. benedicat D. [2] *ychy* D. [3] sign. C .ii.
[4] *Aros ychy* D. [5] ? *gwragath* D. [6] ? *mwy* D.
[7] *pabellter* D. [8] *naw* D. [9] *crosso* D.

A talk in Welsh and English.

Syr, you be hartyly welcome!
Sera, mae yn grosso duw worthy!
Maystres, haue you any good meat and lodgyng?
Vey maistres, oes gennowh whe thin or booyd ta a clettyda?
Syr, I haue good meate and good lodgyng.
Sere, mae gennyf vid ta a cletty da.
Hostes, what is it a clocke?
Veye cleto wraac, beth idioo hy ar i glowh?
Syr, it is .vi. a clock.
Sere, me hy yn wheh ar y glowh.
Hostes, when shall we go to supper?
Vey cleto vraac pamser i cawn̄[1] *ny in supper?*
By and by. *Yn ynian.*
Gyue me some drynke! *Moes imi diod!*
Gyue me some ale! *Moes imi currow!*
Gyue me some bred! *Moes imi*[2] *vara!*
Gyue me some chese! *Moes imi gaws.*
Hostes, geue me a rekening!
Vey leto wraac moes[3] *imi gyfry.*
[4] Syr, ye shall pay thre pens for your supper.
Sere, whe delowgh tair keinowh dio se[5] *ich sopper.*
Hostes, God thanke you!
Voy cleto wraac[6] *dew a thiolchah!*[7]
Much good do it you! *Enwhyn thawen!*
How do you fare? *Par bewiut charuoh*[8] *whe?*
Good morow! *Daws.*[9]
Good nyght to you.[9] *Nos a dawh a whe.*
Farewell! *Yni awn!*[10]
Tary, tary, come hydder! *Arow arow*[11] *therdomma!*
Hold thy peas, hold your peas! *Tau, tau son!*

Thus endeth of Wales.

[1] rawn A. [2] ima A. [3] mee A. [4] sign. C .ii. back.
[5] ? *dros* for *dio se* D. [6] wraas A. [7] thiolphah A.
[8] *arnoch* D.
[9] Upcott's reprint of B leaves out these phrases, though B has them. [10] Yn i awh A. [11] for *Aros, aros* D.

¶ The thyrde Chapter treateth of Irland. And of the naturall disposicion of an Irishe man, & of theyr money and speche.

¶ [1] I am an Iryshe man, in Irland I was borne;
I loue to weare a saffron shert, all though it be to-torne. *I wear a saffron shirt, and am hasty.*
My anger and my hastynes doth hurt me full sore;
I cannot leaue it, it creaseth more and more;
And although I be poore, I haue an angry hart.
I can kepe a Hobby, a gardyn, and a cart;
I can make good mantyls, and good Irysh fryce; *I make frieze and aqua vitæ.*
I can make aqua vite, and good square dyce.
Pediculus other whyle do byte me by the backe, *Lice bite me.*
Wherfore dyvers times I make theyr bones cracke.
I do loue to eate my meate, syttyng vpon the ground, *I squat on the ground, and sleep in straw.*
And do lye in oten strawe, slepyng full sound.
I care not for ryches, but for meate and drynke;
And dyuers tymes I wake, whan other men do wynke.
I do vse no potte to seeth my meate in,
Wherfore I do boyle it in a bestes skyn;

[1] C .iii. not signed.

Than after my meate, the brothe I do drynk vp,
I don't use cups; I care not for my masȝer, neyther cruse nor cup.
I am not new fangled, nor neuer wyll be;
and I live poor. I do lyue in pouerty, in myne owne countre.

Ireland is divided into the English Pale, and the wild Irish.

¶ Irland is a kingdo*m*ship longing to the kyng of England. It is in the west parte of *th*e world, & is deuyded in ii. partes. one is *th*e Engly[sh] pale, & the other, *th*e wyld Irysh. The English pale is a good countrey, plentye of fishe, flesh, wyldfoule, & corne. There be good townes & cities, as Du[b]lyn & Waterford, wher *th*e English fashion is, as in meat, drinke, other fare & lodgi*n*g. The people of *th*e Englyshe pale be metely wel manerd, vsing the Englishe tu*n*ge; but naturally they be testy, specially yf they be vexed; Yet there be many well disposed people, as wel in the Englysh pale as in the wylde Iryshe, & vertuous creatures, whan grace worketh aboue nature. ¶ The other parte of Irland is called the wilde Irysh; and the Redshankes be [1] among them. That countrey is wylde, wast & vast, full of marcyces [2] & mou*n*tayns, & lytle corne; but they haue flesh sufficient, & litle bread or none, and none ale. For *th*e people there be slouthfull, not regarding to sow & tille theyr landes, nor caryng for ryches. For in many places they care not for pot, pan, kettyl, nor for mattrys, fether bed, nor such implementes of houshold. Wherfore it is presuppose *tha*t they lak maners & honesty, & be vntaught & rude; the which rudenes, *with* theyr melo*n*coly complexion, causeth the*m* to be angry & testy wythout a cause. ¶ In those partyes they wyll eate theyr meat syttyng on the ground or erth. And they wyl sethe theyr meat in a beastes skyn. And the skyn shall be set on manye stakes of wood, & than they wyll put in the water and the fleshe. And than they wyl make a great fyre vnder *th*e skyn betwyxt the stakes, & the skyn wyl not greatly

Men of the Pale have English ways, but are testy.

The wild Irish and Redshanks don't sow or till, or care for household goods.

They are rude and wrathful; they boil their meat in a skin.

[1] C .iii. back. [2] marrycces B.

bren. And whan the meate is eaten, they, for theyr drynke, wil drynk vp the brothe. In suche places men and women wyll ly to-gether in mantles and straw. There be many the which be swyft of fote, & can cast a dart perylously. I did neuer finde more amyte and loue than I haue found of Iryshe men the whyche was borne within the English pale. And in my lyfe I dyd neuer know more faythfuller men & parfyt lyuers than I haue knowen of them. ¶ In Irlond there is saynt Partryckes[1] purgatory, the whych, as I haue lerned of men dwellyng there, and of them that hath be there, is not of that effycacyte as is spoken of, nor nothing lyke. Wherfore I do aduertise euery man not haue affyaunce in such matters; yet in Ierland is stupendyous thynges; for there is neyther Pyes nor venymus wormes. There is no Adder, nor Snake, nor Toode, nor Lyzerd, nor no Euyt, nor none suche lyke.

Men and women lie together in straw.

I never knew better men than some of the Pale.

St Patrick's Purgatory isn't much good.

There are no Magpies, Snakes, Toads, or Efts, in Ireland.

[2] I haue sene stones the whiche haue had the forme and shap of a snake and other venimous wormes. And the people of the countre sayth that suche stones were wormes, and they were turned into stones by the power of God and the prayers of saynt Patryk. And Englysh marchauntes of England do fetch of the erth of Irlonde to caste in their gardens, to kepe out and to kyll venimous wormes. ¶ Englysh money goth in Irelond, for Irlond belongeth to England, for the kynge of Englonde is kyng of Irlond. In Irlond they haue Irysh grotes, and harped grotes, & Irysh pens. ¶ If there be any man the which wyll lerne some Irysh, Englysh and Irysh dothe folow[3] here togyther.

I've seen there stones, said to have been once snakes.

Irish earth is bought to kill venomous worms.

They have groats and pence.

One. two. thre. foure. fyue. syx. seuen. eyght.
Hewen. dow. tre. kaar. quiek. seth. showght. howght
nyne. ten. aleuyn. twelue. thirtene. fourtene.
nygh. deh. hewnek. dowek. tredeek. kaardeck.

The Irish numerals.

[1] patriarkes B. [2] C .iv. not signed. [3] fololow A; folowe B.

Irish numerals.

fyuetene. syxtene. seuentene. eyghtene.
quiekdeek.[1] *sehdeek. showghtdeek. howghtdeek.*
nynetene. twenty. one & twenty. ii. & twenty. thre & twenty
nythdek. feh. hewn feet. dowhfeet. trefeet.
Thirty. forty. fyfty. syxty. a hondred.
Dehfeet. eayfeet. dewhegesdayth.[2] *trefeet. keede.*

A talk in Irish and English.

God spede you, syr! *Anoha dewh sor!*
You be welcome to the towne. *De van wely.*
How do you fare? *Kanys stato?*
I do fare well, I thanke you.
Tam agoomawh gramahogood
Syr, can you speke Iryshe? *Sor, woll galow oket?*
[3] ¶ I can speke a lytle. *Tasyn agomee.*
Mayden, come hether, and gyue me som meate!
Kalyn, tarin chowh, toor dewh!
¶ Wyfe, haue you any good meate?
Benitee, wyl beemah hagoot?
¶ Syr, I haue enoughe. *Sor, tha gwyler.*
¶ Wyfe, gyue me bread! *Benytee, toor harun!*
¶ Man, gyue me wine! *Farate, toor fyen!*
¶ Mayden, gyue me chese! *Kalyn, toor case!*
¶ Wyfe, gyue me fleshe! *Benyte, toor foeule!*
Gyue me some fyshe! *Toor yeske!*
¶ Much good do it you! *Teena go sowgh!*
¶ How far is it to Waterford?
Gath haad o showh go port luarg.
It is one an twenty myle. *Myle hewryht.*
¶ What is it a clocke? *Gaued bowleh glog?*
¶ It is .vi. a clocke. *She wylly a glog.*
¶ Whan shal we go to supper?
Gahad rah moyd auer soper?
¶ Giue me a rekenyng, wyfe.
Toor countes doyen, benitee
¶ Ye shall pay .iii. pens. *Yekc ke to tre pyn Iny.*

[1] quiekdeek B. [2] dewhegesnayth B. [3] C .iv. back.

¶ Whan shal I go to slepe, wyfe?
Gah hon rah moyd holowh?
¶ By an by. *Nish feene.*
¶ God night, sir! *Ih may sor!*
Fare wel, fare wel! *Sor doyt, sor doit!*

¶ Thus endeth the maner and speche of[1] Irland.

[2]¶ The fourth[3] chapter treateth of Scotland, and the natural disposycion of a Scotyshe man. And of theyr money, and of theyr speche.[4]

I Am a Scotyshe man, and trew I am to Fraunce;
In euery countrey, myselfe I do auaunce;
I wyll boost myselfe, I wyll crake and face; *I always boast.*
I loue to be exalted, here and in euery place.
an Englyshe man I cannot naturally loue, *I can't like Englishmen.*
Wherfore I offend them, and my lorde aboue;
He that wyll double with any man,
He may spede wel, but I cannot tell whan.
I am a Scotyshe man, and haue dissymbled muche, *I dissemble, and don't keep my promise.*
and in my promyse I haue not kept touche.

[1] of of AB. [2] sign. D .i. [3] fouth A; fourth B.
[4] A note written here in Mr Christie-Miller's copy says, "vid. etiam Jo. Brucrinum in suo lib. de re Cibaria."

Great morder and theft in tymes past I haue vsed;
I trust to God hereafter, such thynges shal be refused.

Whenever I speak I swear by the Foul Evil (see p. 127).

And what worde I do speake, be it in myrth or in borde,
"The foule euyll" shalbe at the end of my worde;
Yet wyl I not chaunge my apparell nor aray,
although the French men go neuer so gay.

Scotland is a kyngdome, the kynge of the whyche [1]hath in olde tyme come to the parliament of the kyng of England, and hath be subiect to England. Scotland is deuyded in two partes; the one part, that is to say, nexte England, is Hayden, Edenborow, Lythko, Sterlynge, Glasco,[2] saynt Androwes, saynt Iohns towne, wyth the countres anexed, and adiacent to the aforesayd cities and townes: [therein] is plenty of fysh and flesh, and euell ale, excepte Leth ale; there is plenty of hauer cakes, whiche is to say, oten cakes: this parte is the hart and the best of the realme. The other parte of Scotlande is a baryn and a waste countrey, full of mores, lyke the lande of the wylde Ireshe. And the people of *that* parte of Scotland be very rude and vnmanered & vntaught; yet that part is somwhat better than the North parte, but yet the Sowth parte wyll gnaw a bone, and cast it into the dish again. Theyr Fyshe and Fleshe, be it rosted or soden, is serued wyth a syrup or a sause in one disshe or platter: of al nacyons they do sethe theyr fysh moste beste. The borders of Scotland toward England,—as they the which doeth dwell by Nycoll forest, and so vpward to Barwyke, by-yonde the water of Twede,—lyueth in much pouertie and penurye, hauynge no howses but suche as a man maye buylde wythin .iii. or .iiii. houres: he and his wyfe and his horse standeth all in one rome. In these partyes be many out-lawes and stronge theues, for muche of theyr

South Scotland has bad ale, but much oat cake.

The Highlands are full of moors.

The Southern Scots will gnaw a bone, and put it back in the dish.

In the Borders they live in penury, in huts;

man, wife, and horse in one room.

[1] D .i. back.

[2] Boorde studied and practised in Glasgow. See the *Forewords*, p. 59.

lyuyng standeth by stelyng and robbyng. Also it is naturally geuen, or els it is of a deuyllyshe dysposicion of a Scotysh ma*n*, not to loue nor fauour an Englyshe ma*n*.[1] And I, beyng there, and dwellynge amonge them, was hated; but my scyences & other polyces dyd kepe me in fauour, that I dyd know theyr secretes.[2] The people of *th*e cou*n*trey be hardy men, and stronge men, and well fauored, & good musycyons; in these .iiii. qualytes they be mooste lyke, [3]aboue all other nacions, to an Englyshe man; but of al nacyons they wyll face, crake, and boost themselfe, theyr frendes, and theyr cou*n*trey, aboue reason; for many wyll make strong lyes. In Scotland a man shall haue good chere —he that can away wyth it after the countrey fashion— for litle money. The most parte of theyr money is bras. In bras they haue pens, and halfe plackes, & plackes: four Scotish pens is a placke, and a placke is almost worth an Englysh peny, for .xviii. Scotish pens is worthe an Englyshe grote: in Scotland they haue Scotysh grotes of syluer, but they be not so good, nor so muche worth, as an Englysh grote. In golde they haue halfe face crownes, worth of our money .ii. shyllynges and .iiii. pens. And they haue crownes of .iiii. shillinges & .viii. pens. if a Scotyshe man do pay .xx. crownes of golde, or a thousande crownes of golde, he doth say, "I haue payde .xx. pound, or a thousande pounde"; for euery crowne of .iiii. shillinges and .viii. pens is a pounde in Scotland. In Scotlande they haue two sondry speches. In the northe parte, and the part ioynyng to Ierland, that speche is muche lyke the Iryshe speche. But the south parte of Scotland, and the vsuall speche of the Peeres of the Realme, is lyke the northen speche of England. Wherfore yf any man

Scotchmen don't like Englishmen.

I was hated by 'em, but still got at their secrets.

They're good musicians,

but the biggest braggers in the world;

they tell strong lies.

Living is cheap.

Scotch placks, pence,

silver grotes, gold ½-face-crowns, and crowns.

4s. 8d. is a Scotch pound.

Northerners talk like Irishmen.

Southerners like North-Englishmen.

[1] See the note from *The Complaynt of Scotland*, p. 59 above.
[2] See Boorde's Letter VI, to Secretary Cromwell, in the *Forewords*, p. 59.
[3] D .ii. not signed.

wyl learne to speake some Scotysh,—Englysh & Scotish doth folow together.

Scotch numerals.

¶ One, two, three, foure, fyue, syx, seuyn, eyght, nyne,
Ene, twe, dre, foore, feue, sax, sauen, awght, neen,
ten, aleuen, twelue, thertene, fourtene, fyftene, syxtene.
tane, alauen, twalue, dertene, fortene, vyuetene, saxtene.
seuentene, eyghtene, nyntene, twenty, one and twentye.
sauentene, awghtene, nyntene, twante, ene and twanty.
two & twenty, a hondred.
twe an twanty, a hondryth.

A talk in Scotch and English.

[1] God morow, syr! *Gewd day, sher!*
Do you know me, good fellow?
Ken ye me, gewd falowh?
Ye syr, wel Inough! *Ye sher, in good fayth!*
What countrey man be you?
What contryth man be ye?
I am a good felow of the Scotyshe bloud.
I es a gewd falow of the Scotland blewd.
Than haue you plenty of sowes and pygges.
Than haue ye fell many of sewes and gryces.
A pygge is good meate. *A gryce is gewd sole.*[2]
Syr, by my fayth you be welcome!
Sher, by my fayth but yows wel come!

Scotch is like Northern English.

For as muche as the Scotysh tongue and the northen Englyshe be lyke of speche, I passe ouer to wryte anye more of Scottyshe speche.

[1] D .ii. back. [2] soul, flavour. See p. 122, l. 16.

¶ The .v. chapytre treateth of Shotland and of Fryceland & of the naturall disposycion of the people of the countrey.

[1] I Was borne in Shotland, my countrey is ful colde;
And I was borne in Friceland, where muche fysh is sold;
For corne and for shoes, our fyshe we do sell;
And symple rayment doth serue us full well;
Wyth dagswaynes and roudges [2] we be content;
And our chiefe fare, in the tyme of Lent,
Fyshe, at any tyme seldome we do lacke.
But I beshrew the louse that pyncheth vs by the back!

In Friesland we sell fish for corn and shoes.

We live on fish.

¶ Shotland is a smale countrey or Ilande, the whyche is a colde countrey and baryn, for there is nothinge the whyche is commodious nor pleasaunt, except fyshe.

In Shetland, nothing is nice but fish.

¶ Fryce is in maner of an Ylande, compassed aboute on the one syde with the occyan sea, hauyng hys begynnyng at the ende of the water of Reene, and doth end towarde Denmarkes sea. And although they be anexed to Germany, yet they do dyffer, for they do vse contrary fashyons, as well in theyr apparel as in

Friesland is nearly an island.

The Frisians differ from the Germans.

[1] D .iii. not signed. [2] coarse cloths and rugs.

theyr maners, for they be rurall and rusticall; they haue no wood there, but turfes and dung of beastes, to make theyr fyre. They wolde not be subiect to no man, although they be vnder the Emperours dominion: they do loue no war, nor bate, nor strife, nor they loue not, nor wyl not haue no greate lordes amonge them; but there be admitted certayn Iustices, And Iustice that loueth, and prayseth, Chastyte. The countrey is could, baryn, and poore, lackyng riches; yet there is plenty of pasture: theyr speche is lyke to base Germanyens spech; it doth dyffer but lyttle. One of the chiefe townes of Fryce land is called Grunnyghen. In golde they haue Ryders, Gylders, and Clemers gylders. In syluer they haue Iochymdalders.

Frislans have no firewood;

and no great Lords, but only Justices.

Friesic is like Low German or Dutch.

Groningen.

Frislan coins.

[1]¶ The .vi. Chapter treateth of Norway & of Islonde, and of the natural disposicion of the people of the countrey, and of theyr money and speche.

[1] D .iii. back. See p. 142 for a note on the cuts.

I Am a poore man, borne in Norway;
Hawkes and fysh of me marchauntes do by all daye.
And I was borne in Islond, as brute as a beest;
Whan I ete candels ends, I am at a feest.
Talow and raw stockfysh, I do loue to ete;
In my countrey it is right good meate;
Raw fysh and flesh I eate whan I haue nede;
Upon such meates I do loue to feed.
Lytle I do care for matyns or masse,[1]
And[2] for any good rayment, I do neuer passe;
Good beastes skyns I do loue for to were,
Be it the skins of a wolfe or of a beare.

In Norway we sell hawks and fish.

In Iceland we eat candle-ends (see *Notes*)

and raw fish and flesh.

We wear wolves' and bears' skins.

[3] ¶ Norway is a great Ilond compassed abowt almost wyth the See; the countre is very colde, wherefore they haue lytle corne, and lytle bread and drynke; the countre is wylde, and there be many rewde people. They do lyue by fysshyng and huntyng. Ther be many castours and whyte beares[4], & other monsterous beastes; there be welles, the whyche doth tourne wood in to Irone. In somer there be many daies that the sunne doth neuer go downe, but is continuallye daye. And in many dayes in wynter it is styll nyght. In Norwaye ther be good hawkes: ther is lytle money, for they do barter there fysh and hawkes for Mele, and shoes, and other marchaundies.

Norway has little corn.

It has Beavers and White Bears, and Petrifying Wells.

It's night all winter.

¶ Iselond is beyond Norway: It is a great Ilond compassed about wyth the Ise See; the countre is wonderful cold, and in dyuers places the see is frosyn, and full of Ise. There is no corne growynge there; nor they haue lytle bread, or none. In stede of bread they do eate stockefyshe; and they wyll eate rawe fyshe and fleshe; they be beastly creatures, vnmanered and vntaughte. They haue no houses, but yet doth lye in

Iceland is very cold,

and grows no corn.

Icelanders eat raw fish, and are beastly creatures.

[1] anye of gods seruasse B. This change implies that Mary's reign was over. *Forewords*, p. 19.
[2] And as B. [3] D .iiii. not signed.
[4] No white bears in Norway.—G. Vigfusson.

Icelanders lie in caves like swine; give away their children, and

caues[1], al together, lyke swyne. They wyll sell there Iselond curres, & gyue a-way their chyldren. They wyll eate talowe candells, and candells endes, and olde grece, and restye tallowe, and other fylthy thinges. They do were wylde beastes skinnes[2] and roudges. They be lyke the people of the newe founde land named Calyco. In Iselond there be many wylde beastes.

are like the people of Calyco.

They barter fish for meal, &c., and use no money.

The people be good fyshers; muche of theyr fyshe they do barter wyth English men, for mele, lases, and shoes, & other pelfery. They do vse no mony in the countre, but they do barter or chaunge one thynge for another. There be som prestes the whych be beggers, yet they wyll haue concubynes. In Sommer tyme they haue, in maner, no nyghte. And in wynter tyme they haue, in lyke maner, [3] fewe howres of dayelyghte. theyr language I can not speke, but here and there a worde or two, wherfore I do passe ouer to wryte of it.

Priests, though beggars, have Concubines.

No night in summer.

I can't speak Icelandic.

[1] In Iceland the subterranean dwelling is a standing phrase.—G. Vigfusson.

[2] No wild beasts in Iceland.—G. V. Skins got from abroad.

[3] D .iiii. back.

Instead of the two cuts at the head of chap. vi., of the Rose-Garland edition (1547 or -8), the Lothbury edition of 1562 or -3 substitutes the two below :

[1] 

¶ The .vii. Chapytre sheweth howe the auctor of thys boke, how he had dwelt in Scotland and other Ilandes, did go thorow and rounde about Christendom, and oute of Christendome; declarynge the properties of al the regions, countreys, and prouynces, the whiche he did trauel thorow.

[2] OF noble England, of Ireland and of Wales,
And also of Scotland, I haue tolde som tales;

[1] On this woodcut the late Mr Dyce remarks in his *Skelton's Works*, i, "the portrait on the title-page of *Dyuers Balettys and Dyties solacyous* (evidently from the press of Pynson; see Appendix II. to this Memoir) is given as a portrait of 'Doctor Boorde' in the *Boke of Knowledge* (see reprint, sig. I)." The pinnacle over the Doctor's head is complete in A, broken in B as in our cut. The cut that Wyer used for Boorde is on the title-page of Barnes's *Treatyse* on Beards below, p. 305. [2] sign. E .i.

And of other Ilondes I haue shewed my mynd;
He that wyl trauell, the truthe he shall fynd.

I write conscientiously.

After my conscyence I do wryte truly,
Although that many men wyl say that I do lye;
But for that matter, I do greatly pas,
But I am as I am, but not as I was.

Tho' my metre is doggrel,

And where [as] my metre is ryme dogrell,
The effect of the whych no wyse man wyll depell,

wise men will take my meaning.

For he wyll take the effect of my mynde,
Although to make meter I am full blynde.

Our royal Realm of England has no equal.

For as muche as the most regall realme of England is cytuated in an angle of the worlde, hauing no region in Chrystendom nor out of Chrystendom equiualent to it,—The co*m*modyties, the qualite, & the qua*n*tyte, wyth other and many thynges considered, within & aboute the sayd noble realme,—Wherefore[1] yf I were a Iewe, a Turke, or a Sarasyn, or any other infidele, I yet must prayse & laud it, and so wold euery man, yf they dyd know of other co*n*trees as well as England. Wherfore, all nacyons aspyeng thys realme to be so co*m*modyous and pleasaunt, they haue a confluence to it more than to anye other regyon. I haue trauayled rownd about Chrystendom, and out of Christe*n*dom, and I dyd neuer se nor know .vii. Englyshe men dwellynge in any towne or cyte in anye regyon byyond the see, excepte marchauntes, students, & brokers, not theyr parmanent[2] nor abydyng, but resorting thyther for a space. In Englande howe manye alyons hath and doth dwell of all maner of nacyons! let euery man Iudge the cause why and wherfore, yf they haue reason to perscrute the mater. I haue also shewed my mynde of the realme of Ierlande,[3] Wales, and Scotland, [4]and other londes; pretendyng to shew of regyons, kyngdoms, countreys, and prouinces, thorow and round about

Were I a Jew or Turk, I yet must praise it.

All nations flow to it.

In all my travels I never knew 7 Englishmen who lived permanently abroad.

Yet how many aliens live here!

I shall now tell you of more lands I've travelled in.

[1] wherof B. [2] permanent B. [3] England B.
[4] E .i. back.

where that I haue traueylyd, specyally aboute Europ, and parte of Affrycke: as for Asia, I was neuer in, yet I do wryte of it by auctours, cronycles, & by the wordes of credyble parsons, the whiche haue trauelled in those partyes. But concernyng my purpose, and for my trauellyng in, thorow, and round about Europ, whiche is all Chrystendom, I dyd wryte a booke of euery region, countre, and prouynce, shewynge the myles, the leeges, and the dystaunce from citye to cytie, and from towne to towne; And the cyties & townes names, wyth notable thynges within the precyncte [of], or about, the sayd cytyes or townes, wyth many other thynges longe to reherse at this tyme, the whiche boke at Byshops-Waltam—.viii. myle from Wynchester in Ha*m*pshyre,—one Thomas Cromwell[1] had it of me. And bycause he had many matters of [state] to dyspache for al England, my boke was loste,[2] *the* which myght at this presente tyme haue holpen me, and set me forward in this matter. But syth *that* I do lacke the aforesayde booke, humbly I desyre all men, of what nacyon soeuer they be of, not to be discontent wyth my playne wryt-yng, & that I do tell the trewth; for I do not wryte ony thynge of a malycious nor of a peruerse mynde, nor for no euyll pretence, but to manyfest thinges *the* whiche be openly knowen, And the thynges that I dyd se in many Regyons, Cytyes, and Countryes, openly vsed.

I've never been in Asia.

I wrote a *Hand-book of Europe*, with distances and descriptions of towns;

but I lent it to Secretary Cromwell at Bishop's-Waltham, and it was lost.

Do not be offended at my telling the truth.

I don't write from malice.

Pascall the playn dyd wryte and preach manifest thinges that were ope*n* in the face of the world to rebuke sin; wyth the which matter I haue nothyng to do, for I doo speke of many countryes & regions, and of

Paschal [? Pope Paschal II, 1099—1118, A.D.] rebuked sin.

[1] Compare this of the dead, "one Thomas Cromwell," with Boorde's letter to the living, "Right Hone*r*able Lorde the Lord of the Pryue Seale," &c. *Forewords*, p. 62.

[2] Boorde's Itinerary of *England*—not Europe—was printed by Hearne in his edition of "Benedictus Abbas Petroburgensis de Vita et Gestis Henrici III. et Ricardi I.," &c., vol. 2, p. 777 (before and after). Hearne's account of Boorde, from Wood's *Athenæ*, and his own knowledge, is in vol. i. of the same book, p. 36-56. *Forewords*, p. 23.

I describe countries and men.

I wish to tell travellers what they're to do;

and about foreign money and speech.

the natural dysposicyon of the inhabitours of the same, with other necessary thynges to be knowen, specially for them the [1] whiche doth pretende to trauayle the countrees, regions, and prouinces, that they may be in a redines to knowe what they should do whan they come there; And also to know the money of the countre, & to speke parte of the language or speache that there is vsed, by the whiche a man may com to a forder knowledge. Also I do not, nor shal not, dispraue no man in this booke perticulerly; but manifest thinges I doo wryte openly, and generally of comin vsages, for a generall commodite and welth.

I went from Calais through Flanders.

¶ And in beyng ouer sea at Calys, I went first thorow Flaunders; wherefore the Flemmyng confesseth him selfe, sayeng:—

The .viii. Chapiter treateth of Flaunders, And of the naturall disposicion of a Fleming, and of their money and of their speche.

[1] sign. E .ii.

[1] ¶ I Am a Flemyng, what for all that,
Although I wyll be dronken other whyles as a rat?
"Buttermouth Flemyng," men doth me call;
Butter is good meate, it doth relent the gall. 4
To my butter I take good bread and drynke;
To quaf to moch of it, it maketh me to wynk.
Great studmares we bryng vp in Flaunders; 7
We sell them into England, wher they get the glaunders.

I get as drunk as a rat, and am called "Buttermouth Fleming."

We sell our brood-mares in England.

Out of England, and out of the aforsayd regyons to come thorowe England, to fetche the course and cyrcuyt of Europ or Chrystendom:—From London, that noble cyte, let a man take his Iorney to Rochester, Cawnterbury and Douer, or to Sandwiche, to take shyppyng to sayle to the welfauered towne of Calys, the which doth stand commodyously for the welth and succor of all Englande; In the whyche towne is good fare and good cheere, and there is good order, & polytike men, great defence, & good ordynaunce for warre. The sayde towne hath anexed to it for defence, Gynes, Hammes, and Rysbanke, Newman[2] brydge, & a blockehowse against Grauelyng, in Flaunders. . From Calys a man must goo thorowe Flaunders. Flaunders is a plentyfull countre of fyshe & fleshe & wyld fowle. There shall a man be clenly serued at his table, & well ordred and vsed for meat, and drynke,[3] & lodgyng. The countre is playn, & somwhat sandy. The people be gentyl, but the men be great drynkers; and many of the women be vertuous and wel dysposyd. In Flaunders there be many fayre townes: as Gawnt, Burges, & Newport, and other. In Flaunders, and in Braban, and other prouinces anexed to the same, the people wil eate the hynder loynes of frogges,[4] & wyll eate tod-

To go from England round Christendom.

Go from London by Dover or Sandwich, take ship to Calais

(which is well fortified),

and then go through Flanders (a rich country, but flat and sandy).

The Flemings are great drinkers.

(Ghent. Bruges.)

They eat frogs' loins and toadstools.

[1] sign. E .ii. back. [2] Newnam B.
[3] meat, drinke B.
[4] See an old recipe for cooking them, in *Quene Elizabethes Achademy, &c.*, Part ii. p. 152, E. E. T. Soc. 1869.

Flemish speech and money are like Low-German or Dutch (p. 151, l. 7, 8).

stooles. As for the speche & the money of Flaunders, [they] doo not dyffer but lytle from Base-Almayne; wherfore loke in the chapiter of Base-Almayn. [Chap. xiii, p. 157-8.]

[1]

¶ The .ix. chapiter tretyth of Selond, and Holond,[2] and of the naturall dysposycyon of a Selondder, and Holander, & of their money and of theyr[3] speche.

Zealand is an island.

¶ I Am a Selondder, and was borne in Selond;
My cuntre is good, it is a propre Ilond.

Hollanders make cloth.

And I am a Holander; good cloth I do make;
To muche of Englyshe bere, dyuers tymes I do take. 4

[1] E .iii. not signed. See the cut again on p. 155.
[2] Selande, Holand, B. [3] & their B.

We lacke no butter that is vnsauery and salt,
Therfore we quaf the beer[1], that causeth vs to halt.
We haue haruest heryng, and good hawkes,
With[2] great elys, and also great walkes:
Wyth such thynges, other londes we help and fede;
Suche marchaundise doth helpe vs at nede;
[3]Yet to vs it shoulde be a great passyon
To chaunge our rayment or our olde fashyon.

We sell butter, herrings, hawks, eels, and whelks, to other lands. We won't change our old fashions.

¶ Seland and Holand be proper and fayre Ilands, and there is plenty of barelled butter, the whych is resty & salt; and there is cheese, & hering, salmons, Elys, & lytle other fysh *that* I did se. ther be many goshawkes, and other hawkes, & wyld foule. Ther be these good townes in Seland: Mydilborow, and Flosshing, & other mo. In Holand is a good towne called Amsterdame; and yet right many of the men of the countres wyll quaf tyl they ben dronk, & wyl pysse vnder the table where as they sit. They be gentyll people, but they do not fauer Skottysh men. The women in the church be deuout, & vsyth oft to be confessed in the church openly, laying theyr heades in the prestes lap; for prestes there do sit whan they do here confessyons, *and* so they do in many other prouynces anexed to the same. The women be modestyouse, & in the townes & church they couer the*m*self, & parte of theyr face and hed, with theyr mantles of say, gadryd and pleted mouch like after nonnes fashyon. theyr la*n*guage, theyr money, theyr maners and fashyons, is lyke Flaunders, Hanaway, and Braban, which be co*m*modyous and plentyfull countreys.

We have butter, cheese, salmon, goshawks. Middleburgh and Flushing. Amsterdam. Dutchmen drink till it runs out of them. They don't like Scotchmen. Women confess openly in church; they are modest, and wear mantles over their heads. Dutch speech and ways are like those of Flanders and Hainault.

[1] Lorde, how the Flemines bragged, and the Hollanders craked, that Calice should be wonne, and all the Englishemen slain; swearyng, and staryng, that they would haue it within thre daies at the moste; thynkyng verely that the toune of Calice could no more resist their puyssaunce then *a potte of double beere, when they fall to quaffyng*.—Hall's Chronicle, p. 181, ed. 1809.

[2] Whan A; with B. [3] E .iii. back.

¶ The .x.[1] Chapiter treatyth of Braban, and of the natural disposicion of a Brabander, of theyr[2] speche and of theyr money.

[3] ¶ I Was borne in Braban, that is both gentil and free;
All nacyons at all tymes be well-come to mee.
I do vse martes, dyuers tymes in the yere;
And of all thynges, I do loue good Englysh beere.
In Anwarpe and in Barow,[4] I do make my martes;
There doth Englysh marchauntes cut out theyr partes.
I haue good sturgyon, and other good fyshe;
I loue euer to haue good meate in my dyshe;
I haue good lodgyng, and also good chere,
I haue good wyne, and good Englyshe bere;
Yet had I rather to be drowned in a beere barell
Than I wolde chaunge the fashion of my olde apparel.

I hold marts often, and love good beer,

and good meat.

I have good wine.

¶ Braban is a comodyous and a pleasaunt countrey, In the whyche is plentyfulnes of meat, drynke, & corne; there is plenty of fysh, and fleshe; there is good

Brabant is a rich country,

with plenty of fish.

[1] tenth B. [2] the A; theyr B. [3] E .iiii. not signed.

[4] Bacow B. ? Breda. Under 'the .XXIII. yere of Kyng Henry the .VIII.' Hall says: 'In this yere [A.D. 1531] was an olde Tolle demaunded in Flaunders of Englyshmen, called the Tolle of the Hounde, which is a Ryuer and a passage: The Tolle is .xii. pence of a Fardell. This Tolle had been often tymes demaunded, but neuer payed: insomoche that Kyng Henry the seuenth, for the demaunde of that Tolle, prohibited all his subiectes *to kepe any Marte at Antwerpe or Barow*, but caused the Martes to be kepte at Calyes.'—*Chronicle*, p. 786, ed. 1809.

Sturgyon, Tu*n*ney, and many other good fysh, and good chepe. The countrey is playn, and ful of fartylyte. God is well serued in theyr churches; and there be manye good and deuout people; and the people be louyng; & there be many good felowes the whyche wyll drynke all out[1]: there be many good craftes men, speciall, good makers of Ares clothe. There a man may by all maner of lynen cloth, & silkes, & implimentes for howsholde, & plate and precious stones, and many other thynges, of a compytent pryce. The speche there is Base-Douche, and the money is the Emperours coine, that is to saye, Douche money, of the whyche I do wryte of whan that I do speke of Base-Almayne. In Brabant be many fayre and goodly townes: the fyrst is Handwarp, a welfauered marchaunt towne; the spyre of the churche is a curyous and a ryght goodly lantren. There is the fayrest flesh sha*m*bles that is in Chri[2]stendome. There is also a goodly commyn place for marchauntes to stand and to walke, to dryue theyr bargyns, called "the Burse." And Englyshe marchauntes haue there a fayre place. There is another towne called Louane, whiche is a good vnyuersyte. There is also Brusels, and Mawghlyn, and other mo. ¶ Here is to be noted that there is another countre ioynyng to Braban, the whych is called Hanawar or Hanago. The countre is like Braban and Flaunders, as well in the fartylyte[3] and plentyfulnes of the countre, as of the money and the conuersacion of the people: howbeit, Hanaway and the Hanawayes do dyffer somwhat in the premysses; for they do speke in diuers places, as well Frenche as Doche; for it lyeth betwyxt[4] Braban, Flaunders, and Fraunce. Theyr money is the Emperours coyne, as the money of Flaunders & Braba*n* is, and all is one coyne: the chefe town of Hanago is saynt Thomas, and Bargen, and dyuers other.

The folk are devout and loving.

They make good Arras cloth.

They talk Dutch.

Antwerp has a fine church-spire

and shambles,

also a Bourse.

Louvain, Brussels, Mechlin.

Hainault is fertile;

they speak French there as well as Dutch.

St Thomas; Bargen.

[1] *gar aus.* [2] E .iiii. back. [3] fertilitie B. [4] betwene B.

¶ The .xi. Chapter treteth of Gelderlond & of Cleue londe, and of the naturall disposicion of the people of those cuntres, & of their money & their speche.

[1] ¶ I Am of Gelderlond, & brought vp in the lond of Cleue;
In many thynges few men wyl me beleue;
I loue brawlyng and war, and also fyghtyng;
Nyght and day do proull, to get me a lyuyng;
Yet for all that, I am euer poore and bare,
Therfore I do lyue styl in penury and care;
For lack of meat, my chyldren do wepe,
Wherfore I do wake whan other men do slepe.
The fashyon of my rayment, chaunge I wyll not;
I am well contented whan I am warme and hot.

Few men believe me. I like fighting,

and am always poor, and my children lack food.

Although that Gylderlond and Cleue-lond be two sondry countrees & dukedoms, yet nowe one duke hathe them both[2]. Cleuelond is better then Gelderlond, for Gelderlond is sandy, and [has] muche waaste and baryn grownd. The Gelders be hardy men, and vse moche fyghtynge, war, and robbyng. The countrees be poore,

Cleveland is richer than Guelderland.

[1] sign. F .i. See the cut in B on the next page.
[2] 'the Duke of Gelders,' *Hall*, p. 743, A.D. 1527.

for Gelderlond hath vsed moche warre. The chyefe townes of Gelder lond is the towne of Gelder[1], & another towne called Nemigyn. And the chefe towne of Cleuelond is *the* towne of Cleue. In Gelder londe and Cleue lond theyr money is base gold, syluer, & brasse. In gold they haue Clemers gylders, and golden gilders, and gelders arerys: a gelder areris is worth .xxiii. steuers: .xxiii. steuers is worth .iii. s. There is another peece of golde called a horne squylyone: a horne squylyone is worthe .xii. steuers .xii. steuers is worthe .xix. d. ob.[2] In Syluer they haue a snappan; a snappa*n* is worth .vi. steuers: .vi. steuers is worth .ix. d. ob. In brasse they haue norkyns and halfe norkyns, & endewtkynge. their speche is Base Douche.

Chief towns: Gelder, Niemeguen (on the Whaal), Cleves.

Gold Coins: gilders, stivers, a horne-squlyone.

A silver Snappan.

A brass Norkyn ($\frac{1}{4}d.$ and $\frac{1}{8}d.$)

[1] Arnhem is the chief town of the present Guelderland. Gelder is now in Kleveberg, Prussia. [2] ob = $\frac{1}{2}d.$

Instead of the cut of the first, or Rose-Garland edition (1547 or -8), at the head of this chapter, the second, or Lothbury one of 1562 or -3, substitutes the cut on the right here:

¶ The .xii. chapyter tretyth of the lond of Gulyk & of Lewke, and of the naturall dysposycion of the people of the countres and of theyr money and of theyr speche.

¶ I Was borne in Gulyke; In Luke I was brought vp;
Euer I loue to drynke of a full cup.
My geese ones a yere I do clyp and pull;
I do sell my fethers as other men doth wull;
If my goos go naked, it is no great matter,
She can shyft for her selfe yf she haue meat & water.
The fashyon of my rayment, be it hot or cold,
I wyl not leue in ony wyse, be it neuer so old.

I pluck my geese once a year, and sell their feathers.

¶ The lond of Gulyk [2] is a dewkedom, and the lond of Lewke is an Archebyshopryche, for Archebyshoppes in Doche lond hathe great lordshyps and domynyons; yet they, and the aforesayd londes rehersed, from Calys, be vnder the domynyon of the Emperour. Gulyk is a fayre countre, not hylly nor watteryshe, but a playne countre. [3] Euery yeare they wyll clyp and pull theyr

Julich is a dukedom,

and is a fair flat land.

[1] sign. F .i. back. [2] Guylk AB. [3] F .ii. not signed.

geese, and the geese shall go naked; and they do sell the fethers to stuffe fether beds. They haue lytle wyne growyng in the co*un*tre. The chief townes of Gulyk is, the towne of Gulyk, and a towne named Durynge. the people be poore of the countre; townes men be ryche; & a man for his money shalbe well orderyd & intreted, as well for meat & drynke as for lodging. The lond of Lewke is a pleasaunt co*un*tre. The cheefe towne is the cytie of Lewke; there is Lewkes veluet made, & cloth of Arys. The speche of Gulyk a*n*d Lewke is Base-Doche. And theyr money is the Emperours coyne; but the Byshop of Lewke doth coyne both gold, syluer, and bras, the whiche is currant there, and in the londes or countres ther about.

Chief towns: Julich, Duren (between Aix and Cologne).

Liège (where velvet and Arras are made).

The speech is Dutch (Low-German).

The .xiii. Chapiter doth speake of base Almayn, and of the disposicion of the people of the countrey; of theyr speche & of theyr money.

[1] ¶ I Am a base Doche man, borne in the Nether-lond;

[1] F .ii. back. The cut has been used before, on p. 148.

I often get drunk,

Diuerse times I am cupshoten,[1] on my feet I cannot stand;
Dyuers tymes I do pysse vnderneth the borde;
My reason is suche, I can not speke a word;
Than am I tonge tayd, my fete doth me fayle,
And than I am harneysed in a cote of mayle;
Than wyl I pysse in my felowes shoes and hose,
Than I am as necessary as a waspe in ones nose.
Now am I harnest, and redy, Doche for to speke;
Vppon the beere van in the cruse my anger I wyl wreek.
A lomp of salt butter for me is good meat;
My knees shall go bare to kepe me out of heat;
Yet my olde cote I wyl not leaue of,
For if I should go naked, I may catche the cof.

can't speak a word,

and leak.

I like salt butter.

Of Base-Almayne,

(Maintz)

the chief city is Cologne on the Rhine, on the banks of which Rhenish wine is made.

Bonn.

The land is rich, and the people kind, but they get drunk, and make a mess.

¶ Base Almayne, or base Doche londe, rechyth from the hydermost place of Flaunders and Hennago, to the cite of Mense, and to Argentyne, as some Doche men holdeth opynyon. The cheef Cyte of Doche land or Almayne is the noble cyty of Colyn, to the whyche cometh the fayre water of Reene; on bothe sydes of the whyche water of Reene doth growe the grapes of the whyche the good Renysh wyne is made of. There is a vyne of grapes at a towne called Bune, of the whyche reed Renysh wine is made of. al Base-Almayne is a plentiful countre of corne and Renysh wyne, and of meat and honest fare, and good lodgyng. The people be gentyll and kynd harted. The worst fawt that they haue: many wyl be dronken; and whan they fall to quaffyng, they wyll haue in dyuerse places a tub or a great vessell standyng vnder the boord, to pysse in, or else they wyl defyle al the howse, for they wyl pysse as

[1] *Yvre:* com. Drunken, *cupshotten*, tipsie, whitled, flusht, mellow. ouerseene, whose cap is set, that hath taken a pot too much, that hath seene the diuell. *Forbeu* ... mellow, fine, cup-taken, pot-shotten, whose fudling or barley Cap is on.—*Cotgrave.*

they doo syt, and other whyle the one wyll pis in a nother[1] shoes. They do loue sault butter that is resty, and bareled butter. In Base Doche land be many [2] vertuous people, and full of almes dedes. In Base Almayn or Doche lond theyr money is gold, tyn, and brasse. In gold they haue crownes, worth four .s. viii. d. of sterlyng money. They haue styuers of tyn and bras: two styuers and a halfe is worth an Englysh grote. they haue crocherdes; .iii. crocherds is les worth than a styuer. they haue mytes; .xxvi. mytes is worthe an Englyshe peny. They haue Negyn manykens; a manyken is worth a fardyng; a Norkyng is worthe a halfpeny. They haue bras pens; a bras peny is .ii. d. fardynge of theyr money. Who so that wyl lerne to speke some Base Doche,—Englysh fyrst, and Doche, doth folowe.

They love salt butter.

Their money is stivers, crocherds (kreutzers?) mytes, manykens, norkyns, and pence.

Dutch numerals.

One. two. thre. foure. fyue. syx. seuyn. eyght. nyne.
Ene. twe. drie. vier. vie. ses. seuen. acht. nughen.
ten. aleuyn. twelue. thyrtene. fowrtene. fyftene.
teene. elue. twaelue. dertyene. vierteene. viefteene.
syxtene. seuentene. eyghtene. nyntene. twenty.
sestyene. seuentyene. achtyene. negentyene. twengtith.
one and twenty. two and twenty. thre and twenty.
en an twentyth. twe an twentyth. dre an twentith.
thyrty. forty. fyfty. syxty. seuenty. eyghty.
derteh. vierteh. vyntith. sesteh. zeuenteh. achtenteh.
nynte. a hondred. a thowsand.
negenteth. hondret. dowsent.

A talk in Dutch and English.

God morow, brother! *Morgen, brore!*
Syr! God gyue you good day!
Heer! God geue v goeden dah!
Syr! how do you fare? *Heer! hoe faerd ghy?*
Ryght well, blessyd be God!
Seer well, God sy ghebenedyt!

[1] another's. [2] F .iii. not signed.

A talk in Dutch and English.

Frend, whyche is the ryght way from hens to Colyn?
Vryent, welk is den rethten weh van hoer te Colyn?
[1] Syr, hold the way on the ryght hand.
Heer, holden den weh aye drechit hand.
Wyfe, God saue you! *Vrow, God gruet v!*
My syr, you be welcome!
Myn heer, yk hiet you welecome!
Haue you any good lodgyng?
Hab v eneh good herberh?
Ye, syr, I haue good lodgyng.
yo, myn heer, I hab goed harberh.
Wyfe of the house, gyue me some bread![2]
Vrow[3] van de hewse, ghewfft[4] me broot!
Mayd, gyue me one pot of beare!
Meskyn, ghewfft me en pot beere!
Brother, gyue me some egges!
Brore, ghewfft me eyeren!
Gyue me fyshe and fleshe!
Ghewft me fis an flees!
What shall I pay, ostes, for my supper?
How veele is to be talen, warden, for meell tyd?
My syr, .vi. d. *Myn heer, ses phenys.*
Hoste, God thanke you! *Warden, God dank ye!*
God gyue you good nyght and good rest!
God ghewfft v goeden naght an goed rust!
God be wyth you! *God sy met v!*
Sonday, *Sondah.* Monday, *Maendah.*
Tewsday, *Dysdah.* Wensday, *Wensdah.*
Thursday, *donnersdah.* Fryday, *Vrydah.*
Saterday, *Saterdah.*
Can you speke Doche? *Can ye Doch sprek?*
I can not speke Doche; I do vnderstond it.
Ik can net Doch spreke; Ik for stow.

[1] F .iii. back. [2] drynke A; bread B.
[3] Brow A; Vrow B. [4] gefft B.

[1]

[2]

¶ The .xiiii. Chapter treateth of hyghe Almayne or hyghe Doch lond, and of the dysposycyon of the people, and of theyr speche and of theyr money.

I Am a hygh Almayne, sturdy and stout,
I laboure but lytle in the world about;
I am a yonker[3]; a fether I wyll were;
Be it of gose or capon, it is ryght good gere.
Wyth symple thynges I am well content;
I lacke good meat, specyally in Lent.
My rayment is wouyn moche lyke a sacke;
Whan I were it, it hangeth lyke a Iack.
Euery man doth knowe my symple intencyon,
That I wyll not chaunge my olde fathers fashyon[4].

I'm a yonker when I wear a feather.

My coat's like a sack.

[1] F .iiii. not signed.

[2] Instead of the 3 cuts above, from the Rose-Garland edition, the Lothbury edition of 1562-3 gives only the centre one, which it has used before for the Norwegian, p. 142 at foot, and which both editions have used before for the Fleming, p. 146 above.

[3] G. *ein juncker*, a younker, younkster or youngster.—*Ludwig.* Dutch *een Ionck-heer* or *Ioncker*, A young Gentleman, or a Joncker.—*Hexham.*

[4] In 1510, Henry VIII made some 'yong Gentelmen' of his court fight together with battle-axes in Greenwich Park, and then gave them 200 marks to have a banquet together: "The whiche banket was made at the Fishemongers Halle in Teames strete, where they all met, to the number of .xxiii, all ap-

High-Almayne goes from Maintz to Trente in the Tyrol.

¶ Hyghe Almayne, or hyghe Dochelond, begynneth at Mens, and some say it begynneth at Wormes, & contayneth Swauerlond or Swechlond, and Barslond, and the hylles or mountayns of *the* most part of Alpes, stretching in length to a town called Trent by-yonde *the* mountayns: half *the* [1] towne is Doche, & the other halfe is Lombardy. There is a greate dyfference betwyxt Hyghe Almayne and Base Almayne, not only in theyr speche and maners, but also in theyr lodgynge, in theyr fare, and in theyr apparell. The people of Hygh Almayne, they be rude and rustycall, and very boystous in theyr speche, and humbly in their apparell; yet yf some of them can get a fox tale or two, or thre fox tayles, standyng vp ryght vpon theyr cappe, set vp with styckes, or that he maye haue a capons feder, or a goose feder, or any long feder on his cap, than he is called a "yonker." they do fede grosly, and they wyll eate magots as fast as we wyll eate comfets. They haue a way to brede them in chese. Maydens there in certayne places shall drynke no other drynke but water, vnto the tyme she be maryed; yf she do, she is taken for a comyn woman. Saruants also do drynke water to theyr meat. the countre is plentyfull of apples and walnuts; the mountayns is very baryn of al maner of vytels; howbeit the good townes be prouyded of vitels. Snowe dothe ly on the mountaynes, wynter and somer; wherfore, the hotter the daye is, the greater is the

High and Low-Germans differ much.

The High-Germans are rude, and badly drest.

One sticks a fox-tail or feather in his cap, and is called a *Yonker*.

Girls drink only water.

Snow lies on the mountains all the year.

pareyled in one sute or liuery, *after Almain* fashion, that is to say, their vtter garmentes all of yealow Satyne, yealow hosen, yealow shoes, gyrdels, scaberdes, and bonettes with yealow fethers, their garmentes & hosen all cutte and lyned with whyte Satyn, and their scaberdes wounde abought with satyne . . After their banket ended, they went by torche light to the Towre, presentinge them selfes before the kynge, who toke pleasure to beholde them."—*Hall's Chronicle*, p. 516. "the kynge, with .xv. other, apparelled in *Almayne* Iackettes of Crymosyne & purple Satyne, with long quartered sleues" . . . "and then folowed .xiiii. persones, Gentelmen, all appareyled in yealow Satyne, *cut like Almaynes*, bearyng torches." *ib.*, ed. 1809.

The third daie of Maie [1512] a gentleman of Flaunders, called Guyot of Guy, came to the kyng [Henry VIII] with .v.C. *Almaines* all in white, whiche was cutte so small that it could scace hold together.—*ib.*, p. 527.

[1] F .iiii. back.

flods, that they renne so swyft that no man can passe for .v. or .vi. howres, and than it is drye agayne. Certayn mountaynes be so hygh that you shal se the hyll tops aboue the cloudes. In the valy it is euer colde. I haue seen snowe in somer on saynct Peters day and the Vysytacion of our Ladye. A man may see the mountaynes fyftene myle of, at a cyte called Ulmes, where fustyan vlmes is made, that we cal holmes. In Hyghe Almayn be good cities and townes, as Oxburdg, Wormes, Spyres, Gyppyng, Gestynge, and Memmyng. In Hygh Almayne theyr money is golde, alkemy, and bras. In gold they haue crownes of .iiii. s. & .viii. d. In alkemy[1] and bras they haue rader Wyesephenyngs worthe [2] almost a styuer; they haue Morkyns[3], Halardes, Phenyngs[4], Crocherds, Stiuers[5], and halfe styuers. Who so wyl lerne Hygh-Doch,—Englysh fyrst, & Doche, followeth.

On the mountains

I've seen snow on June 29 and July 2.

Ulms, where 'holmes' is made.

German towns, Augsburg, &c.

High-German coins, wheel-white-pennies.

One. two. thre. foure. fyue. syx. seuyn. eyght.
Eyne. sway. dre. feer. vof. sys. zeuen. awght.
nyne. ten. aleuyn. twelue. thyrtene. fowrtene.
neegh. zen. elue. zwelue. dersheene. feersheene.
fyftene. syxtene. seuentene. eyghtene. nyntene.
fiftsheene. sissheene. zeulsheene†. *aughtsheene. neeghsheene.* [† for *seebsheene*]
twenty. one & twenty. two and twenty. thre and twenty.
zwense. eyne en zwense. sway en zwense. [6]*dre en zwense, &c.*
thyrty. forty. fyfty. syxty. seuenty. eyghty.
dreshe. feertshe. vofshe. sysshe. zeuenshe. aughtshe.
nynte. a hondred. a thowsand. two thowsand, &c.
neegshe. a hownder. a dowsand. sway dowsand, &c.

High-Dutch or German numerals.

¶ God morow, my master! *Goed morgen, myh* § *hern!*
My master, whyche is the way to the next towne?
Mih leuer hern, weis me de reighten weg to de awnderstot?
My brother, gyue me whyt bread and wyne!
my leuer broder, geue meh wyse brod en wayne!

[§ ? *myn*]
A talk in German and English.

[1] ? tin. [2] sign. G .i. [3] read 'Norkyns,' hapence: p. 157, 153.

[4] *Pfenning*, the 12th part of a groschen and of a Sterling, Flemish and Lübish shilling, a penny or denier.—*Ludwig.*

[5] *Stiver*, a Dutch coin worth $1\frac{1}{5}$ Penny English, of which 20 make a Guilder, and 6 a Flemish Shilling.—*Kersey's Phillips.* [6] ore AB.

A talk in High-German and English.

Hostes, haue you good meate?
Wertyn, hab ye god eften?
ye, I haue enough. *yo, Ik hab gonowgh.*
Hostes, gyue me egges, chese, and walnots!
Wertyn, geue meh ayer, caase, en walshe nots!
mouch good do it you! *Goot go seken eyh esseu!*
I thank yo[u], my mayster!
Ih dank ze, myh[1] *leuer hern!*
What tyme is it of the day? *What hast is gosloken?*
Hostes, God be with you, wyth al my hert!
Wartyn, Goot go seken for harteon!
my master, wyl ye drynk a pot of wyne?
myh leuer hern, wylter drenke a mose wayne?

The .xv. chapter treateth of Denmark and of the natural dysposycion of the people, and of theyr mony and speche.

¶ I Am a Dane, and do dwell in Denmarke,
Seldom I do vse to set my selfe to[3] warke

[1] ? *myn.* [2] sign. G .i. back. B puts the cuts on the right. [3] a B.

I lyue at ese, and therfore I am content;
Of al tymes in the yere I fare best in Lent;
I wyl ete beenes, and good stock fysh—
How say you, is not that a good dysh?—
In my apparel I was neuer nyce,
I am content to were rough fryce;
I care not if euery man I do tel,
Symple rayment shal serue me ful wel;
My old fashion I do vse to kepe,
And in my clothes dyuers tymes I slepe;
Thus I do passe the dayes of my lyfe,
[1] Other whyle in bate, and other whyle in stryfe;
Wysdome it war to lyue in peace and rest;
They that can so do, shal fynd it most best.

I eat beans and stock-fish,

and wear rough frieze.

I often sleep in my clothes.

¶ By cause I do pretend to writ fyrst of all Europ and Christendome, & to fetch *th*e cyrcuyte about Christendome, I must returne from Hygh Almayn, & speke of De*n*marke, the whiche is a very poore cou*n*tre, bare, & ful of penurite[2]; yet ther doth grow goodly trees, of the which be mastes for shyps made, & the marchau*n*tes of *th*e cou*n*tre do sell many masts, ores, & bowe staues. The Danes hath bene good warryers; but for theyr pouerte I do marueyle how they dyd get ones Englonde; they be subtyll wytted, & they do proll muche about to get a pray. They haue fysh and wyldfoule sufficient. Theyr lodgyng and theyr apparel is very symple & bare. These be the best townes in De*n*mark: Ryp, & By borge. In De*n*mark, their mony is gold, and alkemy,[3] and bras. In gold they haue crownes; & al other good gold doth go there. In alkemy and bras they haue Dansk whyten. Theyr speche is Douche.

Denmark's a very poor country, but has fine trees.

The Danes prowl about after prey.

Ribe and Wiborg.

Danish is Dutch.

[1] G .ii. not signed.

[2] Yet in the great Dearth of wheat in England in 1527, wheat was imported from Denmark, among other places: "the gentle marchau*n*tes of *th*e Styliard brought fr*om* DANSKE, Breme, Hamborough, and other places, great ple*n*tie; & so did other marchau*n*tes from Flau*n*ders, Holand, and Frisland, so that wheat was better chepe in Lo*n*don then in all England ouer."—*Hall's Chronicle*, p. 736, ed. 1809.

[3] *Alkani*, tin. Howel (in Halliwell's Glossary).

¶ The .xvi. Chapter treateth of Saxsony, and of the natural disposicion[1] of the Saxsons, and of their mony, and of theyr speche.

I'm a heretic.

Romans cry vengeance on me, and curse me.

[2]¶ I am a Saxson, serching out new thynges[3];
Of me many be glad to here new tidinges.
I do persist in my matters and opinions dayly,
The which maketh *the* Romayns vengians on me to cry;
Yet my opinions I wyl neuer[4] leue;
The cursyng that they gyue me, to them I do bequeue;
The fashion of my rayment I wyl euer[5] vse,
And the Romayns fashion I vtterly refuse.

I wonder how the Saxons conquered England.

Saxony is fertile;

¶ Out of Denmarke a man may go in to Saxsony. Saxsony is [a][6] Dukedom-shyp, And holdeth of hym selfe. I do maruel greatly how the Saxsons should conquere Englonde, for it is but a smalle countre to be compared to Englond; for I think, if al the world were set against Englond, it might neuer be conquerid, they beyng treue within them selfe. And they that would be false, I praye God too manyfest them what they be. The countre of Saxsony is a plentyful[7] countre, and a

[1] dispocion A; disposicion B. [2] G .ii. back. [3] thynkes A. [4] euer A; neuer B. [5] euer A; neuer B. [6] A omits 'a.' [7] plentyfill A; plentyful B.

fartyll; yet there is many greate mountaynes and woddes, in the whyche be Buckes and Does, Hartes, and Hyndes, and Wylde Boores, Beares, and Wolfes, and other wylde beastes. In Saxsony is a greate ryuer called Weser; And there be salte wels of the water, of the whyche is made whyte salt. In the sayd countre doth grow copper. The people of the countre be bold and strong, and be good warriers. They do not regarde the byshoppe of Rome[1] nor the Romayns, for certaine abusions. Martyn Leuter & other of hys factours, in certayne thynges dyd take synistrall opinions, as concernynge prestes to haue wyues, wyth such like matters. The chefe cyte or town of Saxsony is called Witzeburg, which is a vniuersite. In Saxsony theyr monye is golde and brasse. In golde they haue crownes, In brasse thei haue manye smal peces. There speche is Doch speche.

but has many woods, deer, and wild beasts,

the Weser river,

and copper mines.

The Saxons don't mind the Pope.

Martin Luther held heretical opinions.

Wittenburg University.

Saxon money.

[1] Andrew Boorde speaks, I suppose, as a Saxon heretic here (Pope = Bp of Rome), Romanist though he had been, and condemning Luther as he does in the next lines.

The Lothbury edition, 1562-3, substitutes the cut below for the one at the head of this chapter. The Rose-Garland edition uses it for the man of Bayonne, p. 165, below, and both editions use it for the Egyptian, p. 217.

[1]

¶ The .xvii. chapter treateth of the kyngdo*m* of Boeme, and of the dysposycion of the people of the countre, of theyr monye, and speche.

¶ I Am of the kyngdome of Boeme,
I do not tel al men what I do meane;
For the popes curse I do lytle care;
The more the fox is cursed, the better he doth fare.
Euer sens Wyclif dyd dwel wyth me,
I dyd neuer set by the popes auctorite.
In certayn articles Wyclif dyd not wel,
To reherse them now I nede not to tell,
For of other matters I do speke of nowe;
Yf we do not wel, God spede the plow!
Of our apparrel we were neuer nyce;
We be content yf our cotes be of fryce.

I haven't cared for the Pope's curse since Wyclif's time.

I'm content with frieze.

¶ The kyngdome of Boeme is compassed aboute wyth great hygh mountaynes and great thycke wods. In the [2]whyche wods be many wylde beastes; amonges

Bohemia is circled with mountains.

[1] G .iii. not signed. [2] G .iii. back.

al other beastes there be Bugles, that be as bigge as an oxe; and there is a beast called a Bouy, lyke a Bugle, whyche is a vengeable beast. In dyuers places of Boeme there is good fartyl grownd, the whyche doth bryng forth good corne, herbes, frutes, and metals. The people of Boeme be opinionatyue, standyng much in theyr owne conceits. And many of them do erre contrary to vs in the ministracion of the .vii. sacraments, & other approbated thynges, the which we do vse in holy churche. In Boeme is indifferent lodging, and competent of vitels, but they do loue no Duckes nor malardes. theyr condicions and maners be much lyke to the Hygh Almayns, & they do speke Duch. In Boeme is a goodly cyte called Prage, wher the king of Boeme doth ly much whan he is in the countre. In Boeme theyr monye is Golde, Tyn, and Bras. In Golde they haue crownes; In Bras they haue smal peces as in Doch lond; theyr speche is Doch.

Bugles, Bovy. The Bohemians are self-willed, and err from Holy Church. They don't like ducks. Their chief city is Prague.

Instead of the right-hand cut of the Rose-Garland edition, at the head of this chapter, the Lothbury one has another, of a woman without a flower, and with differences in her skirt. It is given on the right here.

The .xviii. chapter treateth of the kyngdome of Poll, and of the naturall dysposicion[1] of the people, and of theyr mony and spech.

[2] I Am a power man of the kyngdom of Pol;
Dyuers tymes I am troubled wyth a heuy nol.
I like bees; — Bees I do loue to haue in euery place,
I sell honey, — The wex and the hony I do sel a pace;
pitch, and tar. — I do sel flex, and also pyche and tar,
Marchaunts cometh to me, fetchyng it a far.
My rayment is not gorgious, but I am content
To were such thynges as God hath me sent.

¶ The kyngdome of Poll is on the Northe syde of the kyngdom of Boeme, strechynge Estwarde to the kyngdom of Hungary. (In Poland are woods and wild beasts,) In Pol be great wods and wyldernes, in the whych be many bees, and wylde beastes of diuers sortes. In manye places the countre is full of fartillite, and there is much pych, and Tar, and Flex. (pitch, tar, and flax.) There be many good townes; the best towne named[3] Cracoue. (Cracow is their chief town.) The people of the countre of Poll be rewde, and homlye in theyr maners and fashions, and many of them haue learned craftines in theyr byeng and sellyng; (They're crafty dealers; but badly off.) and in the co*n*ntre is much pouerte and euyll

[1] dysposion A; dyspocicion B. [2] G .iiii, [3] anmed A; named B.

fare in certayne places. The people do eat much hony in those parties. they be peasible men; they loue no warre, but louyth to[1] rest in a hole skin. Theyr rayment and apparel is made after the High Doche fashion wyth two wrynck-kles and a plyght; theyr spech is corrupt Doche; the mony of Poll is goulde and bras; all maner of gold goth there.

The Poles don't like war,

and they speak bad German.

[1] too A; to B.

The Lothbury edition of 1562 or 1563 gives this woodcut of the Pole, or 'power man of the kyngdom of Pol,' or rather the personage who does duty for him.

[1]

¶ The .xix. chapter treateth of the kyngdome of Hungary, and of the natural dysposision[2] of the people, and of theyr mony & spech.

I do dwel in the kyngdome of Hungary;
I hate the Turks;
Bytwyxt the Turkes and me is lytle marcy;
And although they be strong, proud, and stout,
Other whyle I rap them on the snowt;
they've won much of our land.
Yet haue they gotten many of our towns,
And haue won of our londs and of our bowns;
If we of other nacions might haue any helpe,
We wold make them to fle lyke a dog or a whelp.
Out of my countre I do syldome randge;
The fashion of my apparel I do neuer chaunge.

[1] G .iiii. back. The right-hand cut is from B, and differs a little from that in A, which is the cut of Boorde on the title-page of *Barnes*, p. 305 below, with a different riband over the head.

[2] dysposion A; dysposision B.

[1] The kyngdom of Hungary is beyond the kyngdome of Poll, estward. The lond is deuided into two partes, the whych be called "great Hungary," and the "lesse[2] Hungary." The countres be large & wyde; there is gret mountayns and wildernes, the whych be repleted with manye wylde beeastes. Ther is salte digged out of hylles. And there is found certayne vaynes of gold. In Hungary ther be many Aliens of dyuers nacions, and they be of dyuerce fashions, as wel of maners as of lyuyng, for the lond doth Ioyne to the lond of Grece at the south syde. The great Turke hath got much of Hungary, and hath it in peasable possession. And for as much as there is dyuerce people of diuerce nacions, ther is vsed diuerce speches, & ther is currant diuerce sortes of mony. ther be many good cytyes & townes the which be called "vouen;" Sculwelyng,[3] Warden, Scamemanger, and a noble cytie called Clipron, and a regal castyl called Neselburgh, And a gret citie called Malla vina, the whych is almost the vttermost cytie of Hungary, by the whych cite doth roune the regall flod of Danuby.[4] The spech of Hungary is corrupt Italien, corrupt Greke, & Turkysh. Theyr mony is gold [&] bras[5]: in gold thei haue duccates & sarafes. In bras thei haue myttes, duccates, & soldes, and other smal peses of brasse which I haue for-got.

Great Hungary and Less.

Gold is found there.

Many aliens dwell there.

The chief towns of Hungary: Stuhlweissenburg, Groswardein, Steinamanger.

By Mostalavina runs the Danube.

Hungarian speech and money.

The .xx. chapter treateth of the lond of Grece, & of Constantine-nople, and of the naturall disposicion of the people, and of theyr mony and speche.

[1] sign. H .i. [2] lessee A ; lesse B.
[3] Sculwelrng A ; Sculwelyng B. [4] daunby AB. [5] good bras B.

[1] I Am a Greke, of noble spech and bloud,
Yet the Romayns with me be mervell*ous* wood;
For theyr wodnes and cursyng I do not care;
The more that I am cursyd, the better I do fare.
Al nacions vnder them, they woulde fayne haue;
Yf they so had, yet would they more craue;
Vnder their subiection I would not lyue,
For all the pardons of Rome if they wold me geue.[2]

The seven provinces of Greece.

Constantinople belongs to the Turks.

St Sophia's is the fairest cathedral in the world.

Constantinople is built with two sides to the sea.

By it is St George's Arm, or the Hellespont.

¶ The lond of Grece[3] is by-yonde Hungary; it is a greate region and a large countre. For they haue .vii. prouinces, whyche be to saye: Dalmacye, Epirs, Eladas, Tessaly, Macydony, Acayra, Candy, and Ciclades. The lond of Grece is a ryche countre & a fartyll, and plenty of wine, breade, and other vytels. The chefe cyte of Grece is called Constantinople: in old time it was an Empyre, and ther was good lawes and trwe Iustyce keepe[4]: but nowe the Turke hath it vnder his dominion, howbeit they be styl Chrysten men, and christened; and there is at Constantinople[5] a patriarke: And in Constantinople they haue the fairist cathedral churche in the Worlde: the church is called Saynte Sophyes Churche, in the whyche be a wonder-full syght of preistes: they say that there is a thowsande prestes that doth belo*n*g to the church: before the funt of the church is a pycture of copper and gylt, of Iustinian, that sytteth vpon a horse of coper. Constantinople is one of the greatyst cytes of[6] the world: the cyte is built lyke a triangle; two partes stondeth and abutteth to the watter, and the other parte[7] hath a respect of[8] the londe: the cyte is well walled, and there commeth to it an arme of the See, called Saynct Georges arme or Hellysponte, or the myghte of Constantinople: saynt Luke and saynt Ioha*n*

[1] H .i. back. [2] geue A; gyue B.
[3] *Hidroforbia* in englyshe is "abhorrynge of water," as I lerned in the partes of grece. *Breuiary*, fol. cxxii. *Forewords* p. 74. [4] kepte B. [5] Constanople A; Constantinople B.
[6] citie in, B. [7] partet A; parte B. [8] to B.

Erisemon lyeth there: and they say that there is the holy crosse, and Iesu Chrystes cote that had no seeme. The v[1]niuersitie[2] of Salerne, where physick [is] practysed is not far from Constantinople. the Greciens do erre & swere in mani articles concerning our fayth, The whyche I do thinke better to obmyt, and to leue vnwryten, than to wryte it. In Constantinople theyr money is gold, syluer, & Brasse: in gold they haue sarafes; a saraf is worth .v.s. sterlynge; in syluer they haue aspers; an asper is worth an Englysh peny; in Bras they haue soldes; .v. sold is worth an Asper. they haue myttes; .iiii. myttes is worth a sold.

Relics. University of Salerne. Greek Church is heretical. Greek money: sarafes, aspers, soldes, myttes.

a letter whiche the Greciens sent to the byshop of Rome:—

Grecians' letter to the Bp of Rome.

Parotenciam tuam summam ci[r]ca[3] tuos subiectos firmiter aredimus; superbiam tuam summam[4] tollerare non possumus; Auariciam[5] tuam saciare non intendimus. dominus tecum! quia dominus nobiscum est.

If any man wil learne to speke Greke, such Greke as they do speke at Constantynople and other places in Grece,—Englysh and Greke doth folow.

Modern-Greek numerals.

One. two. thre. foure. fyue. syx. seuyn. eyght.
Ena. dua. trea. tessera. pente. exi. esta. oucto.
nyne. ten. aleuyn. twelue. thyrtene. fowrtene.
enea[6]. *deca. edecaena. edecadna. decatrea. decatessera.*
fyftene. syxtene. seuentene. eyghtene. nyntene.
decapente. decaexi. decaesta. decaoucto. decaenea.
twenty. one and twenty. two and twenty, &c.
cochi. ecochiena. ecochidua,[7] *&c.*
thyrty. forty. fyfty. syxty. seuenty. eyghty.
trienda. serenda. penenda. exininda. estiminda. outoinda.
nynte. a hondred.
eniminda.[8] *ekathoi.*

[1] H .ii. not signed. [2] vniuesitie A. A leaves out too the next 'is' of B. [3] sūnam cica AB. [4] sūnā AB. [5] Anriciam AB. [6] enca AB. [7] dna AB. [8] enimida AB.

A talk in Modern-Greek and English.

God spede you, Ser! *Calaspes, of-ende!*
Ser, you be welcome! *Ofende, calasurtis!*
Syr, from whens do you come? *Offende, apopoarkistis.*
I did come from England.
Ego napurpasse apo to anglia.
How far is it to Constantinople?
Post strat apo to Constantion.
Ser, ye haue .xxti. myle. *Ofende, ekes ecochi mila.*
Mastres, good morow! *Chira, cala mera!*
Mastres, haue you any good meate?
Chira, ekes kepotes calonofy.
Ser, I haue enough. *Ofende, ego expolla.*
Mastres, geue me bread, wyne, and water!
Chira, moo dosso me psome, cresse apo to nero!
Com hyder, and geue me some flesh.
Eila do dosso moo creas.
Bryng hyder to me that dish of flesh!
Ferto to tut obsaria. creas.
Good nyght! *Cale spira!*

The trewe Grek foloweth.

Another talk in true, or Classical, Greek.

Good morow! *Cali himera!*
Good spede! *Calos echois!*
Good euyn! *Cali hespera!*
You be welcome! *Cocharitomenos hikis!*
Syr, whych is the way to Oxford?
Oton poi to Oxonionde?
Syr, you be in the right way. *O outtos orthodromeis.*
Hostiler, set vp my horse, and gyue him meate!
Zene[1], *age ton hippon apon apothes, kae sitison avton.*
Mayd, haue you any good meate? *Eta, echis ti sition?*
Ye, master, enowgh. *Echo dapsilos.*
Geue me some bread, drynke, and meate.
Dos mi ton arton, poton, kae siton.
What is it a clok? *Po sapi hi hora tis himeras?*

[1] *Zene* AB.

Wyfe or woman, geue me a reckenyng! A talk in true-Greek and English.
Gyny[1], *eipe moi ton Analogismon.*
I ame contentyd or plesed. *Arescy moy.*
hostes, fare wel! *Zene*[2], *chere!* or els, *Errosa!*
Syr, you be hertely welcome!
[3]*Kyrie, mala cocharitomenos ilthes.*
Woulde to God that you woulde tary here styl!
Eithe ge to entautha men aei para hymas menois.[4]
O wyfe, I can not speake no Greke!
Ohe gyny[1], *ov dyname calos elinisci legin.*
Syr, by a lytel and a lytyle you shal lerne more.
O outes dia microu mathois an ablinisci lalein.
O hostes, there is no remidy but I must depart.
Zene, anagaeos apieton esci moy!
Syr, than God be your sped in your iorney!
Deospota, theos soi dixios esto metaxi procias!
Fare wel to you al! *Cherete apapapantes!*
God be with you! *Thos meth ymon!*

The .xxi. chapter treateth of the kyngdome of Sicell, and of Calabre, And of the naturall disposicion of the people, and of theyr mony and speche.

I was borne in the kyngdome of Sycel;
I care for no man, so that I do wel.
And I was borne in Calabry, In Calabria, flies bite us.
Where they do pynche[5] vs many a fly. 4

Gyuy AB. [2] Zeue AB. [3] H .iii. not signed. Kyrle AB.
[4] meuois AB. [5] theyr doth bynche B

We be nayboures to the Italyons,
Wherfore we loue no newe fashions;
For wyth vs, except he be a lord or a Grecyon,
[1]Hys rayment he wyl not tourne from the old fashyon.

I shall now come back from Greece, towards Calais,

¶ I haue spokyn of Grece, one of the endes or poynts of Europ; wherfore I pretend to returne, and to come round about, & thorow other regyons of Europ vnto the tyme I do come to Calas agayne,—where that I dyd take my first iorney poynt out of Englond,—& other landes anexed to the same; wherfore in my returnyng

and speak first of Sicily and Calabria.

I wyl speke fyrst of Sicel & Calabry. Sycel is an Ilond, for it is compased wyth water of the see. ther be

In Sicily are mosquitoes (?), like our English flies;

many flyes, the whych wyl styng or byte lyke the flyes of Italy; and loke, where that they do stynge, they wyll bryng the bloud after; and they be such flyes as do set on our table & cup here in England. But they be so eger and so vengeable that a man can not kepe hym selfe from them, specially if he slepe the day tyme. in Sycel

and great storms.

is much thondoryng and lyghtnyng, and great impietouse[2] wyndes. The countrey is fartyl, and there is

Syracuse.

much gold. The chefe towne is Ciracus. & there is a

The river Arethusa.

goodly ryuer called Artuse, where is found whyt corall.

Calabria.

¶ Calabre is a prouince ioyned to Italy; & they do vse the Italion fashion; and theyr mony and spech is muche lyke Italy money and speche.

The .xxii.[3] chapter treateth of the kingdome of Naples, and of the naturall dysposicion of the people and of theyr speche and of there money.

¶ In the kyngdome of Naples I do dwel;
I can nod[4] with my hed, thynkyng euel or well.

I keep my own counsel.

Whan other men do stond in great dout,
I know[5] how my matters shalbe brought about;

[1] H .iii. back. [2] iupietouse A (impetuous); iupirtouse B.
[3] .xx. A. [4] not A; nod B. [5] knew AB.

The fashyon of my rayment I wyl neuer leue ;
Al new fashyons, to Englond I do bequeue ;
I am content with my meane aray,
[1]Although other nacions go neuer so gay.

I keep old fashions, and leave new ones to England.

I must nedes go out of the cyrcuyt, and not dyrectlye go round about Europ & Chrystendom ; for if I should, I shold leue out kyngdomes, cou*n*tres & prouinces ; wherfor, as I went forward, so I wyl come bakeward, and wyll speke of the kyngdom of Naples.[2] The cou*n*tre, & specially the citye of Naples, is a populus cytye & cou*n*tre ; yet I dyd not se nor know that they were men of gret actiuite, for they do liue in peace without warre. The cou*n*trey is ful of fartylite, & plentiful of oyle, wine, bread, corne, fruit, and money. The Napulions do vse great[3] marchaundyse ; & Naples is ioyned to Italy, wherfore they do vse the fashio*n*s and maner of Italyons and Romayns ; and marchauntes passeth from both parties by the watter of Tiber. in Naples ther be welles of water the whych be euer hot, and they be mediscenable[4] for sycke people. the chefe cathedral churche of Naples is called Brunduse. Theyr spech is Italyan corrupted. In Naples theyr money is gold and brasse, lyke money of Italy and Lumberdy ; and they do vse the fashyons of the Italyans.

I (A. Boorde) can't go direct to Calais, but must turn off to Naples.

Naples has many people, who are not active ;

but they're great merchants.

Hot wells in Naples.

Brindisi.

The .xxiii. chapter treateth of Italy and Rome, and of *th*e naturall dysposycyon of the people, and of theyr money & speche.

[1] H .iiii. not signed. [2] Napls AB. [3] gerat A ; great B.
[4] modiscenaple A ; mediscenable B.

My country is fertile.

¶ I am a Romayne, in Italy I was borne;
I lacke no vytayles, nor wyne, breade, nor corne;
All thynges I haue at pleasure and at wyll;
Yf I were wyse, I wolde kepe me so styl;

I want the world to be subject to me.

Yet all the worlde I wolde haue subiecte to me,
[1]But I am a-frayd it wyll neuer be.
Euery nacion haue spyed my fashions out;
To set nowght by me now they haue no dout.

I've let my church fall down.

My church I do let fall; prophanes your[?] is vsed;
Vertu in my countre is greatly abused;
Yet in my apparel I am not mutable,
Althowh in other theynges I am founde variable.

Tiber. Rome. St Peter's Church. Little virtue, and abominable vices in Rome. The Italians, &c., reckon from one to 24 o'clock, which is midnight.

¶ Italy is a noble champion coun*tre*, plesaunt, & plentyfull of breade, wyne, and corne. There be many good pastures & vinyerdes.[2] The noble[3] water of Tyber doth make the countre rych. The people of the coun*tre* be homly and rude. The chefe cytye of Italy is called Rome, the whych is an old cyte, & is greatly decaide; & saint Peters churche, whych is theyr head church & cathedral churche, is fal downe to the grounde, and so hath lyen many yeres wythout reedyfiyng.[4] I dyd se lytle vertue in Rome, and much abhominable vyces, wherfore I dyde not lyke the fashion of the people; such matters I do passe ouer. who so wyl se more of Rome and Italy, let hym loke in the second boke, the .lxvii. chapter.[5] The Latyns or the Italio*ns*, the Lomberdes & the[6] Veneciens, wyth other prouynces anexed to the same, doth vary in dyuers nu*m*bringe or rekanynge of theyr cloke.[7] At mydnyght they doth[8] begyn, and do reken vnto .xxiiii. a cloke,[7] & than[9] it is

[1] H .iiii. back. [2] vniyerdes A; vinyardes B.
[3] nople A; noble B. [4] redyfiyng A; reedifiyng B.
[5] See *The Extrauagantes*, or second Part of *The Breuyary*, fol. v. back, and vi., extracted in the *Forewords* above, p. 77-8. On 'the second boke,' see p. 21. [6] that A; the B.
[7] clocke B. After 'cloke,' A wrongly inserts "and than it is mydnyghte and at one a cloke," which it repeats a line further on. [8] doo B. [9] then B.

mydnyght; and at one a clok[1] thei do begyn agayne. also theyr myles be no longer[2] than[3] our miles be, and they be called Latten miles. Doch myles and French leges[4] maketh .iii. of our myles, and of[5] Latyn myles. In Rome and Italy theyr monye is gold, syluer, & bras. In gold thei haue duccates, in syluer they haue Iulys,—a Iuly is worthe .v.d. sterlynge,—in bras they haue kateryns, and byokes, and denares. who that wyl learne some Italien,[6]—Englyshe and Italyen doth folow.

Latin miles are the same as ours.

Ducats, jules (or juliwaes),

kateryns, baiocchi, denari.

[7] One. two. thre. foure. fyue. syx. seuyn. eyghte. nyne.
Uno. two. tre. quater. sinco. si. serto†. octo. nono.
ten. aleuyn. twelue. thyrtene. fowrtene. fyftene. syxtene.
dees. vnse. duose. trese. quaterse. kynse. sese.
seuentyne. eyghtene. nyntene. twenty. one and twenty.
dessetto. desotto. desnono. vincto. vinto vno.
two and twenty. thre and twenty. foure and twenty.
vincto duo. vincto tre. vincto quater.
therty. forty. fyuete. sexte. seuente.
trento. quaranto. sinquanto. sessento. settanto.
eyghte. nynte. a honderd. a thowsande.
octento. nonanto. cento. milya.

Italian numerals.

[† for *setto*.

Good morow, my syr! *Bonus dies, nu sir!*
Good lyfe be to you, mastres! *Bona vita, ma dona!*
Ys thys, or that, the ryght way to go to Rome?
Est kela, vel kesta, via recta pre andare Rome?
(The true wryting is thus: *Est quela vel questa via;*
But, and[8] I shoulde so write as an Italyan doth, an Englyshman, without teachyng, can not speake nor prelate the wordes of an Italyan.)
¶ How farre is Rome hens? *Sancta de ke est Roma?*
Hit is .xl. myles hence. *Est karenta milia.*[9]
Brother, how farre is it to the nexte lodgyng?
Fradel, kanta de ke ad altera ostelaria?

A talk in Italian and English.

I write phonetically, to enable Englishmen to understand Italian.

[1] clocke B. [2] long or A. [3] then B. [4] leages B.
[5] or AB. [6] Italien and AB. [7] sign. I .i.
[8] an' if. [9] nulia A; milia B.

A talk in Italian and English.

Hit is .iiii. myle. *Sunt kater*[1] *milia.*
May we haue there this nyght good lodgyng?
Podemus auere bonissima loga pro reposar?
My serre, there is good lodgyng.
My ser, se aueryte bonissima.
You be welcome to this count[r]ye! can you speke Italian?
Vene[2] *venuta kesta terra! se parlare Italionna?*
[* vn, un] [3]Ye ser, I can speke a lytle. *My ser, se vin* pauk*
I do thanke[4] you wyth al my hart! *Regracia, bon cor!*
What tydynges is in your countre?
Auete nessona noua de vostra terra?
There is nothing but good, blessed be God!
[† nome] *Nessona noua† salua tota bona, gracia none Deo!*
How do you fare? *Quomodo stat cum vostro corps?*
I do fare wel. *Ge sta beene.*
Wyl you go eate some meate? *volite mangare?*[5]
[§ kantos] What is it a cloke, brother? *kantar§ horas, fardell?*
Hyt is thre and twenty a clock. *sunt vinccitres horas.*
Wyfe, geue me a pot of wyne!
Ma dona, dona[6] *me vn buccal de vyne!*
Much good do hit[7] you! *Mantingat vos Deus!*
Bryng vs a reckenyng, wyfe!
Far tu la counta, madona!
Hostes, pay to this man .iii. kateryng.
Hostessa, paga kesto hominy tres katerinos.
God be wyth you! *Va cum De!*

[1] katet AB. [2] It is *Vene*, not *Bene* in AB. [3] sign. I .i. back.
[4] tanke A; thanke B. [5] maugare A; mangare B.
[6] doua A. [7] good hit A; good do hit B.

The .xxiiii. chapter treateth of Venys, and of the naturall dysposicyon of the people of the country, of ther mony and of theyr spech.[1]

[2]I am a Venesien both sober and sage ;
In all myne actes and doynges I do not outrage ;
Grauite shal be founde euer in me,
Specially yf I be out of my countrey.
My apparell is ryche, very good and fyne.
All my possessyon is not fully myne,
For part of my possession, I am come tributor[3] to *the* Turke.
To lyue in rest and peace, in my cytye I do lourke.
Some men do saye I do smell of the smoke ;
I passe not for that, I haue money in my pooke
To pacyfye the Pope, the Turke, and the Iue :
I say no more, good felow, now adew !

I am always grave.

My dress is rich.

I pay tribute to the Turk.

I have money to pacify my foes.

Yf I should not bryng in & speke of Venes here, I sholde not kepe the circuit of Europe. whosoeuer that hath not seene the noble citie of Venis,[4] he hath not sene the bewtye & ryches of thys worlde.[5] Ther be

Venice is the beauty of the world.

[1] of theyr speche and of there money B. [2] sign. I .ii.
[3] tribut B. [4] venus A ; venis B.
[5] A rare poem in a paper MS of Mr Henry Huth's, about 1500 A.D.,—a poem of which part is printed in Wey's Pilgrimages for the Roxburghe Club—praises Venice as strongly as Andrew Boorde does :

Merchants flow to Venice.

ryche marchauence and[1] marchauntes; for to Venys is a

Here begynnyth the Pilgrymage and the wayes of Ierusalem.

GOd þat made bothe heuen & helle,
To the, lorde, I make my mone,
And geue me grace þe sothe to telle
Of þe pylgrymage þat I haue to gone.
I toke my leue at **Veynes towne**,—
And bade felowes for me to praye,—
That is a cyte of grete Renowne,
And to **Ierusalem** I toke my waye;
But of alle þe Cetys þat I haue seyne,
That maye **Ueynes** kynge been,
That stondith in þe Grikys see alone:
Hit is so stronge alle abowte,
Of enemyes dare hit not drede;
Corsayntes lyen in þe toune abowte;
Who so wylle hym seke, he shal haue mede.
Saynt **Marke**, Saynt **Nicholas**,
Thes two sayntes they loue & drede;
Saynt **Elyne** þat fonde þe Crosse,
And Saynt **Iorge**, oure ladyes knyghte,
Amonge hem beryth grete voyis,
And lythe in golde & syluere I-dyght;
Saynt **Powle**, þe fyrst Eremyght.
And Saynt **Symone** iust, also
Zachare, þe fadre of Iohan baptiste,
Lyeth thense but a lytel therfro;
Saynt **Luce** and saynt **Barbera**
That holy were, bothe olde & younge;
A M[l] Innocentys and moo
Lythe there closyd;
Saynt **Cristofer** lythe in þe Cyte:
Twyes in þe ȝere, who so theder wyll come,
He shal haue playne Remyscioun
Also wel as in the ȝere of grace.
Than passyd we to þe **Iles** of þe see,
Corfe, Medon, and **Candye**;
And some of þe Iles of þe see with-owten dowte
Ben sevyn houndred myle abowte,
And al longyth vnto Venes towne,
Whiche is a Cyte of grete renowne.
And in þe yle of **Rodys**, as we gone,
We fynde Relikis many one:
A Crosse made of a Basyn swete
That Crist wysshe in his Aposteles feete,
And A thorne off þe Crowne
That stake in his hede abouyn,
That blowyth euery good Frydaye,
A fayre myracle hit is to saye.
Ther is Saynt **Loye**, & seint **Blase**;
Ther is þe hande & þe Arme
Of saint **Kateryn**, þe blessyd virgyn. . . .

I started from Venice to Jerusalem.

Venice is the king of all cities.

Saints' corpses lie in it:

St George,

John the Baptist's father,

1000 Innocents, &c.

He who visits it twice in a year gets remission of his sins.

The Isles of the sea belong to Venice.

In Rhodes are many relics:

a thorn of Christ's crown,

St Loye's body, St Katherine's arm, &c.

[1] of B.

great confluence of marchauntes, as well Christians, as all sortes of infydels. The citie of Venis doth stande .vii. myle wythin the sea: *th*e sea is called the gulf; it doth not eb nor flow. Thorow the stretes of Venys ronnyth the water; and euery marchaunt hath a fayre lytle barge standynge at his stayers to rowe thorow and aboute the citie; and at bothe sydes of the water in euery strete a man maye go whyther he wyll in Venys; but he must passe ouer many bredges. The marchauntes of Venys goeth in longe gownes lyke preestes, with close sleues. The Venyscy*on*s wyll not haue no lordes nor knyghtes a-monges theym, but only the Duke. The Duke of Venys is chosen for terme of hys lyfe; he shall not mary, by cause his sonne shall not clayme no inheritaunce of the dukedomshyp, [1] the Duke may haue lemons & concubyns[2] as manye as he wyl.

Water in every street.

Gondolas.

Merchants wear long gowns.

Venetians won't have Lords.

The Duke of Venice mayn't marry, but may have concubines.

[1] sign. I .ii. back.

[2] Thomas does not notice this custom; though he says that younger brothers in Venice do not marry. Of the Venetian young man he says:—

"his greatest exercise is to go, amongest his companyons, to this good womans house and that. Of whiche in Venice are many thousandes of ordinarye, lesse than honest. And no meruaile of the multitude of theyr common women; for amonge the gentilmen is a certeine vse, that if there be diuers brethern, lightlye but one of theim doeth marie: because the number of gentilmen should not so encrease, that at length their common wealth might waxe vile: wherfore the reste of the brethern doe kepe Courtisanes, to the entent they may haue no lawful children. And the bastardes that they begette, become most commonly monkes, friers, or nunnes, who by theyr friendes meanes are preferred to the offices of most profite, as abbottes, priours, and so forth. But specially the Courtisanes are so riche, that in a maske, or at the feast of a mariage, or in the shrouynge tyme, you shal see theim decked with iewelles, as they were Queenes. So that it is thought no one citee againe hable to compare with Venice, for the number of gorgeouse dames. As for theyr beaultie of face; though they be fayre in deede, I woull not highlye commend theim, because there is maner none, old or yong, vnpeincted. In deede of theyr stature, they are of the most parte veraie goodly and bigge women, wel made and stronge."—Thomas's *Historye of Italye*, fol. 84, back (1549 A.D., edit. 1561).

Many thousand courtesans in Venice.

Only one brother of a family marries; the rest keep courtesans, and make their bastards monks or nuns.

The courtesans are deckt out like Queens,

but they paint their faces.

They're well-made.

In an earlier part of his book, Thomas speaks as follows of the Venetian women:—

The Doge mayn't leave Venice.

the Duke shall neuer ryd, nor go, nor sayle out of the cyte as longe as he dothe lyue.[1] The Duke shall rule the

---

"As for the women,
Some be wonders gaie,
And some goe as they maye.
Some at libertee dooe swymme a flote,
And some woulde faine, but they cannot.
Some be meerie, I wote wel why,
And some begyle the housbande, with finger in the eie.
Some be maryed agaynst theyr will,
And therfore some abyde Maydens styll.
In effect, they are women all,
Euer haue been, and euer shall,

The Venetian women are very gay.

Some Venetian women beguile their husbands.

All dress more gorgeously than any other women.

Churchmen keep fine courtesans.

—But in good earnest, the gentilwomen generally, for gorgeouse atyre, apparayle and Iewelles, excede (I thynke) all other women of oure knowen worlde, I meane as well the courtisanes as the maryed women. For in some places of Italye, speciallie where churchemen doe reigne, you shall fynde of that sorte of women in riche apparaile, in furniture of household, in seruice, in horse and hackeney, and in all thinges that apperteyne to a delycate Lady, so well furnysshed, that to see one of theim vnknowynglye, she should seeme rather of the qualitee of a princesse, than of a common woman. But because I haue to speake hereafter in perticuler, I woull forbeare to treate anye further of theym in thys place."—Fol. 6. *The Historye of Italye*, by W. Thomas, 1549, edit. 1561.

The Venetian *Doge* seems grand, but is really an honourable slave.

He can't go out without leave.

But he can make the Council take a ballot on his opinions.

[1] "They haue a duke called after theyr maner *doge*, who onely (amongest al the rest of the nobilitee) hath his office immutable for terme of life, with a certaine yerely prouision of .4000. duckates, or theraboutes. But that is so appoincted vnto him for certaine ordinarie feastes, & other lyke charges, that hys owne aduauntage therof can be but smal. And though in apparaunce he seemeth of great astate, yet in veray deede his power is but small. He kepeth no house, lyueth priuately, & is in so muche seruitude, that I haue hearde some of the Venetians theim selfes cal him an honourable slaue: For he cannot goe a mile out of the towne without the counsails licence, nor in the towne depart extraordinarily out of the palaice, but priuately and secretely: And in his apparaile he is prescribed an ordre: so that, in effect, he hath no maner of preeminence but the bare honour, the gift of a few smal offices, and the libertee *Di mettere vna porta*, which is no more but to propound vnto any of *the* counsailes his opinion, touching the ordre, reformacion, or correcion of anye thyng: and that opinion euery counsaile is bound taccept into a trial of theyr sentences by Ballot: (the maner of the whych ballotting shal hereafter appeare;) and this priuilege, to haue his onely oppinion ballotted, no man hath but he. And wheras many haue reported, that the Duke in ballottyng should haue two voices, it is nothinge so; for in geuyng his voice, he hath but one ballot, as all others haue."—Thomas's *Historye of Italye*, fol. 77 (1549, edit. 1561).

senyorite, and the seniorite shall gouerne and rule the comynalte[1], and depose and put to deth the Duke if thei do fynd a lawful cause. The Duke weryth a coronet ouer a cap of sylke, the whych stondeth vp lyke a podynge or a cokes come, bekyng forward, of .iii. handfoll longe. The Duke do not come to the butyful church of saint Marke but [on] certen hygh feastes in the yere, & the fyrst eyght daies after that he is made Duke, to shew hym selfe. I dyd neer[2] se within the cyte of Venis no pouerte, but al riches. ther be none inhabitours in the cite that is nede & pour. vitelles there is dere. Venys is one of the chefest portes of all the world. the Venyscions hath great prouision of warre, for they haue euer in a redynes tymber readye made to make a hondred gales or more at [a] tyme.[3] they haue all maner of artilery in a redynes. They haue greate possessions; and Candy, and Scio,[4] with other Iles and portes, cites & landes, be vnder ther dominion. Whan they do heare masse, & se the sacrament, they do inclyne, & doth clap theyr hand on theyr mouth, and do not knock them self on the brest. at hygh masse they do vse prycksong & playnsonge, the orgins & the trum-

The Doge wears a coronet over his cap of silk.

St Mark's.

No poverty in Venice.

Victuals dear there.

Great stores for war. (See *Notes* at the end.)

Many islands and lands belong to Venice.

The Venetians' behaviour at Mass

As our rulers are getting honest enough to give poor and squeezeable voters the protection of the Ballot, I add Thomas's further account of the Venetian system :

"This maner of geuyng theyr [the great Council's] voices by ballotte, is one of the laudablest thinges vsed amongest theim. For there is no man can know what an other dooeth.—The boxes are made with an holow place at the top, that a man may put in his hand; and at the ende of that place hange .ii. or .iii. boxes, into whiche, if he wyll, he may let fall his ballot, that no man can perceiue hym. If there be but two boxes (as commonly it is in election) the one saieth yea, and the other sayeth naye: And if there be .iii. boxes (whiche for the most parte hapneth in cases of iudgement) the one saieth yea, thother sayth naye, and the thyrde saieth nothynge: and they are all well enough knowen by theyr dyuers colours. By this order of ballottyng, they procede in iudgement thorough al offices, vpon all maner of causes: beynge reputed a soueraigne preseruation of iustice." —*Ibid.* fol. 79.

The Venetian Ballot.

How the vote by Ballot is taken in the Venetian Council.

It's a sovereign preserver of Justice.

[1] coymnalte A; comenalte B. [2] neuer B.
[3] at tyme A; at a tim B. [4] sco AB.

and when St Mark is named.

pates. if ther be any gospel red, or song of saynt Marke, they wyl say "sequencia santy euangely secundum istum," poyntyng theyr fynger to s. Mark, the whych do ly in the church. the people do pol their heades, and do let ther berdes grow. Theyr spech is Italion, ther money is gold, that is to say, duccates; & bagantins is brasse; .xii. bagantyns is worth a galy halpeny; & there is galy halpens.

The Venetians poll their heads.

Bagantyns.

[1]

The. xxv. Chapter treateth of Lombardye, and of the natural dysposicion of the people, and of theyr speche and of theyr monye.

I am crafty,

I am a Lombort, and subtyl crafft I haue,
To deceyue a gentyl man, a yeman, or a knaue;
I werke by polyse,[2] subtylyte, and craught, [craft]
The whych, other whyle, doth bryng me to nought.
I am the next neyghbour to the Italion;
We do bryng many thynges out of al fashyon;
We care for no man, & no man caryth for vs;
Our proud hartes maketh vs to fare the worse.

and care for no man.

[1] I .iii. not signed. [2] poplyse AB.

In our countre we eate Adders, snayles,[1] and frogges,
And above al thyng we be sure of kur dogges;
For mens shyns they wyl ly in wayte;
It is a good sport to se them so to bayte.

I eat snakes, snails, and frogs. Lombards have many curs

[2]¶ Lombardy is a champion countrey & a fartyl, plentye of wyne and corne. The Lomberd doe[3] set muche by his berd, & he is scorneful of hys speche; he wyl geue an aunswer wyth wryeng his hed at the one side, displaysynge his handes abrode: yf he cast hys head at the one syde, and do[4] shroge vp hys shoulders, speake no more to hym, for you be answered. The Italyons, and some of the Venecyons, be of lyke dysposicion. In Lomberdy ther be many vengable cur dogges, the whyche wyll byte a man by the legges or he be ware. they[5] wyll ete frogges, guttes and all. Adders[6], snayles, and musheroms, be good meate there. In dyuerse places of Italy and Lombardy they wyll put rose-mary into theyr vessels of wine. Florance is the chefe towne of Lomberdy; it is a pleasaunt towne, and a commodiouse, it standeth betwext two hylles. the Lomberdes be so crafty, that one of them in a countrey is enough (as I haue heard many olde & wyse men say) to mar a whole countrey. the maner of the people and the speche be lyke the Italyons; the people of the countrey be very rewde. In Lomberdy and Italy they go to plow but wyth two oxsone, and they be couered with canuas that the flyes shall not byte them. there money is brasse, called katerins and bagantyns; in syluer they haue marketes; a market is a galy halpeny: in gold they haue duccates.

are proud of their beards; shrug their shoulders; eat frogs whole; and put rosemary in wine. Florence. One Lombard is enough to mar a whole country. They cover oxen with canvas. Lombard money: markets (*marchetti*).

[1] See the recipe for dressing them in *Q. Eliz. Achademy, &c.*, Part II. p. 153. [2] I .iii. back. [3] doth B.
[4] to AB. (The prefix *to* is hardly applicable to *shrug*.)
[5] That is, the Lombards, not their curs.
[6] See p. 273, l. 13.

[1]

[2]The .xxvi. chapter treateth[3] of Iene and of the Ianuayes, and of theyr spech, and of their mony.

[B puts this printer's ornament here.]

¶ I am a marchaunt; borne I was in Iene;
Whan I sell my ware, fewe men knoweth what I mene;
I make good treacle, and also fustyan; *(I make Treacle and Fustian;)*
Wyth such thynges I crauft wyth many a poer man;
Other of my marchaundes[4] I do set at a great pryce;
I counsel them be ware lest on them I set the dyce; *(and (?) take-in my customers.)*
I do hyt dyuerce tymes; som men on the thomes.
Wher soeuer I ryde or go, I wyl not lese my cromes.
In my apperel, the old fashyon I do kepe; *(I stick to my old fashions in dress.)*
Yf I should do other wyse, it would cause me to wepe.
Better it is for a man to haue his rayment tore,
Than to runne by-hynd-hande, and not to be before.

[1] This cut is from B. A has the canopy complete, except a third of the top line, and the cape on the right shoulder is complete, as is the cut of Boorde on the title-page of Barnes's *Treatyse* below.

[2] I .iiii. not signed. [3] trateth A; treteth B.

[4] marchauntes A; marchaundes B: merchandise.

Gorgyouse apparell maketh a bare purse;
It bringeth a man by-hynd, & maketh him worse & worse. 14

[1]¶ The noble cyte of Iene is a plesant and a commodyose cyte, And well serued of all maner of vyttells, for it stondeth on the see syd. there is made veluet and other sylkes; and ther is fustyane of Iene mad[e], and triacle of Iene.

Genoa is a well-victualled city, and makes velvet, silks, fustian, &c.

Iene, Prouince, and Langwadock, lyeth on the cost of Barbary, where the whyte and the blacke[2] mores be[3], & so doth Catalony,[4] Aragon, and Cyuel, and parte of Portyngale; of the[5] whych countres I wyl speke of after in this boke. the Ianewayes be sutyl and crafty men in theyr marchaundes[6]; they loue clenlynes; they be hyghe in the instep, and stondeth in theyr owne consayte. to the fayre and commodiouse citie of Iene belongeth gret possessions, the whyche is ful of fartilite, and plentiful of fysh and frut. whan they do make theyr treacle, a man wyll take and eate poysen and than he wyl swel redy to brost[7] and to dye, and as sone as he hath takyn trakle, he is hole agene. theyr spech is Italyon and French; theyr mony is much lyke[8] the Italyons.

It's opposite Barbary, where the White and Black Moors are.

The Genoese are crafty dealers. (See *Notes*.)

Genoese treacle is an antidote to poison.

[1] I .iiii. back.
[2] placke B.
[3] Who come over and rob the Genoese, &c.: see p. 213.
[4] See Boorde's letter in the *Forewords*, p. 56.
[5] of it of the AB.
[6] merchandise, dealing.
[7] borst B.
[8] lyke to B.

[1]

The .xxvii. Chapter treateth of Fraunce, and of our prouences the whyche be vnder Fraunce, and of the natural dysposicyon of the people, and of ther money and of theyr speche.

I am a French man, lusty and stout;
I jag and cut my clothes. My rayment is iagged, and kut round a-bout;
I am ful of new inuencions,
And dayly I do make new toyes and fashions;
All nations follow my fashions. Al nacions of me example do take,
Whan any garment they go about to make.

[2]Fraunce is a noble countre, and plentiful of wyne, bread, corne, fysh, flesh, & whyld[3] foule. there a man

[1] sign. K .i. [2] sign. K .i. back. [3] wild B.

shalbe honestly orderyd for his mony, and shal haue good chere and good lodging. Fraunce is a rych countre & a plesaunt. in Fraunce is many goodly tounes, as[1] Granople, Lyons, and Parys; the which Parres[2] is deuyded in thre partes:—Fyrste is the[3] towne; the citie, & the vniuersite. in Fraunce is also[4] Orlyance, and Puttyors, Tolose, and Mount Pylor, the which .iiii. townes be vniuercites. beyond Fraunce be these great princes, fyrst is Priuinces and Sauoy, Dolphemy & Burgundy; then is the fayer prouynces of Langwhadock & good Aquytany. The other prouynces I wil speke of whan I shal wryt in retornyng home to Calys, where that I toke my first iorny or vyage. the people of Fraunce doo delyte in gorgious apparell, and wyll haue euery daye a new fashion. They haue no greate fantasy to Englyshmen; they do loue syngyng and dansyng, and musicall instrumentes; and they be hyghe mynded and statly people. The money of Fraunce is gold, syluer, and brasse. In gold they haue French crownes of .iiii. s. viii. d.; in syluer they haue testons, which be worth halfe a Frenche crowne; it is worth .ii. s. iiii. d. sterlyng. in bras they haue mictes, halfe pens, pens, dobles, lierdes, halfe karalles & karales,[5] halfe sowses & sowses; a sowse is worth .xii. bras pens[6]; a karoll is worth .x. bras pens, a lier is worth three brasse pens, a double is worthe two brasse pens .xxiiii. Brasse halpens ys a sowese, [and] is almooste worthe thre halpens of our mony; myttes be brasse fardinges: if any man wyll lerne Fraunce[7] and Englyshe,—Englyshe and Fraunce[7] doth folowe.

Grenoble, Lyons, Paris.

Orleans, Poitiers, Montpelier, &c.

Provence, Dauphiny, Languedoc, &c.

New fashions every day.

Dislike Englishmen.

French money: gold crowns, silver testons,

brass Caroluses, sous = 12 brass pence, liers, doubles; 24 brass ha'pence make a sous; nearly 1½d. English; myttes.

One. two. thre. foure. fyue. syx. seuen. eyghte. nyne.
One. deus. trous. cater. cynk. sys. set. huyt. neyf.
ten. aleuyn. twelue. thyrtene. fowrtene. fyftene. sixtene.
[8]*dix. vngse. deuse. treise. katorse. kynse. seise.*

French numerals.

[1] as a A. [2] partes A; parres B. [3] that AB.
[4] fraunce also AB. [5] from Upcott; 'halfe karalles karalle' AB.
[6] cp. 'eyght shyllynges, *huyt sous*,' p. 193. [7] frenche B.
[8] K .ii. not signed.

French numerals.

seuentyne. eyghtene. nyntene. twenty. one and twenty.
desett. deshuit. desneuf. vinct. vinct[1] *ung.*
therty. forty. fyuete. sexte. seuente. eyghte.
trente. katrente. cynkante.[2] *sesante. septante. hytante.*
nynte. a honderd. a thowsand. x. thowsand.
notante. Cent. mille. dix mille.

A talk in French and English.

Good morow, my syr! *bon iour, mon ser!*
God geue you a good day! *Dieu vous dint bon iore!* *
God spede you, my brother! *Dieu vous gard, mon frer!* *
frend, God saue you! *Amy, Dieu vous salue!*
Of whens be you? *Vnde eta*† *vou?*
I am of England. *Ie sues*† *de Angliater.*
You be welcome, gentyl companyon!
Vous etes bien venu, gentyl companyon!
Syr, how do you fare? *Syr, comment vous portes*† *?*
I fare wel. *Ie porta bene*†.
Howe doth my father and mother?
comment se porte mon peer et me mater† *?*
Ryght wel, blessed be God! *Tresbien, benoyst soyt Dieu!* *
I praye you that ye commend me to my father and to all my good frendes.
Ie vous prie que me commendes a mon pere et a tous mes bons amys. *
Whyche is the right way for to go from hens to Parys?
Quele est la droyt† *voye pour alier dicy a Paris?*
Syr, you must hold the way on the ryght hand.
Syr, il vos fault tenyr le chymin a la droit† *mayn.*
Tel me yf ther be any good lodgyng.
Dictes sil y a poynt de bon logis.
There is ryght good lodgyng.
Il i en ya vng tresbon logis.†
My frend, God thanke you! [3] *Mon amy, Dieu marces.*
Syr, God be wyth you! I must depart.

[1] vinci AB; ? for *vingt et.* [2] onkante AB. [3] K .ii. back.

* These seem to me genuine French of Rabelais' time.—C. Cassal.

† These must be by a travelling Brown, Jones, or Robinson.—C. Cassal.

*Syre, Dieu soit auecques vous, car me fault departer.** A talk in French and English.
fare wel! *adewe!*
dame, God saue[1] you! *Dame, Dieu vous salu!*
You be welcome! *Vous estes bien veneu!**
Dame, shall I be here wel logyd?
Dame, seray ie icy bien loge?
ye, syr, ryght wel. *Ouy, syr, tresbien.*
Now geue me som wyne. *Or done moy de† uyn.*
Geue me bred. *done moy de† pane.*
Dame, is al redy to supper?
[*Dame, est tout pret a souper †?*][2]
Ye, syr, whan it pleaseth you.
Ouy, syr, quant il vous plaira.
Syr, much good do it you! *Syr, bon preu vous face!**
I pray you, mak good chere!
Ie vous prye, factes bon chere!
Now tell me what I shall pay.
Or me dictes combien Ie[3] payera.†
Ye haue in all eyght shyllynges.
*Vous aues en tout huyt sous.**
Syr, God geue you a good nyght, and good rest!
*Syr, Dieu vous doynt bon nuy et bon repose!**
My frend, if you do speke, take hede to thy selfe!
Mon amy, si tu parles, gard a toy!
To speke to much is a dangerous[4] thynge.
Le trop parler est dangereus.[5]

Aquitaine

¶ Here is to be noted, that I, in al the countres that euer I dyd trauyl in, Aquitany,—the whyche is wyth-in the precynt of Fraunce, and on of the vttermost prouinces of [6]Fraunce, Langadok except, the which Aquytany pertainth by ryght to the crowne of Englond, as Gascony and Bion and Normandy doth,—whych is the most plentifullist country for good bred & wyne, consideryng

is the most plentiful and cheap country for bread and wine.

* † See notes on last page.
[1] same A. [2] not in A, but in B. [3] ye AB.
[4] dargerous A; dangerous B. [5] daugereus A; dangereus B.
[6] K .iii. not signed.

A pen'orth of cakes lasted me 9 days in Aquitaine.

the good chep,[1] that I was euer in;[2] a peny worth of whyte bread in Aquitany[3] may serue an honest man a hoole weke; for he shall[4] haue, whan I was ther, .ix. kakys for a peny; and a kake serued me a daye, & so it wyll any man, excepte he be a rauenner. the bred is not so good[5] chepe, but the wyne & other vittels is in lyke maner good chepe. Aquytany ioyneth to Langwadock, the whych Langwadock is a noble country, and plentyful, as Aquytany is: ther is muche wode growyng, specially from Tolose to Mount-piliour. Tolose & Mount-pyliour be vniuersites. in Tolose regneth treue Iustyce & equitie: of al the places that euer I dyd com in, Munpilior is the most nobilist vniuersite of the world for phisicions and surgions. I can not geue to greate a prayse to Aquitane and Langwadock,[6] to Tolose and Mountpiliour.

Languedoc is a noble country.

Toulouse.

Montpelier is the noblest Medical University in the world.

[7] The xxviii. chapter treateth of Catalony and of the kyngedome of Aragon, and of the naturall dysposycyon of the people, and of theyr money and[8] of theyr spech.

The Emperor of Austria dwells in Catalonia.

¶ I am borne in Catalony; the Emproure dwelleth
wyth mee;[9]
Why he so doth, I can not tel the.

[1] chepe B (bargain, cheapness).
[2] Compare the end of Chapter xxxii. p. 206, "Aquitany hath no felow for good wyne & bred."
[3] Aquiany A; aquiani B.
[4] for "should."
[5] god A; good B.
[6] langadwoen AB.
[7] B has for this cut, the king's head on p. 175.
[8] and of A.
[9] "mee" is not in A, but is in B.

[1] Whan I fayght[2] with the Mors, I set al at sixt or seuyn;
He that is in hel thynketh no other heuen.
And I was borne[3] in Aragon, where that I do dwel.
Masyl[4] baken, and sardyns, I do eate and sel,
The whych doth make Englyshe mens chykes lene,
That neuer after to me they wyll come agene:
Thus may you know howe that we do fare,
The countres next vs al be very bare;
We haue no chere but by the se syd,
Although our countres be both large and wyde.
Castyll, and Spane, and we, kepe on vse;
They that leke not vs, let them vs refuce[5];
And playnly now I tell you my intencyon,
My rayment I chaunge not from the olde fashion.

In Aragon we eat measly bacon and sardines, to Englishmen's disgust.

We're like Castille and Spain.

¶ Catalony, whych is a prouince, and Aragon whych is a kyngdome, be anexed to gider.[6] the Emproure doth ly much in Catalony, for in those partes he hath not only Catalony vnder hys dominion, but also he hath the kyngdom of Aragon, the kyngdom of Spayne, the kyngdome of Castil, and Biscay, and part of the kingdom of Nauer. The countres of Catalony and Aragon, except it be by the see syde and great townes, is poer & euyl fare, & worse lodgyng; yet ther is plenty of fruit, as fygges, Poudganades,[7] Orenges, & such lyke. the chefe townes of Catalony is called Barsalone, and Tarragon, and Newe Cartage. in Aragon the chefe towne is called Cesor Augusta[8]; nowe it is called Sarragose. thorowe Aragon doth rone a noble ryuer called Iber. the spech of Catalony & Aragon is Castilion; how be it they dyffer in certene wordes, theyr vsage, theyr maner & fashyons, is much after the Spainierdes fashions; theyr mony is diuerce coynes of the Emperour, for all maner coynes of the Emperour goeth ther.

The Emperor lives in Catalonia.

It and Aragon are poor, but have much fruit, Pomegranates, &c.

Barcelona, Tarragona, Cartagena.

Sarragossa.

Ebro river.

Folks' ways like the Spaniards'.

[1] K .ii. back. [2] faught B.
[3] brone A; borne B. [4] Mesyl B. [5] refuse B.
[6] githcr B. [7] pomgranates. [8] angusta A.

The xxix. Chapter treateth of Andalase, of Cyuel, and of the kyngedome of Portyngale, and of the natural dysposicyon of the people, and of ther speche, and of theyr money.

Andalusia.

I was borne in Andalase
Wher many marchantes commeth to me,
Some to bay,[2] and some to sel;
In our marchantes [3] we sped ful wel.

Seville.

And I was borne in Cyuel, lackyng nothyng;
Al nacions, marchauntes to me doth bryng.
And I was borne in the kyngdome of Portyngale;
Of spices & of Wyne I do make great sale.

Portugal sells spices and wine.

By marchauntes, al my country doth stond
Or els had I [4] very poer land.
Yf any man for marchauntes [5] wyl come to vs,
Let hym bryng with hym a good fat purse,
Than shal they haue of vs theyr full intencion,

[1] K .iiii. not signed. [2] bey B. [3] marchandes B.
[4] I a B. [5] merchaundices B.

[1] And know that in our rayment we kepe the olde fashion.

Portyngale is a rych angle, specially by the Sec side, for the comon corse of marchaunte straungers. the kyng of Portyngale is a marchaunte, & doth vse marchauntes.[2] Lustborne and Acobrynge be the chefe townes of Portyngale. The countre stondeth much by spyces, fruites, and wyne. The Portingales seketh theyr lyuynge fare by the see, theyr money is brasse and fyne golde. In bras they haue mariuades[3] and myttes and other smale peces; in gold they haue cursados worth .v. s. a pece; they haue also portingalus, the whych be worth .x. crownes a pece. the spech of Portingale is Castilyone; how be it in some certen wordes they doth swerue from the true Castilion speche. The men and the women and the maydens doth vse theyr rament after the fashion of the Spainierdes, the men hauyng pold hedes, or els her handgyng one there[4] shoulders; and the[5] maydens be poled, hauynge a[6] garlond about the lower part lyke a Barfote Frier.

Portugal is used by merchants.

Lisbon and Alcoutrin (?).

Portuguese money: maravedies, gold crusados, and portingales.

Portuguese speech is nearly Castilian.

The folk dress like Spaniards.

Girls crop their crowns, and leave a rim like a friar's.

[1] K .iiii. back. [2] marchaundes B. [3] marmades AB.
[4] out that A; one there B. [5] that A; ther B.
[6] at A; a B.

The .xxx. chapter treateth of the natural disposicion[2] of Spanyardes, of the countrey, of the money, and of the speche.

I am a Spaynyard, and Castylyon I can speke;
In dyuers countreys I do wander and peke;
I do take great labour, and also great payne;
To get a poore lyuyng I am glad and fayne;
In my countrey I haue very poore fare,
And my house and my lodgyng is very bare.
A Spanyshe cloke I do vse for to were,
To hyde mine olde cote and myn other broken gere.

I wander about, to pick up a poor living.

I have very poor fare.

Spain inland is very poor.

Biscay and Castille are very barren.

Sardines.

¶ Spayne is a very poore countrey within the realme, & ple*n*tyful by the sea syde; for al theyr riches & marchauntes[3] they bryng to the sea syde. I know nothing, wit*h*in the countre, of ryches, but corne. Byskay & Castyle is vnder Spayne; these countreys be baryn of wine and corne, and skarse of vitels; a ma*n* shall not get mete in many places for no mony; other whyle you shall get kyd, and mesell bakyn, and salt sardyns, which is a lytle fyshe as bydg[4] as a pylcherd,

[1] sign. L .i. [2] dispocion A; disposicion B.
[3] merchandise. [4] bydge B.

& they be rosty. al your wyne shalbe kepte [1] and caryed in gote skyns, & the here syde shalbe inwarde, and you shall draw your wyne[2] out of one of the legges of the skyne. whan you go to dyner & to supper, you must fetch your bread in one place, and your wyne in a nother place, and your meate in a nother place; & hogges *in* many places shalbe vnder your feete at *the* table, and lice in your bed. The cheife cities and townes in Spayne is Burges & Co*m*postel. many of the people doth go barlegged. the maydens be polyd lyke freers; the women haue siluer ringes on theyr eres, & coppyd thinges standeth vpon theyr hed, wi*th*in ther kerchers, lyke a codpece or a gose podynge.[3] In Spayne there money is brasse, siluer, & gold; in brasse they haue marivades[4]; .xxv. marivades[4] is worth an Englyshe grote: they haue there styuers. In siluer they haue ryals & halfe ryalles; a ryal is worth .v.d.ob. in golde they haue duccates and doble duccates. there speche is Castylyon.

Wine kept in goat skins.

Hogs under the table, and lice in beds.

Burgos. Compostella.

Women's head-dress.

Spanish money: maravedies, stivers, reals, 5½d.

[5]

The .xxxi. chapter tretyth of the kyngdome of Castyle, & of Byscay[6], and of the natural disposicion of the people, and of there money & of theyr speche.

¶ In the kyngdome of Castell borne I was,
And though I be poer, on it I do not passe;

I am poor.

[1] L .i. back. [2] wynde A; wyne B.
[3] Cp p. 185, and in chap. xxxiii. p. 207. [4] marmades AB.
[5] B has for this cut, the king's head on p. 175. See too p. 194. [6] byscat AB.

Where so euer I do goe or ryde,
My cloke I wyl haue, and my skayne by my syde.

but wear a skean.

And I was borne in the prouince of Byscay[1];
My countrey is poer; who can say nay?
And though we haue no pastor nor grandge,
Yet our olde fashyon we do not chaunge.

Biscay is a poor country.

[2]¶ Castyle is a kyngdome lyinge bytwyxte Spayne and Byscay; it is a very baron countrey, ful of pouerte. there be many fayre and proper Castels, plenty of aples & of sider, and there be great water mylles to forge yrone, & theyr be great mountaynes & hilles, and euill fare, [and] lodgyng; the best fare is in prestes houses, for they do kepe typlynge houses. and loke, how you be serued in Spayne and Neuer, shal you be serued in Castyle. the chief towne of Castile is called Tolet. Palphans made the tables of astronimye. In all these countreys, yf any man, or woman, or chylde, do dye; at theyr burying, and many other tymes after that they be buryed, they wyl make an exclamacyon[3] saying, "why dydest thou dye? haddest not thou good freendes? myghtyst not thou haue had gold and syluer, & ryches and good clothynge? for why diddest thou die?" crying and clatryng many suche folysh wordes; and commonly euery day they wyll bryng to church a cloth, or a pilo carpit, and cast ouer the graue, and set ouer it, bread, wyne & candyllyght; and than they wyll pray, and make suche a folyshe exclamacion, as I sayd afore, that al the churche shall rynge; this wyll they doe although theyr freendes dyed .vii. yere before; & thys folysh vse is vsyd in Bisca, Castyle, Spayne, Aragon & Nauerre. their money is golde and brasse: in golde they haue single and duble duccates; and all good gold goeth there. in brasse they haue marivades,[4] and stiuers, & other brasse money of the Emperours

Castille is very barren.
Castles;
mills to forge iron.
Priests keep tippling houses.
Toledo.
When any one dies, others cry out,
'Why did you die? You had friends and gold.'
They put a cloth and food over the grave, and cry thus.
Castilian money:
ducats,
maravedies,
stivers.

[1] vyscay A; byscay B. [2] L .ii. not signed.
[3] Compare the Welsh, p. 126.
[4] marmades *or* marinades A; marmades B.

coyne. who so that will learne to speake some Castilion,—Englishe and Castilion doth folowe.

Castilian (or Spanish) numerals.

One. two. thre. foure. fyue. syx. seuen. eyght. nyne.
vna. dos. tros. quarter. sinco. sisse. saeto. ocho. nowe.
tene. aleuen. twelue. thertene. fouertene. fyftene.
diece. onze. dose. treerse. quartorse. quynse.
syxtene. seuentene. eyghtene. nyntene. twenty.
dezisys. dezisyeto. desyocho. desinoue. veynto.
therty. forty. fyfty. syxte. seuente.
[1] *trenta. quarenta. cynquenta. sesenta. setenta.*
eyghte. nynte. a hondred. a thousand.
ochenta. noventa. cyento. mylyes.

A talk in Castilian (or Spanish) and English.

Syr, God geue you a good day!
senyor, Dios os be[2] *bonas dias!*
God saue you, syr! *Dios vos salue, senyor!*
How do you fare? *quomodo stat cum vostro corps*[3]*?*
I do well, thankes be to God!
Ie sta[4] *ben, gracyas a Deos!*
What wold you[5] haue, syr? *ke keris, senyor?*
I would haue some meate. *kero comer.*
Come wyth me, I am hungre.
Veni connigo[6]*, tengo appetito de comer.*
Much good do it you! *bona pro os haga.*
you be welcome, wyth all my harte
Seas been venedo, com todo el corason.
Wyll you drynke, syr? *kerys beuer, senyor?*
It pleaseth me well. *byen me pleze.*
Speke that I may vnderstand you. *halla ke tu entende*[7].
I do not vnderstand you, syr! *non entiende, senyor.*
I do vnderstande Castylion, but I cannot speke it.
Io lo entendo Castyliona; Io no saue hablar.
I do thank you! *mochos mecedo!*

[1] L .ii. back.
[2] *de.*—H. H. Gibbs.
[3] Dog-Latin, not Spanish.—F. W. Cosens.
[4] For *Io sto.*—H. H. Gibbs.
[5] ye B.
[6] For *Ven* or *ben conmigo.*—F. W. C.
[7] For '*habla que tu* entiende.'—F. W. C.

[1]

The .xxxii. chapter treteth of the kingdome of Nauer, and of the naturall disposicyon of the people, and of theyr mone[y] and of theyr speche.

[2] In the kyn[g]dome of Nauer I was brought vp,
Where there is lytle meate to dyne or suppe ;
Sardyns and bacon shall fynde the Spanyard and me, — We eat Sardines and Bacon.
Wyth suche meate we be contente in all our countre :
What wolde other men, other meate craue ?
Such meate as we do eate, such shall they haue.
In my apparell I do kepe the olde raate ;
The Fraunch [3] men with me preforse be at baate, — We're now friends with our old foes the French.
Not now, but in olde tymes past ;
For now our amyte is full fast.

The kyndome [4] of Nauer is ioynynge [5] to Spayne and to Fraunce, & to Catalony, and to Castyle, for it dothe stand in the midle of these [6] iiii. countres. The people be rude and poore, and many theues, and they dothe liue in much pouerte and penury ; the countrey is barayn, for it is ful of mountayns And weldernes ; yet haue they much corne. The chiefe towne is Pampilona, and there is a nother towne called saynt Domyngo, in the whyche towne there is a churche, in the whyche is kept a whit cock and a hene. And euery pilgreme that goeth or commyth that way to saynct

The people of Navarre are poor and thievish.

Pampeluna.

St Domingo has a church with a white cock and hen.

[1] The corner is not broken in A. [2] L .iii. not signed.
[3] frenche B. [4] kingdome B. [5] iunynge AB. [6] the B.

Iames in Compostell, hath a whit feder to set on hys hat. The cocke and the hen is kepte there for this intent:[1]— There was a yonge man hanged in that towne that wolde haue gone to saynct Iames in Co*m*postell; he was hanged vniustly; for ther[2] was a wenche the whych wolde haue had hym to medyll with her carnally; the yonge man refraynyng from hyr desyre, and the whenche repletyd with malyce for the sayd cause, of an euyll pretence conueyed a syluer peece into the bottom of the yonge mans skrip. he, wyth his father & mother, & other pylgrems, going forthe in theyr Iurney, the sayde whenche raysed offycers of the towne to persew after [3] the pylgryms,[4] and toke them, fyndynge the aforesayd peace in the younge mannes scryp: Wherfore they brought to the towne the yong man; and [he] was condemned to be ha*n*ged, and was hanged vppon a payre of galowes,—Whosoeuer that is hanged by-yonde see, shall neuer be cutte nor pulled downe, but shall hange styll on the galowes or Iebet.—the father and the mother of the younge manne, with other of the pylgryms, went forthe in theyr pilgrymage. And whan they returned agayne, they went to the sayd galows to pray for the yong mans soule. wha*n* they dyd come to the place, The yonge man did speke, & sayd "I am not ded; God and his seruaunte saynt Iames hathe here[5] preserued me a lyue. Therfore go you to the iustis of the towne, & byd him come hyther and let me down." vpo*n* the which wordes they went to the Iustice, he syttyng at supper, hauyng in his dyshe two greate chykens; the one was a hen chik, and the other a cock chyk. the messe*n*gers shewyng him this wonder, & what he should do, the iustice sayd to them, "This tale that you haue shewed me is as treue as these two chekenes before

The story of the white Cock and Hen of St Domingo:

A wench wanted to have a young pilgrim.

He refused her.

She put a silver coin in his scrip,

and sent officers after him.

The pilgrim was hanged for robbery,

but, though on the gallows,

St James kept him alive, and he sent for the Justice to let him down.

The Justice, on hearing the story, said,

'It's as true as that my 2 cookt

[1] intentent A; intent B. [2] that A; ther B. [3] L .iii. back.
[4] A wrongly repeats "goyonge forthe in theyr Iorney, the sayde Wenche raysed offycers of the towne to persuc after the pylgryms." [5] ther A; here B.

chickens will crow.' On which the chickens did crow; and the hanged pilgrim was taken off the gallows.

mee in thys dysshe doth stonde vp and crowe." & as sone as the wordes ware spoken, they stode in the platter, & dyd crowe; wher vpon the Iustyce, wyth processyon, dyd fetche in, a lyue frome the galows, that sayd yong man. & for a remembraunce of this stupendyouse thynges, the prestes and other credyble persons shewed me that they do kepe styl in a kaig[1] in the churche a white cocke and a hen. I did se a cock and a hen ther in the churche, and do tell the fable as it was tolde me, not of three or .iiii. parsons, but of many; but for [2] all this, take thys tale folowyng for a suerte. I dyd dwel in Compostell, as I did dwell in many partes of the world, to se & to know the trewth of many thynges, & I assure you that there is not one heare nor one bone of saint Iames in Spayne in Compostell, but only, as they say, his stafe, and the chayne the whyche he was bounde wyth all in prison, and the syckel or hooke,[3] the whyche doth lye vpon the myddell of the hyghe aulter, the whych (they sayd) dyd saw and cutte of the head of saint Iames the more, for whome the confluence of pylgrims resorteth to the said place. I, beynge longe there, and illudyd, was shreuen of an auncyent doctor of dyuynite, the which was blear yed,—and, whether it was to haue my counsell in physycke or no, I passe ouer, but I was shreuen of hym,—and after my absolucion he sayd to me, "I do maruaile greatly that our nation, specially our clergy and they, and the cardynalles of Compostell" (they be called 'cardynalles' there, the whyche be head prestes; and there they haue a cardynall that is called "cardinal[i]s maior," the great cardynal, and he but a prest, and goeth lyke a prest, and not lyke the cardinalles of Rome,) "doth illude, mocke, and skorne, the people, to do Idolatry, making ygnorant people to worship the thyng that is not here. we haue not one heare nor bone of saynct Iames; for

This is why the white cock and hen are kept.

I dwelt in Compostella to get at the truth of things;

and there's no hair or bone of St James, in Compostella.

I was shriven by an old blear-eyed Doctor of Divinity there,

and he told me how the clergy deceived the people, as none of St James's hairs or bones were there.

[1] kaige B. [2] L .iiii. not signed. [3] booke A; hooke B.

saynct Iames the more, and saynct Iames the lesse, sainct Bartilmew, &[1] sainct Philyp, saynt Symond and Iude, saynt Barnarde & sanct George, with dyuerse other saynctes, Carolus magnus brought theym to Tolose, pretending to haue had al the appostels bodies or bones to be congregated & brought together into one place in saynt Seuerins church in Tolose, a citie in Langawdocke." therefor I did go to the citie & vniuersite of Tolose, & [2] there dwelt to knowe the trueth ; & there it is known by olde autentyck wryttinges & seales, the premyses to be of treuth ; but thes words can not be beleued of incipient parsons,[3] specially of some Englyshe men and Skotyshe men ; for whan I dyd dwell in the vniuersite of Orlyance, casually going ouer the bredge into the towne, I dyd mete with .ix. Englyshe and Skotyshe parsons goyng to saynt Compostell, a pylgrymage to saynt Iames. I, knowyng theyr pretence, aduertysed them to returne home to England, saying that "I had rather to goo .v. tymes out of England to Rome,—and so I had in dede,—than ons[4] to go from Orlyance to Compostel;" saying also that "if I had byn worthy to be of the kyng of Englandes counsel, such parsons [5] as wolde take such iornes [6] on them wythout his lycences, I wold set them by the fete.[7] And that I had rather they[8] should dye in England thorowe my industry, than they[9] to kyll them selfe by the way :" wyth other wordes I had to them of exasperacyon. They, not regardyng my wordes nor sayinges, sayd that they wolde go forth in theyr iourney, and wolde dye by the way rather than to returne home. I, hauynge pitie they should be cast a way, poynted them to my hostage, and went to dispache my busines in the vniuersyte of Orliaunce. And after that I went wyth them in theyr iur-

Charlemagne took all the bones to Toloue,

to St Severin's Church :

I went there to know the truth, and saw the writings.

When I was at Orleans, I met 9 English and Scotch men going to Compostella.

I told them how hard a journey it was,

and that it would kill them.

But they *would* go;

so I went with them,

[1] to AB. [2] L .iiii. back.
[3] insipient (unwise, foolish) persons B. [4] then once B.
[5] persons B. [6] iorneys B. [7] In the stocks or prison ?
[8] that thei B. [9] then thei B.

ney thorow Fraunce, and so to Burdious & Byon; & than we entred into the baryn countrey[1] of Byskay and Castyle, wher we coulde get no meate for money; yet wyth great honger we dyd come to Compostell, where we had plentye of meate and wyne; but in the retornyng thorow Spayn, for all the crafte of Physycke that I coulde do, they dyed, all by eatynge of frutes and drynkynge of water, the whych I dyd euer refrayne my selfe.[2] And I assure all the worlde, that I had rather goe .v. times to Rome oute of [3]Englond, than ons to Compostel: by water it is no pain, but by land it is the greatest iurney that an Englyshman may go. and whan I returnyd, and did come into Aquitany, I dyd kis the ground for ioy, surrendring thankes to God that I was deliuered out of greate daungers, as well from many theues, as from honger and colde, and *that* I was come into a plentiful country; for Aquitany hath no felow for good wyne & bred.[4] in Nauerne theyr spech is Castilion: theyr money is gold and brasse; in golde they haue crownes; in brasse they haue Frenche money, and the Emprours money.

and, after nearly starving in Biscay, we got to Compostella.

But, in their return, all 9 Pilgrims died.

I'd rather go 5 times to Rome than once to Compostella by land.

I kist the ground for joy when I got back to Aquitaine

Money of Navarre.

¶ The .xxxiii. chapter treateth of Bion, and of Gascony, and of Lytle Briten, and of the natural disposicion of the people,[5] and of theyr money and of theyr speche.

[1] countres B. See pp. 199, 200, above.
[2] See Boorde's *Breuyary*, ch. C.xxii., extracted in the *Forewords*, p. 74, as to his hydrophobia, or dislike of water.
[3] sign. M .i. [4] See chapter xxvii. p. 193-4.
[5] treateth of the natural disposicion of the people of Bion and of Gascony, and of lytle briten — B.

I was borne in Bion ; ens[1] English I was ;
if I had be so styl, I wold not gretly pas.
And I was brought vp in gentyl Gascony ;
For my good wyne I get money.
And I was borne in Litle Britten ;
Of al nacions, I [hate] free Englyshe men :
Whan they be angry, lyke bees they do swarme ;
I be-shromp them, they haue don me much harme.
Although I iag my hosen & my garment rounde aboute,
[2] Yet it is a vantage to pick pendiculus owt.

Bayonne, once English. *Gascony.* *Brittany.* *I jag my clothes to pick out lice.*

¶ As tochinge Byon, the towne is commodiouse, but the country is poer and barin, in the whiche be many theues. ther is a place calyd the hyue ; it is fyuete or .lx. myle ouer ; there is nothynge but heth, and there is no place to haue succour with-in vii. or eyght myles ; and than a man shal haue but a typling house. The women of Byon be dysgysed as players in enterludes be, with long raiment ; the sayd clokes hath hodes sewed[3] to them, and on the toppe of the hod is a thyng like a poding bekyng forward.[4]

The Hive. *Women of Bayonne ; their cloaks and hoods.*

Gascony is a commodiouse country, for ther is plenty of wyne, bred, & corne, and other vytells, and good lodgyng and good chere, and gentle people. The chefe towne of Gascony is Burdiouse, and in the cathedrall Churche of saint Andreus is the fairist and the gretest payer of Orgyns in al Crystendome, in the whyche Orgins be many instrumentes and vyces, as Giants[5] heds and sterres, the whych doth moue and wagge with their iawes and eyes as fast as the player playeth. Lytle Brytane is a proper and a commodiouse countre, of Wyne, corne, fysh, fleshe ; & the people be hygh mynded & stubborne. These .iii. countres speketh French, and vseth euery thyng, as wel in ther mony &

Gascony. *Bordeaux.* *Grand pair of Organs in St Andrew's Church, with figures that wag their jaws.* *Brittany is a fruitful country.*

[1] once (before 1451-2). [2] sign. M .i. back. [3] swed A : sewed B.
[4] Compare the description of the Spanish women's heads in chapter xxx. p. 199, and the Venetian Doge's cap, p. 185.
[5] Gians A ; Giants B.

Rochelle, Morlaix.

fashions, as French men doth. Rochel & Morles is praysed in Briten to be the best townes.

¶ The .xxxiiii. chapter treateth of Normandy & Picarde, and of the natural disposicion of the people, and of theyr spech and mony.[1]

Normandy, Picardy: we wish we were further from English invasions.

[2] ¶ I was borne and brougt vp in gentyl Normandy;
And I am a man dwellyng in Pycardy;
We border vpon England; I wolde we war forder of;
For whan warre is, they maketh vs take the cof;
For than we do watche both nyght and day,
To prepare ordynaunce to kepe them away.
Yet we wyl kepe new fashyons of Fraunce,
Much lyke to players that is redy to daunce.

Normandy. Rouen; Caen and Sens, where canvas is made. All France belongs to England, by rights. Picardy.

¶ Normandy is a pleasaunt and a comodiouse countrey, in the whiche be many good Cities & townes, specyallye be these, which is to say, Rone[3], Cane, and Seno, withe many other. in Cane and Seno is good Canuis made. the people be after a gentil sort. Normandy doth partaine to England, and so doth al Fraunce by right many wayes, amonge the whyche I wyll resyte one thynge, that yf Fraunce ware not England, king Henry the sixt should not haue ben crowned kinge of Fraunce in Parys, he being in his cunables[4], and an infant. Pycardy is a good countrey ioynyng to

[1] B has no wood-cut. The one above is the upper part of the right-hand cut that Wynkyn de Worde uses for Robert the Devil in his *Robert the Deuyll*, sign. C .ij. back, and D .iv. back. [2] sign. M .ii.

[3] Rome AB, for Rouen; Caen and Sens.

[4] tunables B. *cunables* is cradle, no doubt.

Calys. The countrey is plentyfull of wood, wyne, and corne; how be it naturally they be aduersaries to Cales. Bolyn, in my mynde, is the best town of Pycardy. [1] Boleyn is now ours by conquest of Ryall kyng Henry the eyght.[1]

Picardy.

Boulogne is ours. Henry VIII won it.

¶ Here is to be noted, that in thys matter partrattyng of Europ, I shew at the begynnyng of this boke: If a man wolde go out of England, or other landes anexed to the same, he[2] should go to Calis;[3] and from Calys I haue set the cyrcuyte or the cercumferens of Europ, whyche is al Chrystendome, and am come to Calys agayn, wherfore I wyll speke no more of Europe, but only a chapter of Latyne, and than I wyll speke of other countreys of Affryck and Asya.

I've now treated of all Europe, from Calais, and back to Calais.

[4] 

1—1 This passage is omitted in the Lothbury edition of 1562 or 1563, Boulogne having been restored to France by Edward VI in 1550. See *Forewords*, p. 18.

2 AB have no "he."

3 See the end of Chapter vii, and Chapter viii above, p. 146.

4 sign. M .ii. back.

¶ The .xxxv. chapter treateth of the Latyn man and the Englysh man, & where Laten is most vsed.

I can show my face all over Europe.

Italy has corrupted my speech, and I shall leave her.

¶ I am a Latyn man, and do dwel in euery place;
Thorow al Europ[1] I dare shew my face;
Wyth the Romans and Italyon I haue dwelled longe;
I wyl seke other nacions, for they haue done wronge
In corruptyng my tonge and my ryalte,
Wherfore in other nacyons I loue to dwel and be,
And wher I shalbe dayly accept and vsed,
Regardyng not them where I am abused.

A responcion of the Englysh man.

To England I am welcome. They know Latin well.

I am an Englyshman; Latyn, welcome to me!
In thy tounge I am wel sped, & neuer was in thy countre;
[2] For thou arte indyfferent here and in[3] euery place,
If a man wyll study, and lerne the bokes a pace;
Wherfore bitwixt thee & me we wyl haue some alteracion,
That vnlerned men may know parte of our intencion.

Englyshe, and some Latyne, doth folowe.

A talk in English and Latin.

¶ Helth be to the, now and euer!
Salus tibi, nunc et in euum![4]
I thanke the hartly, and thou art welcome!
Immortalem habeo tibi graciam, & gratissime aduenisti!
What countrey man art thou? *Cuias es?*
I was borne in England, and brought vp at Oxforde.
Natus eram in Anglia, et educatus Oxoni.
Doest not thou know me? *noscis ne me?*
I know thee not *Minime te nosco.*[5]
What is thy name? *Cuius nominis es?*
My name is Andrew Borde.
Andreas parforatus est meum nomen.

[1] Erop AB. [2] M .iii. not signed. [3] A leaves out B's "in."
[4] enum A; et enum B. [5] nosca AB.

A talk in Latin and English.

How haue you fared many a day?
Qua valitudine fuisti longo iam tempore?
I haue faryd very wel, thankes be to God!
Optime me habui; graciarum acciones sunt Deo.
I am very glad of it. *Plurimum gaudio inde.*
Whyther dost thou go now? *Quous tendis modo?*
I go towerd London. *Versus Londinum lustro.*
What hast thou to do ther? *Quid illic tibi negoci est?*
I shal ease my mynd ther?
Animo meo morem gessero illic.[1]
Helth be to you al! *Salus sit omnibus!*
Thou art welcome! *Saluum te aduenisse*[2] *gaudeo!*
[3] I thanke you. *Habeo vobis graciam.*[4]
Hostes, how do you fare? *Hospica, vt tecum est?*
I haue fared wel, yf you haue bene well.
Multa melius me habeo si bene vale.
Hostes, haue you good meate?
Hospita, est ne hic cibus tantus?
Ye, I haue many good dyshes of meate.
Etiam, sana[5] *multa que sunt mihi fercula.*
Geue me drynke, and also bread.
Potum da mihi, Insuper et panem.
I drynke to you all! *propino vobis omnibus!*
Much good do it you! *prosit vobis!*
Farewel, & God be with you al!
Valetote, et Deus vobiscum!
Go[o]d night! *Optata requies!*
Farewel, & let them go *that* wolde any stryfe be-twyxt vs!
Vale! et valeant qui inter nos dissidium volunt!

[1] illis AB. [2] aduinesse AB. [3] M .iii. back.
[4] Habio vobis gracia A; Habo vobis gracia B. [5] santa AB.

¶ The .xxxvi. chapter treteth of the Mores whyche do dwel in Barbary.

Christian men buy me as a slave.

I Am a blake More borne in Barbary;[1]
Chrysten men for money oft doth me bye;
Yf I be vnchristend, marchauntes do not care,
They by me in markets, be I neuer so bare.
Yet wyll I be a good dylygent slaue,
Although I do stand in sted of a knaue;
I do gather fygges, and with some I whype my tayle:
To be angry wyth me, what shal it a-vayle?

I gather figs.

¶ Barbary is a great countrey, and plentyfull of frute, wine, & corne. The inhabytours be Called *the* Mores: ther be whyte mores and black moors; they be Infydels and vnchristened. There be manye Moores brought into [2]Christendome, in to great cytes & townes, to be sold; and Christenmen do by the*m*, and they wilbe diligent, and wyll do al maner of seruice; but thei be set most co*m*onli to vile thynges. they be called slaues; they do gader[3] grapes and fygges, and with some of the fygges they wyl wyp ther tayle, & put them in the frayle. they haue gret lyppes, and nottyd[4] heare,[5] black and curled; there[6] skyn is soft; and ther is nothing white but their teth and the white of the eye. Whan a Marchaunt or anye other man do by them, they be not al of one pryce, for some bee better cheepe then some; they be solde after as they can werke and do there busines. whan they do dye, they be caste in to the watter, or on a dounge hyll, that dogges and pyes and crowes may eate the*m*, except some of them that be christened; they be buried. they

White Moors and Black Moors;

are bought as slaves,

some cheaper than others;

are not buried when they die,

unless they are christened,

[1] Barby A; Barbary B.
[2] M .iiii. not signed.
[3] gader do A; do gader B.
[4] polled, clipt.
[5] heare is AB.
[6] the there A; there B.

do kepe muche of Macomites[1] lawe, as the Turkes do. they haue now a gret captyn called Barbarerouse,[2] whiche is a great warrier. thei doth harme, diuerce tymes, to the Ianues, & to Prouynce and Langewadocke, and other countres that do border on them, & for they wyl come ouer the straytes, &[3] stele pygges, and gese, and other thynges.

are Mahometans; are led by Barbarossa; plunder the Genoese, &c. (See p. 189.)

¶ Who so wyl speke any Moryshe, Englyshe and Morysh[4] doth folow.

Moorish numerals.

One. two. thre. foure. fyue. syx. seuen.
Wada. attennin. talate. arba. camata. sette. saba.
eyght. nyne. tene. aleuyn. twelue. thertene.
tamene. tessa. asshera. hadasshe. atanasshe. telatasshe.
fortene. fyuetene. syxtene. seuenten.
arbatasshe. camatasshe. setatasshe. sabatashe.
eyghtene.[5] nyntene. twente. one and twenty, &c.
tematasshe. tyssatasshe. essherte. wahadaessherte, &c.

A talk in Moorish and English.

Good morow! *sabalkyr!*
Geue me some bread and mylke and chese.
[6]*Atteyne gobbis, leben, iuben.*
Geue me wyne, water, flesh, fysh, and egges.
Atteyne nebet, moy, laghe, semek, beyet.
Much good do it you! *sahagh!*
You be welcome! *Marrehababack!*
I thanke you! *Erthar lake heracke!*
Good nyght! *Mesalkyr!*

[1] Maconites A (Mahomet's). See next chapter.

[2] Heyradin Barbarossa, a Corsair king of Algiers, born about 1467, died 1547.—*Hale.* See *Forewords*, p. 55.

[3] A has not B's "&."

[4] This 'Morysh' is undoubted Arabic, but in a very corrupt state. . . For instance, 'one' in Arabic is *ahad* or *wahid:* what are we to do with Boorde's *wada?* 'Five' is *khamsa* or *khamsat:* how correct Boorde's *camata?* I shall therefore correct only a few glaring errors, where one letter has been mistaken for another, *attennin, arba, tamene, hadasshe, sabalkyr*, for Boorde's, or his printer's wrong *m, o, c, b, s*, in these words.—Ch. Rieu.

[5] eyghtent A.

[6] M .iiii. back.

¶ The .xxxvii. Chapter tretyth of the natural disposicion of the Turkes, and of Turkey, and of theyr money and theyr spech.

I keep Mahomet's laws, and don't eat pork.

¶ I am a Turk, and Machamytes law do kepe;
I do proll for my pray whan other be a slepe;
My law wyllith me no swynes flesh to eate;
It shal not greatly forse, for I haue other meate.
In vsyng my rayment[1] I am not varyable,
Nor of promis I am not mutable.

The Great Turk has conquered many lands.

¶ In Turky be many regions & prouynces, for the great Turke, whyche is an Emproure, hath, besyd hys owne [2]possessyons, conqueryd the Sarsons londe, and hath obtayned the Sophyes lond, and the ylond of the Roodes,[3] with many other preuynces, hauyng it in pes-

[1] On Shrove Sunday in Henry VIII's first year, 1509-10, at his banquet in the Parliament Chamber at Westminster, "his grace, with the Erle of Essex, came in appareled after *Turkey fasshion*, in long robes of Bawdkin, powdered with gold, hattes on their heddes of Crimosyn Veluet, with greate rolles of Gold, girded with two swordes called Cimeteries [scimetars], hangyng by greate bawderikes of gold."—*Hall's Chronicle*, p. 513, ed. 1809. [2] sign. N .i.

[3] See Hall's account of its siege and capture in 1522.—*Chronicle*, ed. 1809, p. 653-5.

able possession. he doth conquere and subdue, as wel by polyce and gentylnes, as by hys fettes of warė. in Turkey is cheppe of vittyls, & plenty of wyne & corne. The Turkes hath a law called Macomites law, and the booke that there lawe is wrytten in, is called the Alkaron. Macomyt, a false felow, made it[1]; he sedused the people vnder thys maner: he dyd bryng vp a doue, and would put .ii. or thre pesen in his eare, & she would euery day come to his eare and eate the peason, and then the people would thynke the holy goost, or an Angell, did come & teache him what the people should do. And then he made hys booke, and vsyd to feede a tame Camel in his lappe; and euery daye he wolde feede *th*e Camel, *th*e which he taught to set downe on his knees when he did eate his meate. And whan he had broken the Camel to thys vsage, he monisshed *th*e people. saying, that God wolde sende them a law written in a booke, and to whome soeuer the booke was brought vnto, he should be the prophit of God, & conductor of the people. The*n* Macomit did poynt a day, And did conuocate the people together at a place where he was vsyd to feede a camel, by the whych place was a greate wood or wyldernes full of wylde beastes. The aforesayd day appoynted, yerly in *th*e morninge, Macomit sent one of hys seruau*n*tes to the wood with the Camel, bindi*n*g the booke a-boute the Camelles necke, *th*e whych[2] he had made before, chargyng his seruaunte, that whan all the people war gathered about him, to heare him make an exortacion, *tha*t he should let the Camell go, and that he shoulde preuely thorow the wood get himselfe home. Macomyte & the people beyng gathered together at the aforesayde place [3]appoyntcd, and makyng an exortation of the people, had his face to the

Turkey is a cheap and fertile country.

Alcoran.

Mahomet and his tricks: his Dove,

his Koran and his Camel.

He taught his Camel to kneel and feed out of his lap; and told the people God would send their Law to their Prophet and Ruler.

On a set day he sent his Camel with his book round its neck to a wood,

and told his man to let it go when the people were round him.

[1] See Sir John Mandeville's *Voiage*, ch. xii, on the Sarasines and Machomete, p. 131, ed. 1839.

[2] which book. [3] sign. N .i. back.

wood to looke whan the camel woldé *come*; and spyeng the camel, he dyd fynysh his exhortacion, and dyd couet of the prayse of the people, [and] stoude before the people. the Camel, seing his mayster, did come to him, and kneeled downe to haue eaten hys prouender. and Macomit sayd: "this Camell hath brought our law that we must keepe, to me;" and tooke of the booke from the Camels necke, and did reede it to the people; the whiche they did, and dothe, take it for a law. And they do take Macomite for a prophit. by thys, euery man may perceyue many subtyll and crafty castes be played in certeyn regions, long to reherse at this time, as it appered by the mayde of Kent[1], & other. The money the which is in Turke[2] is Golde and Siluer and Brasse: there be so many coynes, that it war long to reherce. in brasse they haue Torneys. In syluer they haue Aspers and Souldes; & ther be som Souldes that be brasse, *that* v. is worthe an Englishe peny. In golde they haue saraffes. A saraf is worth an Englysh crowne. In Turky is vsed diuers speches and langweges: some dothe speake Greeke, & some doth speake corrupt Caldy, and some dothe speake Moryske speche; wherfore I do now shew but litle of Turkey speche, the whych doth folow.

Mahomet, seeing the camel, finisht his speech; the Camel came and knelt to him, and Mahomet took his book off its neck, as the people's Law. The Turks think him a prophet. Turkish money: Torneys, Aspers, Souldes, Saraffes. Languages in Turkey. The Turkish numerals.

One. two. three. foure. fyue. syx. seuen. eyght. nyne.
bir[3]. *equi*. *vg*. *dort*[4]. *bex*. *alti*. *ȝedi*. *zaquis*. *dogus*.
tenne. aleuyne. twelue. thirten. fouertene. fyftene.
on. *onbir*[3]. *on equi*. *on vg*. *ondort*[4]. *on bex*.
sixtene. seuyntene. ayghtene. nynetene. twenty.
on alti. *onȝedi*. *onzaquis*. *on dogus*[5]. *on ygrimi*.
One and twenty. two and twenty. thre & twenty. &c.
ygrimi bir[6]. *ygrimi esqui*. *ygrim vg*, *&c*.
Bellahay.[7]

[1] Elizabeth Barton, the Holy Maid of Kent, executed April 21, 1534. See Hall's *Chronicle*, p. 814, ed. 1809.
[2] Turkye B. [3] bix A. [4] doit A. [5] doguc A.
[6] big A. [7] ? meaning. Both A and B have it.

[1]

¶ The .xxxviii. Chapter treteth of Egypt, and of theyr mony and of theyr speche.

¶ Egipt is a countrey ioyned to Iury ;
The countrey is plentyfull of wine, corne, and Hony.

Egypt is next to Judæa, and has deserts where holy Fathers lived.

Ther be many great wyldernes, in the which be many great wylde beastes. In the which wildernes liuid many holy fathers, as it apperyth in *vitas patrum.*[2] The people of the country be swarte, and doth go disgisyd in theyr apparel[3], contrary to other nacyons : they be lyght fyngerd, and vse pyking[4]; they haue litle maner, and euyl loggyng, & yet they be pleas[a]unt daunsers. Ther be few or none of the Egipcions *that* doth dwel in Egipt, for Egipt is repleted now *with* infydele alyons. There mony is brasse and golde. yf there be any man *that* wyl learne parte of theyr speche, Englyshe and Egipt speche foloweth.

The Egyptians steal, but dance well.

Few live in Egypt.

[1] sign. N .ii. See this cut before, p. 165, 206.
[2] The great mediæval storehouse of pious and lying legends.
[3] The other two ladies [A.D. 1510] . . . Their heades roulded in pleasauntes and typpers, *lyke the Egipcians,* enbroudered with gold. Their faces, neckes, armes & handes, couered with fine pleasaunce blacke : Some call it Lumberdynes ; which is merueylous thine ; so that the same ladies semed to be nygrost or blacke Mores.—*Hall's Chronicle,* p. 514 (see also p. 597), ed. 1809.
[4] cp. 'picking and stealing.'

A talk in Egyptian and English.

¶ Good morow! *Lach ittur ydyues!*
How farre is it to the next towne? *Cater myla barforas?*
[1] You be welcome to the towne *Maysta ves barforas*
Wyl you drynke some wine? *Mole pis lauena?*
I wyl go wyth you. *A vauatosa*
Sit you downe, and dryncke. *Hyste len pee*
Drynke, drynke! for God sake! *pe, pe, deue lasse!*
Mayde, geue me bread and wyne!
Achae, da mai manor la veue!
Geue me fleshe! *Da mai masse!*
Mayde, come hyther, harke a worde!
Achae, a wordey susse!
Geue me aples and peeres! *Da mai paba la ambrell!*
Much good do it you! *Iche misto!*
Good nyght! *Lachira tut!*

The .xxxix. Chapter treateth of the naturall disposicion of the Iues, and of Iury, and of theyr mony and of theyr speche.

I'm a Hebrew or Jew,

¶ I am an Hebrycyon; some call me a Iew;
To Iesu Chryst I was neuer trew.
I should kepe Moyses olde lawe;
I feare at length I shall proue a daw;
Many thynges of Moyses lawes do I not keepe;
I beleue not the prophetes; I lye to longe a sleepe.

and don't believe the prophets.

Judæa is a noble country.

[2] Iury is called the lande of Iude; it is a noble countre of ryches, plenty of wine and Corne, Olyues, ponegarnardes, Milke & Hony, Figges and Raysins, and all other fruites: ther be great trees of Cipres, palme

[1] sign. N .ii. back. [2] sign. N .iii.

trees, & Ceders. the chief towne of[1] Iury is Ierusalem, which was a noble citie, but now it is destroyed, and there doth neuer a Iue dwell in al Iury; for it was prophised *No Jews dwell in Judæa;* to theym by theyr lawe, that yf they woulde not beleue in Messias, whych is Chryst, they should be expelled out of their countrey; & so they were, and theyr citie destroyed by Vaspacion and Tytus; and the Iewes do *but all among Christian folk.* dwell amonge Christian people in diuers cities & townes, as in Rome, Naples, Venis, and diuerce other places. and forasmuche as our Lorde did suffer death at Ierusalem, And that there is a great confluence of pylgrims *As pilgrims go to the Holy Places, I'll tell you what I saw there.* to the holy Sepulcre and to many holy places, I wyl wryte[2] somwhat that I doo know and haue sene in *that* place. Who so euer that dothe pretende to go to Ierusalem, let him prepare himselfe to set forth of England *To make a pilgrimage to Jerusalem,* after Ester .vii. or .viii. dayes, and let him take his waye to London, to make his banke, or exchaunge of his mony, with some marchaunt, to be payd at Venis; and than let him go or ride to Douer or Sandwich, to take *start from Dover or Calais,* shypping to Calys; from Calis let him goe to Grauelyng, to Nuporte, to Burges, to Anwarpe, to Mastryt, to *go through Antwerp,* Acon, to During, to Colyn, to Boune, to Coualence, to *Coblentz,* Mense, to Wormes, to Spyres, to Gypping, to Geslyng, *Spiers,* to Memmyng, to Kempton, to the .vii. Kirkes, to *Kempten,* Trent, to Venis. Whan you be there, you must make *to Venice.* your bargen wyth the patrone of the Galy that you shall *Get the galley-captain to supply you with food,* go with-all, for your meate and drinke, & other costes. you must bye a bed, to haue into the Galy; you must *buy a bed, and a chest to keep wine, &c., in.* bye a bygge cheste with a locke and kaye to kepe-in wyne, and water, and spices, and other necessary thynges. [3] one Corp[u]s Christy daye[4] you shal be hous- *Be shriven on ship to Rhodes,* elled, and within two or three dayes you shall take your shyppyng, and you shall come to many fayrer portes, as

[1] A puts "of" after "is." [2] wyshe A; wishe B. [3] sign. N .iii. back.
[4] Corpus Christi is a festival of the Church of Rome, kept on the next Thursday after Trinity Sunday [a moveable summer feast-day] in honour of the eucharist.—*Webster.*

Candy, *the* Rodes, and dyuers other, longe to wryte; than, when you come to porte Iaffe, you shal go a foote to Ierusalem, except you be sycke, for at port Iaffe you enter in to the Holy Land. when you come to Ierusalem, the friers which be called Cordaline,[1]—they be of saynct Fraunces order,—they wyl receaue you with deuocion, & brynge you to the sepulcre. the holy sepulcre is wythin the church, and so is the mount of Caluery, where Iesu Chryst did suffer his passions. The churche is rounde, lyke a temple; it is more larger then anye temple that I haue sene amonges the Iues. The sepulcre is grated rounde about wyth yrone, that no man shall graet[2] or pycke out any stones. The sepulcre is lyke a lytle house, *the* which by masons was dydgyd[3] out of a rocke of stone. There maye stonde wythin the sepulcre a .x. or a .xii. parsons; but few or none dothe go into the sepulcre, except they be singulerly beloued, & than they go in by night, wyth great feare and reuerence. And forasmuch as ther be many[4] that hath wrytten of the Holy Lande, of the stacyons, & of the Iurney or way, I doo passe ouer to speake forther of this matter. wherfore yf any man wyll learne to speake some Hebrew,—Englyshe and Hebrew foloweth.

Joppa.

At Jerusalem the Cordaline Friars will lodge you.

The Holy Sepulchre

is railed round with iron,

but few are allowed to go into it.

The Hebrew numerals

¶ One. two. thre. fouer. fyue. syx.
Aleph. beth. gymel. daleth. he. vauf.
seuyn. eyght. nyne. tenne. aleuyne.
zain. heth. theth. Iod. Iod aleph
twelue. thertene. fouertene. fyftene. sixtene.
Iod beth. Iodgymel. Iod daleth. Iod he. Iod vauf.
seuentene. eyghtene. nintene. twenty. therty.
Iod zain. Iod heth. Iod teth. Chaph. lamed.

[1] *Cordeliers*, from the rope they wore as a girdle. [2] grate B. [3] diggyd B.

[4] It is curious how few early writers in English there are on Jerusalem and its Stations, &c. Except Sir John Maundevile (*Voiage*, ch. 7—11, p. 73—130, ed. 1839), Mr Huth's late MS poem quoted above, p. 182, of which the handwriting is about 1500 A.D., the less complete copy, &c., in Wey's Pilgrimages, the old printed tract reprinted for the Roxburghe Club, and I do not know any.

forty. fyfty. sixte. seuynte. eyghte. nynte. a hunderd.
[1]*mem. vn. sameth. yami. pee.*[2] *phe. zade.*

¶ The Hebrew the whych the Iues doth speak now, these dayes, doth alter from that[3] trew Hebrew tongue, (except the Iues be clerkes,) as barbarouse Latin doth alter from trew Latins, as I haue knowen the trueth whan *that* I dyd dwel amonges them, as it shall appere to them *that* doth vnderstande the tounge or speche folowynge.

Modern Hebrew is corrupt.

A talk in corrupt Hebrew.

God speede, god speed, syr! *Hosca, hosca, adonai!*
You be welcome, master! *Baroh haba, rabbi!*

Thys aforesayde Hebrew is corrupt, and not good Hebrew; but thys Hebrew that foloweth, is perfyt:

A talk in good Hebrew and English.

You be welcome, syr! *Eth borachah, adonai!*
(Or els you may say) *Im borachah, adonai!*
Wenche, or gyrle, geue me meate!
Alma, ten lii schaar!
Mayde, geue me drynke! *Bethela, ten lii mashkeh!*
Woman, geue me bread! *Nekeua, ten lii hallechem!*
Woman, geue me[4] egges! *Ischa, ten lii baet sim!*
Man,[5] geue me wyne! *Isch, ten lii iaiiu!*
Master, geue me flesh! *Rauf, ten lii basar!*
Geue me fyshe! *Ten lii daga!*
Fare wel, wife! *Schasom lecha nekeua!*
God nyght, syr! *Iailah tof, adonai!*
God be wyth you, master! *Leschalom rauf!*
Iesus of Nazareth, kyng of Iues! The son of God haue mercy on me! Amen.
Iesuch Natzori, melech Iuedim. Ben Elohim conueni! Amen[6]*!*

[1] M .iiii. not signed.
[2] A little bit of the last leaf of A, with *i*, *pee*, and part of *phe* on it, has been torn out.
[3] ye B. [4] mo A. [5] Mam A; man B.
[6] In B, the colophon follows, and is: "¶ Imprented at London in Lothbury ouer agaynste Sainct Margarytes church, by me Wyllyam Copland." Upcott's reprint was printed by Richard and Arthur Taylor, Shoe Lane.

¶ Imprinted at London in Fleetestrete, at the Signe of the Rose Garland, by me William Copland.

(.·.)

¶ Hereafter folo

weth a compendyous Regyment or a dyetary of Helth, made in Mountpyllier, compyled by Andrew Boorde of Physycke doctour, dedycated to the armypotent Prynce, and valyaunt Lorde Thomas Duke of Northfolche.

Galyen prynce,

of Physycke.

[*Beside the Preface of the first edition of* 1542 *is set that of Powell's edition of* 1547, *in order that readers may see the differences between the two, and judge whether any one but Andrew Boorde himself could have made the alterations.*]

[ed. 1542.]

¶ The preface.

¶ To the precellent and armypotent prynce, lorde Thomas, duke of Northfolch,[1] Andrew Borde, of Physycke doctour, doth surronder humyle commendacyon.

FOrasmoch as it pleased your grace to send for me (to syr Robert Drewry, knyght,)—whiche was the yeare in the whiche lorde Thomas, cardynal, bishop of york, was commaunded to go to his see of york,[2]—to haue my counceyll in Physycke, in certayne vrgent causes requyryng to the sauyte of your body: at that tyme I, beyng but a yonge doctour in my scyence or faculte, durst not

[1] Thomas Howard, 8th Duke, inherited the dukedom on his father's death in 1524, was attainted in 1546, when his honours became forfeited; they were restored in 1553, and the Duke died in 1554.—*Nicolas's English Peerage*, ii. 473.

[2] A.D. 1530.

[ed. 1547.]

¶ The preface or the proheme.

To the armypotent Prynce and valyent lorde Thomas Duke of Northfolke Andrewe Boorde of physycke doctor: dothe surrender humyle commendacyon with immortall thankes.

AFter the tyme that I had trauelled for to haue the notycyon & practes of Physycke in diuers regyons & countres, & returned into Englande, and [was] requyred to tary and to remayne and to contynue with syr Robert Drewry, knyght, for many vrgent causes, Your grace, heryng of me, dyd sende syr Iohan Garnyngham—nowe beynge knyght[3]—to me, to come to youre grace, to haue my counsell in physycke for your infyrmytes. The mesage done, I with festynacyon & dylygence dyd nat prolonge the tyme, but dyd come to your grace accordynge to my deuty. The whiche was in the tyme whan lorde Thomas Cardynall Archebysshop of Yorke was commaunded to go to his

[3] No doubt Sir R. Drury's son-in-law. "Edward Jernegan, Esq., his son and heir, who was afterwards knighted. He had two wives, first, Margaret, daughter of Sir Edmund Bedingfield, of Oxborough, in Norfolk, Knt., by whom he had *Sir John Jernegan*, of Somerleytown, in Suffolk, Knt., *who married, first, Bridget, daughter of Sir Robert Drury*, of Hawsted, in Suffolk, Knt., from whom the Jernegans of Somerleytown, in Suffolk, descended."—*The English Baronetage*, 1741, vol. i. p. 455, 'Jernegan or Jerningham, of Cossey, Norfolk.' 'From this house (Drury) branched off the Drurys of Hawsted, Suffolk, who built Drury house in London, temp. Elizabeth, the road leading to which has ever since retained the name of Drury Lane. It stood a little behind the site of the present Olympic Theatre.'

to presume to mynyster any medysone to you with*h*out the counceyl of mayster doctour Butte, whiche had a longe continuau*n*ce with you, & a [1 sign. A .ij.] great cognys[1]cyon, not onely of your infyrmyte, but also of your complexyon & dyet. But he not co*m*myng to your grace, thankes be to God, your grace recuperatyng your helth, And conuocated thorowe the kynges goodnes to wayte on his prepotent mageste, I than dyd passe ouer the sees agayne, And dyd go to all the vnyuersyties and scoles approbated, and beynge within the precinct of chrystendome. And all was done for to haue a trewe cognyscyon of the practis of Physycke; the whiche obtayned, I than, cotydyally remembryng your bountyfull goodnes shewed to me, & also beynge at the well-hed of Physycke, dyd consult with many egregyous Doctours of Physycke / what matter I shuld wryte, the whiche myght be acceptable, and profitable for the sauyte of your body. The sayde

see of Yorke. And after my co*m*mynge to you, and felynge the pulses of your herte, the pulses of your brayne, and the pulses of your lyuer, and that I had sene your vryne & your egestyon, I durste nat to enterpryse or medyll with out the counsell of Mayster doctor [Sign. + ii.] Buttes, the which dyd know, nat onely your co*m*plexcion & infyrmite, but also he dyd know the vsage of your dyete, And the imbecyllyte and strength of your body, with other qualytes expedyent & necessary to be knowen: but brefely to conclude, [for] your recuperatyng or recouering your health, And for synguler trust and hygh fauour, the which the kyng had to you, [I] was compocated[2] to be in the presence of his magesty. I than dyd passe ouer the sees agayne, and dyd go to all the vnyuersytes and great Scloles,[2] the whiche be approbated with in the precynct of Chrystendome, for to haue the practes of physycke. I seynge many expedyent thynges in dyuers regyons, at the last I dyd staye my selfe at Mountp[y]llyoure, which is the hed vniuersite in al Europe for the practes of physycke & surgery or chyrmi*n*g. I beinge there, And hauyng a cotydyal reme*m*brance vpon youre bountyfull goodnes, dyd consulte with many egregyous

[2] so in the original.

doctours, knowynge my trewe intencyon, dyd aduertyse me to compyle and make some boke of dyete, the which, not onely shuld do your grace pleasure, but also it [1]shuld be necessary & profytable for your noble posterite, & for many other men the whiche wolde folowe the effycayte of this boke / the whiche is called the Regyment or dietary of helth. And where that I do speake in this boke but of dietes, and other thynges concernynge the same, If any man therfore wolde haue remedy for any syckenes or dyseases, let hym loke in a boke of my makynge, named *the* Breuyare of helth. But yf it shall please your grace to loke on a boke, the which I dyd make in Mountpyller, named *the* Introductory of knowlege, there shall you se many new matters / the whiche I haue no doubte but that your grace wyl accept and lyke the boke, the whiche is a pryntynge besyde saynt Dunstons churche within Temple barre ouer agaynst *the* Temple.[2] And where I haue dedycated this

[1] [sign. A .ij. back]

Doctours of physycke what maner that I myghte wryte the whiche myght be acceptable for the conseruacyon of the health of youre body. The sayde doctors, knowynge my zele and true intencyon had to you, dyd aduertyse me to make a boke of dyete, nat only for your grace, but also for your noble posteryte, and for all men lyuynge: wherfore I do nomynate thys boke The Dyetary of health, the which doth pertract howe a man shuld order him selfe in all [Sign. ✝ .ii. back.] maner of causes partenynge to the health of his body: yf your grace or any man wyl haue forther knowledge for dyuers infyrmites, let him loke in a boke of my makynge named *the* Breuyary of health. And where I haue dedycated this boke

[2] There is no early edition of this book in the British Museum. The reprint of 1814 says, 'The rarity of this Tract is such, that Mr West was induced to believe that no other copy existed than the one in his collection; after his death it passed into the hands of Major Pearson; and at the sale of his library, in 1788, Mr Bindley became the possessor.' This is the only copy 'known of the edition printed by *Copland in Fletestrete, at the signe of the Rose Garland.* Of the edition *printed by him in Lothbury* a copy is in the Bodleian Library, among Selden's books, B. 5, 6, [another in the Chetham Library at Manchester,] and from one in the publishers' hands [? now Mr Christie-Miller's copy] the present reprint has been executed.'

boke to your grace, and haue not ornated and floryssheḍ it with eloquent speche and rethorycke termes, *th*e which [1] [sign. A .iij.] in all wry[1]tynges is vsed these modernall dayes, I do submyt me to your bou*n*tyful goodnes. And also dyuers tymes in my wrytynges I do wryte wordes of myrth / truely it is for no other intencyon but to make your grace mery,—for myrth is one of the chefest thynges of Physycke, the which doth aduertyse euery man to be mery, and to beware of pencyfulnes,—trustynge to your affluent goodnesse to take no displeasure with any contentes of this boke, but to accept my good wyl and dylygent labour. And furthermore I do trust to your superabundau*n*t gracyousnes, that you wyll consydre *th*e loue and zeale, the which I haue to your prosperyte, and that I do it for a co*m*mon weele, the whiche I beseche Iesu chryst longe to contynew, to his wyll and pleasure in this lyfe, And after this transytory lyfe remunerate you with celestyal ioy and eternall glorye. From Mountpyllier. The .v. day of May. The yere of our Lorde Iesu Chryste .M.v.C.xlij.

to your grace, And haue nat ornated hit with eloquence & retorycke termes, the whiche in all maner of bokes and wryttynges is vsed these modernall dayes, I do submytte me to your bountefull goodnes. And also dyuers tymes in my wrytynges I do wryte wordes of myrth: truely it is for no other inte*n*cion, but to make your grace mery;—for myrth is one of *th*e chefest thynges of physycke,[2] the which doth aduertise euery man to be mery, and to beware of pencyfulnes;—trustynge to youre affluent goodnes to take no displeasure with any of the co*n*tentes of this boke, but to accept my good wyll & dylygent labour. And, forthermore, I do truste to your superabundaunt gracyousnes, that you wyll consyder the loue and zele, the which I haue to your prosperyte, and that I do it for a common weale; the which I beseche Iesu chryst longe to continue, to his wyll and pleasure in this lyfe; And after this transytory lyfe, to remunerate you with celestyall ioye and eternal glorye. From Mountpyller. The fyft daye of Maye. The yere of our Lorde Iesu Chryste. M. CCCCC.XLVII.[3]

[2] See *Forewords*, p. 89, and *Dyetary*, p. 244.

[3] Powell's title is: "A com-/pendyous Regyment or a Dyetary of healthe made in Mount-pyllyer by Andrewe Boorde of phy-/sycke Doctour newly corrected / and imprynted with dyuers ad-/dycyons Dedycated to the / Armypotent Prynce and / valyent Lorde Tho-/mas Duke of / Northfolke. : " ✠ A B C D E F G H in fours, I in six. For Colophon, see p. 304.

[1] ¶ Here foloweth [2] the Table of the Chapytres.

THe fyrste Chapytre doth shewe where a man shuld cytuat or set his mancyon place or howse, for the helth of his body. (p. 232)

¶ The seconde Chapytre doth shewe a man howe he shulde buylde his howse, and that the prospect be good for *the* conseruacion of helth. (p. 234)

¶ The thyrde Chapitre doth shewe a man to buylde his howse in a pure and[3] fresshe ayre, for to lengthen his lyfe. (p. 235)

¶ The .iiii. Chapytre doth shewe vnder what maner a man shuld buylde his howse or mansyon, in eschewynge thynges that shuld shorten his[4] lyfe. (p. 237)

¶ The .v. Chapytre doth shewe howe a man shuld ordre his howse concernyng the implementes to comforte the spyrytes of man. (p. 240)

¶ The .vi. Chapytre doth shewe a man howe he shulde ordre his howse and howsholde. and[5] to lyue in quyetnes. (p. 241)

¶ The .vii. Chapytre doth shew howe the hed of a[6] howse, or a howseholder,[7] shulde exercyse hym selfe for the helth of the[8] soule and body. (p. 242)

¶ The .viii. Chapytre doth shew howe a man shulde order hym selfe in slepynge, and wat[9]chynge,[10] and in his apparell wearynge. (p. 244)

¶ The .ix. Chapitre doth shew that replecion or surfetynge doth moche harme to nature, and that abstynence is the chyfest medyson of all medysons.[11] (p. 250)

¶ The .x. Chapytre treateth of all maner of drynkes, as of water, of wyne, of ale, of bere, of cyder, of meade, of metheglyn, & of whay.[12] (p. 252)

[1] sign. A .iij. back.
[2] Wyer's undated edition (A), and Colwel's of 1562 (B) read: "¶ The Table. ¶ The Table of the Chapters foloweth." Powell's edition of 1547 (P) has: "Here foloweth the Table of the Chapiters."
[3] and a P. [4] the AB. [5] AB omit 'and.'
[6] the B; A reads 'of house.' [7] householde P. [8] his AB.
[9] leaf A. 4, not signed. [10] watche AB. [11] medyson P. [12] AB add '&c.'

¶ The .xi. Chapytre treateth of breade. (p. 258)

¶ The .xii. Chapytre of potage, of sewe, of stew pottes, of grewell, of fyrmente, of pease potage, of almon[1] mylke, of ryce potage, of cawdels, of culleses, of alebrues, of hony soppes, and of all other maner of brothes. (p. 262)

¶ The .xiii. Chapitre treateth of whyt meate, as of egges, butter, chese, mylke, crayme, posettes; of almon[1] butter, and of beane butter. (p. 264)

¶ The .xiiii. Chapytre treateth of fysshe. (p. 268)

¶ The .xv. Chapytre treateth of wyld fowle, of[2] tame fowle, and of byrdes.[3] (p. 269)

¶ The .xvi. Chapytre treateth of flesshe, wylde and domestycall. (p. 271)

The .xvii. Chapytre treateth of partyculer thynges of fysshe and flesshe. (p. 276)

¶ The .xviii. Chapitre treateth of rost meate, of fryde meate, of soden or boyled meate, of bruled meate, and of baken meate. (p. 277)

[4]¶ The .xix. Chapytre treateth of rootes. (p. 278)

¶ The .xx. Chapytre treateth of certayne vsuall herbes.[5] (p. 280)

¶ The .xxi. Chapytre treateth of fruytes. (p. 282)

¶ The .xxii. Chapytre treateth of spyces. (p. 286)

¶ The .xxiii. Chapytre sheweth a dyate for sanguyne men. (p. 287)

¶ The .xxiiii. Chapytre sheweth a dyate for flematycke men. (p. 288)

¶ The .xxv. Chapytre sheweth a dyate for colorycke men. (p. 288)

¶ The .xxvi. Chapytre doth shewe a dyate for melancoly men. (p. 289)

¶ The .xxvii. Chapytre treateth of a dyate and of an order to be vsed in the pestyferous tyme of *the* pestilence & the swetyng syckenes. (p. 289)

¶ The .xxviii. Chapytre treateth of a dyate for them the whiche be in an agew or a feuer. (p. 291)

¶ The .xxix. Chapitre treateth of a dyate for them the whiche haue the Ilyacke, or the colycke, and the stone. (p. 292)

[1] almonde AB. [2] and AB. [3] and byrdes AB.
[4] A 4, back. [5] of herbs P.

¶ The .xxx. Chapytre treateth of a dyate for theym the whiche haue any of the kyndes of the gowtes. (p. 293)

¶ The .xxxi. Chapitre treateth of a dyate for them the which haue [1] any kyndes of[1] lepored. (p. 293)

¶ The .xxxii. Chapytre treateth of a dyate [2]for theym the whiche haue any of the kyndes of the fallynge syckenes. (p. 294)

¶ The .xxxiii. Chapytre treateth of a dyate for them[3] whiche haue any payne in theyr hed. (p. 295)

¶ The .xxxiiii. Chapytre treateth of a dyate for them the whiche be in a consumpcyon. (p. 296)

¶ The .xxxv. Chapytre treatheth of a dyate for them the which be asmatycke men, beynge short-wynded, or lackynge breath. (p. 296)

¶ The .xxxvi. Chapytre doth shewe a dyate for them the whiche hath[4] the palsy. (p. 297)

¶ The .xxxvii. Chapitre doth shew an order & a dyate for them *that*[5] be mad & out of their wyt. (p. 298)

¶ The .xxxviii. Chapytre treateth of a dyate for them[3] which haue any[6] kynde of the dropsy.[7] (p. 299)

¶ The .xxxix. Chapytre treateth of a general dyate for all maner of men or[8] women[9] beynge sycke or whole. (p. 300)

¶ The .xl. Chapytre doth shew an order or a fasshyon, howe a sycke man shall[10] be ordered in his syckenes. And how a sycke man shuld be vsed that is lykly to dye. (p. 301)

¶ Here endeth[11] the Table.

¶ Here foloweth the dyetary or the[12] regyment[13] of helth.

[1–1] any of the kyndes of the AB. [2] sign. B .i. [3] them the AB. [4] haue AB. [5] the whiche AB. [6] any of the AB. [7] of dropsy P. [8] and AB. [9] woman B. [10] shulde A ; shoulde B. [11] The ende of AB. [12] "or the" is repeated in B, the 1562 edition. [13] And here foloweth the Dyetary.

[*In the Text, the small initials of some proper names have been made Capitals; and the stops have been often altered.*

In the Notes, "*A*" *stands for Wyer's undated edition* (Forewords, p. 13); *B for Colwel's edition with the Dedication dated* 5 *May*, 1562; *and P for Powell's edition, dated* 5 *May*, 1547, *in the Dedication, and* 1567 *in the Colophon. Powell prints* nat *for* not. *Differences of spelling, and printers' mistakes, are seldom noted.*

In Wyer's original of 1542, *the Galien cut on the next page stands by itself, and* '*the fyrst Chapytre*' *begins on the page after.*]

[2]¶ The fyrst Chapytre doth shew where a man shulde cytuate or[3] sette his mancyon place or howse for the health of his body.

Whoever means to build

WHat man of honour or worshyp, or other estate, the whiche doth pretende to buylde a howse or any mancyon place to inhabyte hym selfe, Or elles doth pretende to alter his howse, or to

or alter a house,

[1] sign. B .i. back. No cut in ABP. [2] sign. B .ii.
[3] for P.

alter olde buyldyng in-to co*m*modyous and pleasaunt buyldynge, not onely for his owne proper co*m*modite, welth, & helth, but also for other men the whiche wyll resort to hym, hauyng also a respect to his posterite,—

¶ Fyrste, it is necessarye and expedyent for hym to take hede what counceyll God dyd gyue to Abraham; and after that to take hede what counceyll God dyd gyue to Moyses, and to the chyldren of Israell, as it appereth in the .xiii. chapytre of Exodi, and the .xx. chapytre of Numeri, & the .vi. chapytre of Deutronomii[1]; and also in the boke of Leuites, saying fyrste to Abraham: "Go thou forth of [2]thy countre, & from thy cognacion or kynred, And come thou in to the countrey the whiche I wyll shew to the, a countrey abundynge, or plentyfull, of mylke and hu*n*ny." ¶ Here is to be noted, that where there is plenty of mylke there is plenty of pasture, and no skarsyte of water; & where there is plenty of hu*n*ny there is no skarsyte, but plentyfulnesse, of woddes, for there be mo bees in woddes (and so consequently abundaunce of hu*n*ny,) than there be bees, or hu*n*ny, or waxe, in *th*e hyues in gardyns or orchardes; wherfore it appereth that whosoeuer[3] wyl buylde a mancyon place or a house, he must cytuat and set it there where he must be sure to haue both water and woode, except for pleasure he wyll buylde a howse in or by some cytie or great towne, the whiche be not destitude of such co*m*modytes. But he the whiche wyll dwell at pleasure, and for proffyte and helth of his body, he must dwell at elbowe-rome, hauyng water and woode anexed to his place or howse; for yf he be destytuted of any of the pryncypalles, that is to say, fyrst, of water for to wasshe and to wrynge, to bake and to brewe, and dyuers other causes, specyally for parrell[4], the whiche myghte fall by fyre, [it][5]

must first heed how God told Abraham

to go to a country of milk and honey;

one with pasture, water, woods, and gardens.

A man must dwell at elbow-room,

and look ... for water.

[1] Deutro. P. [2] sign. B .ii. back. [3] euer that AB.
[4] peryll AB. [5] it AB.

were a great dyscommodyous thynge. And better it [1]were to lacke woode than to lacke water, the premysses consydered, althoughe that woode is a necessarye thynge, not onely for fewell, but also for other vrgent causes, specyally concernynge buyldynge and reperacyons.

2. for wood.

¶ The seconde Chapytre doth shewe a man howe he shuld buylde his house or mansyon, that the prospect be fayre & good for the conseruacyon of helth.[2]

Next to the soil and place,

After that a man haue chosen a conuenyent soyle and place accordynge to his mynde and purpose to buylde his howse or mansyon on, he must haue afore cast in his mynde, that the prospect to and fro the place be pleasaunt, fayre, and good to the eye, to beholde the woodes, the waters, the feldes, the vales, the hylles, & the playne grounde, And that euery thynge be desent and fayre to the eye, not onely within the precyncte of the place appoynted to buylde a mansyon or a howse, to se the commodyties aboute it, but also [that] it may be placable to the eyes of all men to [3]se & to beholde whan they be a good dystaunce of[4] from the place, that it do[5] stande commodyously. For the commodyous buyldyng of a place doth not onely satysfye the mynde of the inhabytour, but also it doth comforte and reioyseth a mannes herte to se it, specyally the pulcruse prospect. For my consayte is suche, that I had rather not to buyld a mansyon or a howse, than to buylde one

you must see that the prospect be good,

so that it may please people far off.

The sight of a well-placed house rejoices a man's heart.

[1] sign. B .iii.
[2] As to the building and pitching of houses, see Burton's *Anatomy*, Part ii., sect 2.—W. C. H.
[3] B .iii. back. [4] of = off. [5] doth A; doeth B.

without a good respecte[1] in it, to it, & from it. For and the eye be not satysfyed, *the* mynde can not be contented. And the mynde can not be contented, the herte can not be[2] pleased: yf the herte & mynde be not pleased, nature doth abhorre. And yf nature do abhorre, mortyfycacyon of the vytall, and anymall, and spyrytuall powers, do consequently folowe.

The eye must be satisfied, or the heart 'll not be pleased.

¶ The thyrde Chapytre doth shewe a man to buylde his howse in a pure & a fresshe ayre, to lengthen his lyfe.

THere is nothynge, except poyson, that doth putryfye or doth corrupt the blode of man, and also doth mortyfye the spyrytes of man, as doth a corrupt and a conta[3]gyous ayre. For Galyen, *terapentice*[4] *nono*, sayeth, "whyther we wyll or wyll not, we must grau*n*t vnto euery man ayre; for without the ayre, no man can lyue." The ayre can not be to clene and pure: consyderynge it doth[5] compasse vs rounde aboute, and we do receyue it in to vs, we can not be without it, for we lyue by it as the fysshe lyueth by the water. Good ayre, therfore, is to be praysed. For yf the ayre be fryske,[6] pure, and clene, about the mansyon or howse, it doth conserue the lyfe of man, it doth comfort the brayne, And the powers naturall, anymall, and spyrytuall, ingendrynge and makynge good blode, in the whiche consysteth the lyfe of man. And contraryly, euyl and corrupt ayres doth infecte the blode, and doth ingendre many corrupte humours, and doth putryfye the brayne, and doth corrupte the herte; & therfore it doth brede many dyseases & infyrmytyes, thorowe the which, mans

Bad air corrupts the blood and spirits of man.

Air can't be too pure.

Bright air comforts the brain, and makes good blood.

Bad air corrupts the heart, and

[1] prospecte AP; prospect B.
[2] A omits "be."
[3] B .iv. not signed.
[4] terapentico AB.
[5] close and doth AB.
[6] fresshe AB.

shortens man's life.

lyfe is abreuyated and shortned. Many thynges doth infect, putryfye, and corrupteth the ayre, as[1] the influence of sondry sterres, and standyng waters, stynkyng mystes, and marshes, caryn lyinge longe aboue the grounde, moche people in a smal rome lying vnclenly, and beyng fylthe and sluttysshe; wherfore he [2]that doth pretende to buylde his mansyon or house, he must prouyde that he do nat cytuat hys howse nyghe to any marsshe or marysshe growndo; that[3] there be nat, nygh to the place, stynkynge and putryfyed standyng waters, pooles, pondes, nor myers,[4] but at lestwyse that such waters do stande vpon a stony or a grauayle growndo myxt with claye, and that some fresshe sprynge haue a recourse to nourysshe and to refresshe the sayd standyng waters. Also there must be circumspection had that there be not aboute *th*e howse or mansyon no stynkynge dyches, gutters, nor canelles, nor corrupt dunghylles, nor synkes, excepte they be oft and dyuers tymes mu*n*dyfyed and made clene. Swepyng of howses and chambres ought nat to be done as long as any honest man is within the precynct of the howse, for the dust doth putryfy the ayre, makynge it dence. Also, nygh to the place let nother[5] flaxe nor hempe[6] be watered; & beware of the snoffe of candelles, and of the sauour of apples, for these thynges be contagyous and infectyue. Also, mysty & clowdy dayes, impetous and vehement wyndes, troublous and vaporous wether is nat good to labour in it, to open the pores[7] to let in infectious ayre. Furthermore, [8]beware of pyssynge in drawghtes; & permyt no co*m*mon pyssyng place be aboute the howse or mansyon; & let the co*m*mon howse of easement be ouer some water, or elles elongated from the howse. And beware of emptynge of pysse-pottes,

As standing waters, &c., putrefy the air,

take care that you don't build your house near stinking ponds, &c.;

or near any stinking ditches, channels, or sinks,

or where flax is steept;

and don't have a urinal or privy near your house.

[1] The fyrst is AB. [2] B. 4, back. [3] And that AB.
[4] meeres AB. [5] nat her P. [6] hempe nor flaxe AB.
[7] powers AB. [8] sign. C.

and pyssing in chymnes, so that all euyll and contagyous ayres may be expelled, and clene ayre kept vnputryfyed. And of all thynges let the buttery, the celler, the kytchen, the larder-howse, with all other howses of offyces, be kept clene, that there be no fylth in them, but good & odyferous sauours: and, to expell & expulse all corrupt & contagyous ayre, loke in the .xxvii. Chapytre of this boke. [p. 289.]

Mind that your kitchen and offices are kept clean.

¶ The .iiij. Chapytre doth shew vnder what maner & fasshyon a ma*n* shuld buylde his howse or mansyon, in exchewynge thynges that shortneth mans lyfe.[1]

WHan a man doth begyn to bylde his hous or ma*n*syon place, he must prouyde (sayth Jesus Chryst), before *that* he begyn to buylde, for all thyng*es* necessary for the performacyon of it, lest that whan he [2]hath made his fou*n*dacion, & can not fynysshe his worke that he hath begon, euery man wyl deryde hym, saying: "This man dyd begyn to buylde, but he can not fynysshe or make an end of his purpose:" for a man must consyder the exspence before he do begynne to buylde; for there goeth to buyldynge, many a nayle, many pynnes, many lathes, and many tyles, or slates, or strawes, besyde other greater charges, as tymber, bordes, lyme, sand, stones, or brycke, besyde the workmanshyp and the implementes. But a man the whiche haue puruyd,[3] or hath in store, to accomplysshe his purpose, and hath chosen a good soyle and place to cytuat

When you begin to build, provide beforehand enough to finish, as Christ tells you.

Many a nail, pin, straw, and board will be needed.

[1] thynges the whiche shulde shorten the lyfe of man AB.
[2] C .i. back. [3] prouyded AB.

hys howse or mansyon, and that the prospecte be good, and that the ayre be pure, fryske, and clene, Then he that wyll buylde, let hym make his fundacyon vpon a graualy grownde myxt with clay, or els let hym buylde vpon a roche of stone, or els vpon an hyll or a hylles syde, And ordre & edyfy the howse so that the pryncypall and chefe prospectes may be Eest and weest, specyally North-eest, Sowth-eest, and South-weest, for the merydyal wynde, of al wyndes is the moste worst, for the South wynde doth corrupt and doth make euyl vapours. The Eest wynde is tem[1]perate, fryske, and fragraunt.[2] The weest wynde is[3] mutable. The North wynde purgeth yll vapours; wherfore, better it is, of *the* two worst, that the wyndowes do open playne North than playne Sowth, althoughe that Jeremy sayth, "from the North depe*n*deth all euyl[4];" and also it is wryten in Cantica cant[ic]or*um*[5]: "Ryse vp, North wynde, and come, thou Sowth wynde, and parfyat[6] my gardayne." Make the hall vnder such a fasshyon, that the parler be anexed to the heade of the hall. And the buttery and pantry be at the lower ende of the hall, the seller vnder the pantry, sette somwhat abase; the kychen set somwhat[7] a base from *the* buttry and pantry, co*m*myng with an entry by the wall of the buttry, the pastry-howse & the larder-howse anexed to the kychen. Than deuyde the lodgynges by the cyrcuyte of the quadryuyall courte, and let the gate-howse be opposyt or agaynst the hall-dore (not dyrectly) but *the* hall-dore standynge a base, and the gate-howse in the mydle of the front entrynge in to the place: let the pryue chambre be anaxed to *the*[8] chambre of astate, with other chambres necessarye for the buyldynge, so that many of the chambres maye haue a prospecte in to the Chapell. If

Lay your foundation on gravel and clay, rock, or a hill,

facing East and West, or that by South; but not full South.

North is better than South.

Parlour at top of the Hall; Pantry at bottom;

Kitchen next,

with a Larder.

Lodgings on another side of the Quadrangle;

Gate in middle of front; Privy-chamber next State-chamber;

all looking into the Chapel.

[1] sign. C .ii. [2] Compare Charles Kingsley's poem on the East Wind. [3] AB omit "is." [4] euyll AB.
[5] canticorum AB. [6] perfecte A; perfect B.
[7] AB omit "somewhat." [8] *the* great AB.

there be an vtter courte made, make it qua[1]dryuyal, with howses of easementes, and but one stable for horses of pleasure; & se no fylth nor dong be within the courte, nor cast at the backe-syde, but se the donge to be caryed farre from the mansyon. Also, the stables and the slaughter-howse, [and] a dyery[2] (yf any be kept) shulde be elongated the space of a quarter of a myle from the place. And also the backe-howse and brew-howse shuld be a dystaunce from the place and from other buyldyng. whan all the mansyon is edyfyed and buylte, yf there be a moote made aboute it, there shulde some fresshe sprynge come to it; and dyuers tymes the moote ought to be skowered, and kept clene from mudde and wedes. And in no wyse let not the fylth of the kychen descende in to the moote. Furthermore, it is a com-modyous and a pleasaunt thynge to a mansyon to haue an orcherd of soundry fruytes; but it is more commo-diouse[3] to haue a fayre gardain repleted wyth herbes of aromatyck & redolent sauours. In the gardayne maye be a poole or two for fysshe, yf the pooles be clene kept. Also, a parke repleted with dere & conyes is a necessarye and a pleasaunt thyng to be anexed to a mansyon. A doue howse also is a necessary thyng aboute a mansyon-place. And amonge other [4]thynges, a payre of buttes is a decent thynge aboute a mansyon; & other whyle, for a great man, necessary it is for[5] to passe his tyme with bowles in an aly: whan all this is fynysshed, and the mansyon replenysshed with Implementes, There must be a fyre kept contynually for a space to drye vp the contagyous moysters of the walles, & the sauour of the lyme and sande. And after that a man may ly and dwell in the sayd mansyon without takynge any incon-uenyence of syckenes.

Have an outer Quadrangle; with privies, and one stable for riding horses.

Other stables, slaughter-house and dairy, half a mile off.

The moat must be kept fresh and clean;

no kitchen filth in it.

Fruit-orchard.

Garden of sweet herbs.

Fish-pool.

Park with deer and conies.

A pair of Butts;

a Bowling alley.

Fire to dry the walls.

[1] sign. C .ii. back. [2] dayery A; dayerye B; dery P.
[3] more commodyouser AB. [4] sign. C .iii. [5] AB omit "for."

¶ The .v. Chapytre doth shewe howe a man shulde ordre his howse conser-nynge the Implementes to comforte the spyrytes of man. ☞

When you've built your house,

if you can't furnish it,

but must borrow salt here, a sheep's head there,

you'll be put to a shift, and never be at peace,

and men'll call you a fool.

Look ere you leap!

WHen a man hath buylt[1] his mansyon, and hath his howses necessary aboute his place, yf he haue not howsholde stuffe or implementes the whiche be nedefull, but muste borowe of his nayghbours, he than is put to a shefte [2]and to a great after deale; for 'these men the which do brew in a botyl and bake in a walet, it wyll be long or he can by Iacke a[3] salet'; yet euery thynge must haue a begynnynge, and euery man must do after his possessyons or abylyte: this notwithstanding, better it is not to set vp a howseholde or hospytalyte, than to set vp housholde, lackynge the performacyon of[4] it, as nowe to ron[5] for malt, and by-and-by for salt; nowe to sende for breade, and by-and-by to sende for a shepes-heade; and nowe to sende for this, & nowe to sende for that; and by-&-by he doth send he can not tell for what: such thynges is no prouysion, but it is a great abusyon. Thus a man shall lese his thryfte, and be put to a shefte; his goodes shall neuer increase, and he shall not be in rest nor peace, but euer in carcke and care, for his purse wyll euer be bare; wherfore I do counceyll euery man to prouyde for hym selfe as soone as he can; for yf of implementes he be destytuted, men wyll call hym lyght-wytted, to set vp a great howse, and[6] is not able to kepe man nor mowse: wherfore, let euery man loke or he lepe, for many cornes maketh a great hepe.

[1] buylded AB. [2] C .iii. back. [3] & A; and B. [4] on B. [5] come AB. The rest of this chapter runs into rude rimes. [6] & he P.

[1]¶ The .vi. Chapytre doth shewe howe a man shuld ordre his howse and howseholde, and to lyue quyetly.

Vho soeuer he be that wyll kepe an howse, he must ordre the expenses of his howse according to the rent of his laudes. And yf he haue no landes, he must ordre his howse after his lucre wynnynge or gaynes. For he that wyll spende more in his howse than the rentes[2] of his landes, or his gaynes, doth attayn to, he shal fal to pouerte, and necessite wyl vrge, cause, and compel hym to sel his lande, or to waste his stocke; as it is dayly sene by experye*n*ce of many men; wherfore they the whiche wyll exchewe such prodygalyte and inconuenyence, must deuyde his rentes, porcyon, & exspences, wherby that he doth lyue, in to .iii.[3] equal porcyons or partes. ¶ The fyrst parte must serue to prouyde for meate and drynke, & all[4] other necessary thynges for the sustencyon[5] of the howseholde. ¶ The seconde porcyon or parte must be re-[6]serued for apparell, not onely for a mannes owne selfe, but for all his howseholde, & for his[7] seruauntes wages, deductynge somwhat of this porcyon in almes dede to pore neyghbours and pore people, fulfyllynge [one or] other of[8] the .vii.[9] werkes of mercy. ¶ The .iii.[10] porcyon or parte must be reserued for vrge*n*t causes in tyme of nede, as in syckenesse, reparacyon of howses, with many other cotydyall exspences, besyde rewardes, & the charges of a mans[11] last end. If a ma*n* do exsyde[12] this

Order your house according to your rents.

Divide your income into 3 parts:

1 for food, &c.;

1 for dress, liveries, wages, alms;

1 for urgent cases, as sickness, repairs, your funeral, &c.

[1] C .iv. not signed. [2] rent A; rente B. [3] the three AB.
[4] also AB. [5] sustentacion A; sustentation B. [6] C .iv. back.
[7] AB omit "his." [8] P omits "other of." [9] seuen AB.
[10] thyrde AB. [11] of mans B. [12] excede AB.

ordre, he may soone fall in det, the whiche is a daungerous thynge many wayes, besyde the bryngynge a man to trouble. And he that is ones behynde hande and in trouble, he can not be in quyetnesse of mynde, the whiche doth perturbe the herte, & so consequently doth shorten a mannes lyfe; wherfore there is no wyse man but he wyll exchewe[1] this inconuenyence, & wyll caste before what shal folowe after. And in no wyse to sette vp a howseholde, before he hath made prouysyon to kepe a howse. For yf a man shall bye euery thynge that belongeth to the keping of his[2] howse with his peny, it wyl be longe or he be ryche, and longe or that he can kepe a good howse. But he is wyse, in my conceyte, that wyll haue, or he do sette vp his [3]howseholde, .ii. or .iii.[4] yeares rent in his cofer. And yf he haue no landes, than he must prouyde for necessarye thynges or that he begyn howseholde, leest that he repent hymselfe after, through the whiche he do[5] fall in to pencyfulnes, and after that in to syckenes & dyseases, lyuyng not quyetly, wherby he shal abreuyate his lyfe.

Once get behindhand, and you'll never be in peace.

Before you set up housekeeping, have 3 years' rent in your coffer.

¶ The .vii. Chapytre doth shewe howe the hed of a howse, or a howseholder shulde exercyse hym selfe, for the helth of the[6] soule & body.

AFter that a man hath prouyded all thynges necessary for his howse and for his howseholde, expedyent it is for hym to knowe howe he shuld exercyse hym selfe both bodely and ghostly. For there is no catholycke

How to take care of body and soul.

[1] eschewe AB. [2] a AB. [3] sign. D .i.
[4] two or thre B. [5] doth AB. [6] his AB.

or chrysten man lyuyng, but he is bounde in conscyence to be more circum*m*specter aboute the welth of his soule then the helth of his body. Our Sauyour Iesus Chryst sayth, "what shall it profyte vnto [1] man yf he geat all the worlde, and lese hym selfe, and bryng hym [2] selfe to a detryme*nt*?" wherfore it appereth that a man ought to be circumspecte for the helth and welth of his soule; For he is bounde so to lyue, that nyght and day, and at all houres, he shulde be redy; than [3] whan he is called for to departe out of this worlde, he shuld nat feare to dye, saying these wordes with saynt Ambrose: "I feare not to dye, bycause we haue a good God." whan a man hath prepared [4] for his soule, and hath subdued sensualyte, and that he hath brought hym selfe in a trade, or a vsage of a ghostly or a catholycke lyuynge in obseruyng the co*m*maundementes of God, than he must study to rule and to gouern them the whiche be in his howseholde,[5] or vnder his custody or domynyon, to se *that* they be not ydle; for kynge Henry the eyght sayd, when he was yong, "ydlenes is chefe maistres[6] of vyces all." And also the heade of a howse must ouer-se that they the which be vnder his tuyssyon serue God the holy dayes as dylygently, yee, and more dylygentler[7] than to do theyr worke the feryall dayes, refraynynge them from vyce and synne, compellynge them to obserue the co*m*maundementes of God, specyally to punysshe swearers, for in all the worlde there is not suche odyble swearyng as is vsed in En[8]glande,[9] specyally amonge youth & chyldren, which is a detestable thyng to here it, and no man doth go aboute to punysshe it. Suche thynges reformed, than may an howseholder be glad, not cessynge to instruct them the whiche be ygnorant; but

Care more for the well-being of your soul than the health of your body.

Be always ready to die.

And when you're trained yourself to godliness,

see that your household are not idle;

make them serve God on Holy-Days, keep them from vice, and punish swearers,

for there's more swearing in England than anywhere else in the world.

[1] to AB. [2] sign. D .i. [3] and P. [4] prouyded AB.
[5] Compare Hugh Rhodes in *The Babees Book*, p. 64.
[6] maisters P. [7] diligentlyer A; dylygentlyer B.
[8] sign. D .ii. [9] See *Forewords*, p. 82.

Set your people a good example; and then be merry.

also he must contynewe in shewynge good example of lyuynge; than may he reioyse in God, and be mery, the whiche myrth & reioysyng doth[1] lengthen a mans lyfe, and doth expell syckenes.[2]

¶ The .viij. Chapytre doth shewe howe a man shulde ordre hym selfe in slepynge and watchynge,[3] and in weryng his apparell.

WHan a man hath exercysed hym selfe in the daye tyme as is rehersed, he may slepe soundly and surely in God, what chaunce so euer do fortune in the nyght. Moderate slepe is moste praysed, for it doth make parfyte[4] degestyon; it doth nourysshe the blode, and doth qualyfye the heate of the lyuer; it doth acuate, quycken, & refressheth the memory; it doth restore nature, and [5]doth quyet all the humours & pulses in man, and doth anymate and doth comforte all the naturall, and anymall, and spyrytuall powers of man. And suche moderate slepe is acceptable in the syght of God, the premysses in the aforesayd Chapytre obserued and kept. And contraryly, *im*moderate slepe and sluggyshnes doth humecte and maketh lyght the brayne; it doth ingendre rewme and impostumes; it is euyll for the palsy, whyther it be vnyuersall or partyculer; it is euyll for *th*e fallynge syckenes[6] called Epilencia, Analencia, & Cathalencia, Appoplesia, Soda, with all other infyrmytyes in the heade; for it induceth and causeth oblyuyousnes; for it doth obfuske and doth obnebulate the memorye and the quyckenes of wyt.

Moderate sleep quickens all man's powers, and pleases God. Immoderate sleep breeds boils, and apoplexy (see *The Breuyary*), and dulls the wits.

[1] do A; doe B. [2] See *Forewords*, p. 88-9; and p. 228.
[3] slepe and watche AB; P leaves out "and watchynge."
[4] perfecte AB. [5] D .ii. back. [6] syckenesses B.

And shortly, to conclude, it doth perturbe the naturall, and anymall, and spyrytuall powers of man. And specyally it doth instygate and lede a man to synne, and doth induce and infer breuyte of lyfe, & detestably it displeaseth God. Oure lorde Iesu Chryste dyd not onely byd or *commaunde* his dyscyples to watche, but dyd anymat them and al other so to do, saying: "I say not onely to you, watche, but to all men I say, watche." And to Peter he said, "myghtest not thou one houre wat[1]che with me:" althoughe these holy scryptures, with many other mo, the whiche I myght allygate for me, althoughe they be not greatly referred to this se*n*ce, yet it may stande here with my purpose & matter without reprehensyon. These matters here nede not[2] to be rehersed; wherfore I do returne to my purpose, and do say that the moderacyon of slepe shulde be mesured accordyng to the natural co*m*plexyon of man, and in any wyse to haue a respect to the strength and the debylyte, to age & youth, and to syckenes & helth of man. ¶ Fyrste, as concernynge *the* naturall complexyon of man, as[3] sanguyne and colorycke men, .vii.[4] houres[5] is suffycyent for them. And nowe, consyderynge the imbecyllyte and wekenes of nature, a flemytycke man may slepe .ix. houres or more. Melancoly[6] men may take theyr pleasure, for they be [the][7] receptacle and the dragges of all the other humoures. ¶ Secondaryly, youth and age wolde haue temporau*n*ce in slepynge. ¶ Thyrdly, strength maye suffre a brount in watche, the whiche debylytye and wekenes can not. As I wyl show by a famylyor example. There were two men set at the dyce togither a day and a nyght, & more; the weke man said to hym, "I can playe no longer." The stronge [8]man sayde to hym, "fye on thè, benche-

Excessive sleep leads a man to sin, and is detestable to God.

Christ bade all men watch.

Sleep moderately,

according to your state:

Sanguine men for 7 hours;

Phlegmatic men 9 hours;

Melancholy men, as long as they like.

Weak men can't sit up so long as strong ones.

[1] sign. D .iii. [2] not greatly AB. [3] AB omit "as." [4] seuen AB. [5] howres of slepe AB. [6] Melancolycke AB. [7] be the AB. [8] D .iii. back.

whystler! wylt thou sterte away nowe?" The weke man, to satysfye the stronge mannes mynde, appetyte,[1] & desyre, playeth with hys felow; throughe *the* which he doth kyl hym selfe. The stronge man doth hym selfe lytel pleasure, all thynges consydered; the whiche I do passe ouer. wherfore I wyll retourne to the sycke man, whiche maye slepe at all tymes whan that he maye get it; but yf he maye slepe at any tyme, best it is for hym to refrayn from slepe in the day, & to take his naturall rest at nyght, whan all thynges is, or shulde be, at rest and peace; but he must do as his infyrmyte wyll permyt and suffre. whole men, of what age or complexyon soeuer they be of, shuld take theyr natural rest and slepe in the nyght, & to exchew merydyall slepe. But, an [2] nede shall compell a man to slepe after his meate, let hym make a pause, and than let hym stand, and leane and slepe agaynst a cupborde, or els let hym sytte vpryght in a chayre, & slepe. Slepynge after a full stomacke doth ingendre dyuerse infyrmyties; it doth hurte the splen, it relaxeth the synewes, it doth ingendre the dropsyes and the gowte, and doth make a man loke euyll coloured. Beware of Veneryous actes before [3]the fyrste slepe, and specyally beware of such thynges after dyner, or after a full stomacke, for it doth ingendre the crampe, the[4] gowte, and other displeasures. To bedwarde be you mery, or haue mery company aboute you, so that, to bedwarde, no anger nor heuynes, sorowe nor pencyfulnes, do trouble or disquyet you. To bedwarde, and also in the mornyng, vse to haue a fyre *in* your chambre, to wast and consume the euyll vapours within the chambre, for the breath of man maye putryfye the ayre within the chambre. I do aduertyse you not to stande nor to syt by the fyre, but stand or syt a good waye of from the fyre, takynge the

A sick man may sleep whenever he can,

though night is best.

Healthy men shouldn't sleep in the day.

If they must, they should do it standing against a cupboard, or in a chair.

No venery early at night or on a full stomach.

Before bed time be merry,

and have a fire in your room,

but don't stand or sit by the fire.

[1] appyted, *orig.* [2] and AB (if). [3] D .iv. not signed. [4] and the AB.

flauour of it; for fyre doth aryfye & doth drye vp a ma*n*nes blode, and doth make sterke the synewes & ioyntes of man. In the nyght, let the wyndowes of youre howse, specially of your chambre, be closed; whan you be in your bed, lye a lytel whyle on your left syde, & slepe on your ryght syde. And whan you do wake of your fyrste slepe, make water yf you fele your bladder charged, and than slepe on the lefte syde; and loke, as ofte as you do wake, so ofte tourne yourselfe in the bed from the[1] one syde to the other. To slepe grouelynge[2] vpon the stomacke and belly is not good, oneles [3]the stomacke be slow and tarde of digestyon; but better it is to lay your hande, or your bed-felowes hande, ouer your stomacke, than to lye grouelyng. To slepe on *th*e backe vpryght is vtterly to be abhorred. when *tha*t you do slepe, let not your necke, nother your shoulders, nother your handes, nor fete, nor no other place of your body, lye bare vndyscouered. Slepe not with an empty stomacke, nor slepe not after that you haue eaten meate, one houre or two after. In your beed, lye with your heed somwhat hygh, lest that the meate which is in your stomacke, thorow eructuacyons, or some other cause, ascend to *th*e gryfe[4] of the stomacke. Let your nyght-cap be of skarlet; & this I do aduertyse you, for[5] to cause to be made a good thycke quylt of cotton, or els of pure flockes, or of clene woull, and let the couerynge of it be of whyte fustyan, and laye it on the fether-beed that you do lye on; and in your beed lye not to hote nor to colde, but in a temporaunce. Olde auncyent doctours of Physycke sayth, .viii. houres of slepe in So*m*mer, & .ix. houres of slepe[6] in wynter, is suffycyent for any man, but I do thynke

Shut your bedroom windows at night.

Sleep on your right side.

Don't sleep on your belly,

or flat on your back.

Cover up all your body.

Lie with your head high.

Have a scarlet nightcap,

a good thick quilt, covered with fustian,

and a feather bed.

[1] AB omit "the."

[2] The adverb in *-lynge* (A.Sax. *-linga*, *-lunga*).—R. Morris, *Phil. Soc. Trans.*

[3] D .iv. back.

[4] oryfe AB; oryfice P (see p. 265, note [11]).

[5] you to AB.

[6] AB omit "houres of slepe."

that slepe ought to be taken as the complexyon of man is. whan you do ryse in the morenynge, ryse with myrth, [1]and remembre God. Let your hosen be brusshed within and without, and flauour the insyde of them agaynst the fyre; vse lynnen sockes or lynnen hosen next your legges. whan you be out of your bedde, stretche forth your legges and armes, and your body; coughe and spyt, and than go to your stole to make your egestyon; and exonerate your selfe at all tymes that nature wold expell. For yf you do make any restryction in kepynge your egestion, or your vryne or ventosyte, it maye put you to dyspleasure in bredyng dyuers infyrmyties. After you haue euacuated your body, & trussed your poyntes, kayme your heade oft; and so do dyuerse tymes in the daye. And wasshe your handes and wrestes, your face and eyes, and your tethe, with colde water. & after that you be apparelled, walke in your gardayne or parke a thousande pace or two; & than great and noble men doth vse to here masse, & other men that can not do so, but must applye theyr busynes, doth serue God with some prayers, surrendrynge thankes to hym for his manyfolde goodnes, with askyng mercye for theyr offences. & before you go to your refection, moderatly exercyse your body with some labour, or playing at the tennys, or castyng a [2]bowle, or paysyng wayghtes or plomettes of ledde in your handes, or some other thynge, to open your poores, and to augment naturall hete. At dyner & supper vse not to drynke of[3] sondry drynkes; & eate not of dyuers meates, but fede of two or thre[4] dysshes at the moste. After that you haue dyned & supped,[5] laboure not by-and-by after, but make a pause, syttynge or standyng vpright the space of an houre or more, with some pastyme; drynke not moch after dyner. At

Rise with mirth.
Brush and air your breeches.
Wear linen hose.
Stretch your legs,
go to stool,
truss your points, and comb your head.
Wash in cold water.
Walk a mile or two.
Hear mass, or pray to God.
Play tennis, or work your dumb-bells.
Eat of 2 or 3 dishes only,
and then amuse yourself for an hour.

[1] sign. E .i. [2] E .i. back. [3] AB omit "of."
[4] .ij. or .iij. A. [5] and supte.

your supper, vse light meates of digestyon, & refrayne from grose meates; go not vnto bedde [1] with a ful nor [2] emptye stomacke. And after your supper, make a pause or you go to bedde; and go to bed, as I sayde, with myrth. Furthermore, as co*n*cernynge your apparell: in wynter, next your sherte vse [3] to were a petycote of skarlet; your doublet vse at plesure; but I do aduertyse you to lyne your Iacket vnder this fasshyon or maner: by you fyne skyn*n*es of whyte lambe & blacke lambe, and let your skynner cut both the sort*es* of the skynnes in smale peces tryangle wyse, lyke halfe a quarel of a glase wyndow. And than sewe togyther a whyte pece and a blacke, lyke a whole quarel of a glasse wyndowe; & so sewe vp togyther quarell-wyse as moche as wyll [4] lyne your Iacket; this fur, for holsomnes, is praysed aboue sables or any other furre: your exteryall apparell vse accordynge to your honour. In so*m*mer, vse to were a skarlet petycote made of stamele or lynsye [5]-wolsye. In wynter and so*m*mer, kepe not [6] your hed to hote, nor bynde it to strayte; kepe euer youre necke warme. In so*m*mer, kepe your necke and face from the so*n*ne; vse to were gloues made of goote-skynnes, [7] perfumed with amber-degrece. And beware in standynge or lying on the grownde in the reflyxyon of the so*n*ne, but be mouable. If you [8] shall co*m*mon or talke with any man, stande not styll in one place yf it be on [9] the bare grownde, or grasse, or stones, but be moueable in such places: stande nor syt vpon no stone nor [10] stones; stand nor syt long bareheed vnder a vawte of stone. Also beware that you do not lye in olde chambres whiche be not occupyed, specyally such chambres as myse, rattes, and snayles resorteth vnto. lye not in suche chambres the whiche be depryued clene from the

Eat a light supper; then rest, and go to bed merry.

In winter, line your jacket with black and white lambskin

sown in triangles.

In summer, wear a red linsey petticoat,

and good skin gloves.

Don't stand or lie in the sun,

or sit on a stone.

Don't lie in ratty and snaily rooms.

[1] to bed AB. [2] nor an AB. [3] vse you AB.
[4] sign. E .ii. [5] lynsyn P. [6] not AB; nor *orig.*
[7] skyn AB. [8] thou AB. [9] vpon A; vppon B.
[10] or AB.

sonne & open ayre; nor lye in no lowe chambre except it be borded. Beware that you take no colde on your feete and legges; and of all wether, beware that you do not ryde nor go in great and impyteous wyndes.

Don't take cold in your feet.

[1]¶ The .ix. Chapytre doth shewe that replecyon[2] or surfetynge doth moche harme to nature / and that abstynence is the chefyst medyson of all medysons.

GAlen, declaryng Hypocrates sentence vpon eatynge to moche meate, saith: "More meate than accordeth with nature, is named replecyon,[2] or a surfete." Replecyon[2] or a surfet is taken as well by gurgytacyons, or to moche drynkynge, as it is taken by epulacyon,[3] of eatynge of crude meate, or eatynge more meate than doth suffyce, or can be truely dygested. Or els replecyon[2] or a surfyt is whan the stomacke is farced or stuft,[4] or repleted with to moche drynke & meate, that the lyuer, whiche is the fyre vnder the potte, is subpressed,[5] that he can not naturally nor truely decocte, defye, ne dygest, the superabundaunce of meate & drynke the whiche is in the potte or stomacke; wherfore dyuers tymes these impedymentes doth folowe: the tounge is depryued of his offyce to speke, the wyttes or sensys be dull & obnebulated from reason. Slouth [6]and sluggyshnes consequently foloweth; the appetyde is withdrawen. The heade is lyght, and doth ake, and [is] full of fantasyes; & dyuers tymes some be so sopytyd, *that* the malt worme playeth the deuyll so fast in the heade, that all the worlde ronneth rownde aboute on

Repletion or surfeit comes from drinking as well as eating.

The liver, or fire under the pot, is so prest that it can't cook the meat;

the senses get dull,

the head aches, and the malt-corn plays the devil in it.

[1] sign. E .ii. back. [2] replexion AB. [3] *epulatio*, feasting.
[4] stufted AB. [5] suppressed AB. [6] sign. E .iii.

wheles; then both the pryncepall membres & the offycyall membres doth fayle of theyr strength, yet the pulsys be full of agylyte. Such replecyon,[1] specyally suche gurgytacyons, doth ingender dyuers infyrmytes, thorowe the whiche, breuite and shortnes of lyfe doth folowe. For the wyse man sayth, that "surfetes do kyll many men, and temporaunce doth prolonge the lyfe." And also it is wrytten, Eccle. xxxvii.,[2] That "there doth dye many mo by surfette, than there doth by the sworde;" for, as I sayde, surfetynge ingendreth many infyrmytes, as the Idropyses,[3] the gowt*es*, lepored, sawsfleme & pymples in the face, vehemen*t* impressyons, vndygest humours, opylacyons, feuers, and putryfaccyons. And also it doth perturbate the heade, the eyes, the tounge, and the stomacke, with many other infyrmyties. For, as[4] Galen sayth, "ouer moche replecyon[1] or surfeting causeth strangulacion and soden death;" for, as I sayde, the stomacke is so inferced[5], [6]and the lyuer is so sore obpressed,[7] that naturall heate and the poores[8] be extyncted; wherfore abstynence for this matter is the moste best and the parfytest medysone that can be. And in no wyse eate no meate vnto *th*e tyme the stomacke be euacuated of all yll[9] humours by vomet or other conuenyent wayes; for els, crude and rawe humours vndygested wyll multiply in the body to *th*e detryment of man. Two meales a daye is suffycyent for a rest man; and a labourer maye eate thre tymes a day; & he that doth eate ofter, lyueth a beestly lyfe. And he that doth eate more than ones in a day, I aduertyse hym that the fyrste refeccyon or meale be dygested or that he do eate the seconde refeccyon or meale. For there is nothynge more hurtfull for mans body than to eate meate vpon meate vndy-

Repletion shortens a man's life,

and breeds dropsy, sawsfleme (p. 101-2), gout, and fevers.

Abstinence is the best medicine for it.

Two meals a day are enough for a resting man; 3 for a labouring one.

[1] replexion AB. [2] 37 A. [3] dropses AB. [4] AB omit "as."
[5] enforced AB. [6] sign. E .iii. back. [7] oppressed AB.
[8] powers AB. [9] euyll AB

gested. For the last refeccyon or meale wyll let the dygestyon of the fyrste refeccyon or meale. Also sondry meates of dyuers operacyons eaten at one refeccion or meale, is not laudable; nor it is not good to syt longe at dyner and supper. An houre is suffycyent to syt at dynner; and not so longe at supper. Englande hath an euyll vse in syttynge longe at dyner and at supper. And Englysshe men hath an euyll [1] vse; for, at the begynnynge at dyner and supper he wyll fede on grose meates, And *th*e best meates which[2] be holsome and nutratyue, and lyeth[3] of dygestion, is kept for seruauntes; for whan the good meate doth come to the table, thorowe fedynge vpon grose meate, the appetyde is extynct whan *th*e good meet doth come to the table; but ma*n*nes mynde is so auydous, althoughe he haue eate ynoughe, whan he seth[4] better meate come before hym, agaynst his appetyde he wyll eate; wherupon doth[5] come replecyon[6] and surfetes.

Don't eat several meats at a meal.

Sit only an hour at dinner.

Englishmen sit too long at it,

and stupidly eat gross meat first,

leaving the best for the servants.

Men are so greedy.

¶ The .x. Chapytre treateth of al maner of drynkes, as of water, of wyne, of ale, of bere, of cyder, of meade, of metheglyn, and of whay.

WAter is one of the foure Elemente*s*, of the whiche dyuers lycours or drynkes for ma*n*nes sustynaunce be made of, takyng theyr orygynall and substaunce of it, as ale, bere, meade, and metheglyn. water is not holsome,[7] sole by it selfe, for an Englysshe man, consyde[8]rynge the contrarye vsage, whiche is not concurraunt with nature: water is

Water is not wholesome by itself.

Water is bad for an Englishman.

[1] E .iv. not signed. [2] the whiche AB; meate which P.
[3] lyght BP. ? *Lyeth* is A.Sax. *líð*, mild. [4] seeth AB.
[5] do AB. [6] replexion AB. [7] See *Forewords*, p. 74.
[8] E .iv. back.

colde, slowe, and slacke of dygestyon. The best water is rayne-water, so be it that it be clene and purely taken. Nexte to it is ronnyng water, the whiche doth swyftly ronne from the Eest in to the west vpon stones or pybles. The thyrde water to be praysed, is ryuer or broke water, the which is clere, ronnyng on pibles and grauayl. Standynge waters, the whiche be refresshed with a fresshe sprynge, is commendable; but standyng waters, and well-waters, to the whiche the sonne hath no reflyxyon, althoughe they be lyghter than other ronnyng waters be, yet they be not so[1] commendable. And let euery man be ware of all waters the whiche be standynge, and be putryfyed with froth, duckemet,[2] and mudde; for yf they bake, or brewe, or dresse meate with it, it shall ingender many infyrmytes. The water *the* which euery man ought to dresse his meate with all, or shall vse bakynge or bruyng, let it be ronnyng; and put it in vesselles *that* it may stande there .ii. or .iii.[3] houres or it be occupyed; than strayne the vpper parte[4] thoroughe a thycke lynnyn cloth, and cast the inferyall parte awaye. If any man do vse to drynke water with wyne, let it be purely [5]strayned; and than seth it, and after it be cold, let hym put it to his wyne: but better it is to drynke with wyne, stylled waters, specyally the water of strawberes, or *the* water of buglos, [or the water of borage,][6] or the water of endyue, or *the* water of cycory, or the waters of southystell and daundelyon. And yf any man be combred with the stone, or doth burne in the pudibunde[7] places, vse to drynke with whyte wyne the water of hawes and the water of mylke: loke for this water in a boke of my makynge, named "the breuyary of health".[8]

Rain-water is best; running-water next; river-water third.

Well-water isn't so good.

Standing water is bad.

For cooking, use running-water, strained.

Water drunk with wine must be boiled or distilled with herbs.

For stone, drink water of haws, with white wine. See my *Breuary*.

[1] AB omit "so." [2] docknet AB; duckemeat P.
[3] two or three B. [4] parte that B. [5] sign. F .i.
[6] AB put in "or the water of borage" (not P).
[7] pubibnude, *orig.* [8] Chapter 207, Fol. lxxii; p. 80, above.

¶ Of[1] wyne.

¶ All maner of wynes be made of grapes, excepte respyse,[2] the whiche is made of a bery. Chose your wyne after this sorte: it muste be fyne, fayre, & clere to the eye; it must be fragraunt and redolent, hauynge a good odour and flauour in the nose; it must sprynkle in the cup whan it is drawne or put out of the pot in to the cup; it must be colde & pleasaunt in the mouth; and it must be strong and subtyll of substaunce: And than, moderatly dronken, it doth acuate and doth quycken a mans wyttes, it doth co*m*fort the hert, it doth scowre the lyuer; specyally, yf it be whyte wyn, it doth reioyce all the powers of man, and doth now[3]rysshe them; it doth ingender good blode, it doth comforte and doth nourysshe the brayne and all the body, and it resolueth fleume; it ingendreth heate, and it is good agaynst heuynes and pencyfulnes; it is ful of agylyte; wherfore it is medsonable, specyally whyte wyne, for it doth mundyfye and clense wou*n*des & sores. Furthermore, the better the wyne is, the better humours it doth ingen*der*. wyne must not be to newe nor to olde; but hyghe wynes, as malmyse, maye be kep[t]e[4] longe. And bycause wyne is full of fumosyte, it is good, therfore, to alaye it with water. wynes hyghe and hote[5] of operacyo*n* doth comfort olde men and women, but there is no wyne good for chyldren & maydens; for in hyghe Almayne, there is no mayde shall drynke no wyne, but styl she shal drynke water vnto[6] she be maried. the vsuall drynke, there & in other hyghe countres, for youth, is fountayn water; for in euery towne is a fountayne or a shalowe wel, to the which all people

Respyse is raspberry wine.

The qualities of good wine.

Good wine comforts the heart and scours the liver.

White wine nourishes the brain and

cleanses sores.

Wine mustn't be too old.

Mix it with water.

In Germany, maidens mustn't drink wine.

Abroad, there's a water-fountain in every town.

[1] AB omit "Of."

[2] See *Babees Book*, 125/118; p. 204; 267/21.

[3] sign. F .i. back. [4] kepte ABP.

[5] hyghe and hote. Wynes AB.

[6] vnto the time AB: *vnto* = until. See ch. xiv, p. 159, on Hygho Almayne, in the *Introduction*.

that be yonge, and seruauntes, hath a confluence and a recourse to drynke. Meane wynes, as wynes of Gascony, Frenche wynes, & specyally Raynysshe wyne that is fyned, is good with meate, specyally claret wyne. It is not good to drynke nother wyne [1] nor ale before a man doth eate somwhat, althoughe there be olde fantastycall sayinges to the contrarye. Also these hote wynes, as malmesye, wyne course, wyne greke, romanysk, romny, secke, alygaunt, basterde, tyre, osay, Muscadell, caprycke, tynt, roberdany,[2] with other hote wynes, be not good to drynke with meate; but after mete, & with oysters, with saledes, with fruyte, a draught or two may be suffered. Olde men maye drynke, as I sayde, hyghe wynes at theyr pleasure. Furthermore, all swete wynes and grose wynes doth make a man fatte.

Light wines, specially claret, are good with meat.

Hot wines are not good

with meat, but may be drunk after it.

[1] sign. F .ii.

[2] See *The Babees Book*, p. 202-7, with extracts from Henderson's *History of Ancient and Modern Wines*, 1824, p. 75, above, and *Notes*. Of the wines mentioned above, but not in *B. B.*,

Course is the Italian '*Córso*, wine of Corsica.' (Florio.)

Alygaunt is 'Alicant, a Spanish wine . . said to be made near Alicant, and of mulberries.' (Nares.)

Tynt is the modern *Tent* used in the Sacrament, 'a kind of wine of a deep red colour, chiefly from Galicia or Malaga in Spain.' (Webster.)

At Alicant, in the province of Valencia, a *vino tinto* is procured from the *tintilla* grape, which resembles the Rota wine, and contains a large quantity of tannin, holding in solution the colouring matter, and precipitating animal gelatin. It is sweet and spirituous, having a reddish orange colour, and a bitter and somewhat rough after-taste. Like the Rota, it is chiefly used for medicinal purposes.—*Henderson*, p. 193-4; and see p. 251.

Neither *Roberdany* nor *Romanyske* is mentioned by Henderson.

Sack. See *Henderson*, p. 298-309, and his quotation, p. 315, of Markham, "Your best *Sacks* are of Xeres in Spain; your smaller, of Gallicia and Portugall; your strong Sacks are of the islands of the Canaries and of Malligo . ." Also from the *Discovery of a London Monster called the Black Dog of Newgate*, printed in 1612, "There wanted neither *Sherry Sack*, nor Charneco, Maligo, nor amber-coloured Candy, nor liquorish Ipocras, brown beloved *Bastard*, fat *Aligant*, nor any quick-spirited liquor."

¶ Of[1] ale.

Ale comes naturally to an Englishman. Properties of Ale. It should be 5 days old, and makes a man strong.

¶ Ale is made of malte and water; and they the which do put any other thynge to ale then[2] is rehersed, except yest, barme, or godesgood, doth sofystical[3] theyr ale. Ale for an Englysshe man is a naturall drynke. Ale must haue these propertyes: it must be fresshe and cleare, it muste not be ropy nor smoky, nor it must haue no weft nor tayle. Ale shuld not be dronke vnder .v. dayes olde. Newe ale is vnholsome for all men. And sowre ale, and deade ale[4] the which doth stande a tylt, is good for no man. Barly malte maketh better ale then oten malte or any other corne doth: it doth ingendre[5] grose humoures; but yette[6] it maketh a man stronge.

¶ Of[1] bere.

Beer is a Dutch drink, but has lately come into England. It blows out the belly.

¶ Bere is made of malte, of hoppes, and water: it is a naturall drynke for a Dutche man. And nowe of late dayes it is moche vsed in Englande to the detryment of many Englysshe men; specyally it kylleth them the which be troubled with the colycke, and the stone, & the strangulion;[7] for the drynke is a colde drynke; yet it doth make a man fat, and doth inflate the bely, as it doth appere by the Dutche mens faces & belyes. If the bere be well serued, and be fyned, & not new,[8] it[9] doth qualyfy *the* heat of the lyuer.

¶ Of[1] cyder.

The best Cider is made of Pears.

¶ Cyder is made of the iuce of peeres, or of[1] the iuce of aples; & other whyle cyder is made of both; but the best cyder is made of cleane peeres, the which be dulcet; but the beest[10] is not praysed in physycke, for

[1] AB omit "Of." [2] than AB. [3] sophysticat P.
[4] AB insert "and ale." [5] sign. F .ii. back.
[6] AB omit "yette;" P has "yet."
[7] strayne coylyon AB. [8] be wel brude and fyned P
[9] newi, t *orig.* [10] best AP; beste B.

cyder is colde of operacyon, and is full of ventosyte, wherfore it doth ingendre euyll humours, and doth swage to moche the naturall heate of man, & doth let dygestyon, and doth hurte the stomacke; but they the which be vsed to it, yf it be dronken in haruyst, it doth lytell harme.

Cider breeds evil humours,

but may be drunk at harvest.

¶ Of[1] meade.

[2] ¶ Meade is made of ho*n*ny and water boyled both togyther; yf it be fyned and pure, it preserueth helth; but it is not good for them the whiche haue the Ilyacke or the colycke.

Mead is bad for the colic.

¶ Of[1] metheglyn.

¶ Metheglyn is made of ho*n*ny & water, and herbes, boyled and soden togyther; yf it be fyned & stale, it is better in the regyment of helth than meade.

Metheglyn is wholesomer than Mead.

¶ Of[1] whay.[3]

¶ whay, yf it be wel ordered, specyally that whay the which doth come of butter, is a temporate drynke, and is moyst; and it doth nourysshe, it doth clense the brest, and doth purge redde colour, and [is] good for sausfleme faces.

Whey from butter is nourishing.

¶ Of[1] poset ale.

¶ Poset ale is made with hote mylke & colde ale; it is a temporate drynke, and is good for a hote lyuer, and for hote feuers, specyally yf colde herbes be soden in it.

Posset ale is good for a hot liver.

[1] AB omit "Of."

[2] sign. F .iii.

[3] Pover cilly shepperd*es* they gett/
Whome into their farmes they sett/
Lyvynge on mylke / whyg / and *whey* [whyg = butter-milk, or sour whey].—Roy's *Satire*, Pt II, p. 111, of Pickering's reprint, p. 17 of my *Ballads from MSS*, 1868.

We tourmoyle oure selfes nyght and daye,
And are fayne to drycke whygge and *wheye*,
For to maynteyne the clurgyes facciones.
1530, *A Proper Dyaloge*, fol. C; *Ballads from MSS*, p. 22.

¶ Of[1] coyte.

Coyte is a usual drink in Holland, &c.

¶ Coyte is a drynke made of water, in the whiche is layde a sowre and a salt leuyn .iii. or .iiii. houres; then[2] it is dronke. it is a vsual drynke in Pycardy, in Flaundres, in Holande, in Brabant, and Selande; [3]hit dothe but quench the thyrste.[3]

For a Ptisane, Hippocras, &c., see my *Breuyary*.

¶ To speake of a ptysan, or of oxymel, or of [4]aqua vite, or of Ipocras, I do passe ouer at this tyme; for I do make mensyon of it in the Breuyary of health.[5]

¶ The .xi. Chapytre treateth of breade.

Wheat bread makes a man fat.

Unleavened bread is better than leavened.

In Rome, loaves are only as big as a walnut, and are saffron'd.

Manchet, with no bran, I like.

AVycen sayth, that breed made of whete maketh a man fatte, specyally when the breade is made of newe whete; and it doth set a man in temporaunce. Breade made of fyne flower without leuyn is slowe of dygestyon, but it doth nouryсshe moche yf it be truely ordered and well baken. whan the breade is leuened, it is soone dygested, as some olde Aucthours sayth; but these dayes is proued the contrary by *the* stomacke of men, for leuyn is heuy and ponderous. Breade hauynge to moche brande in it is not laudable. In Rome, and other hyghe countres, theyr loues of breade be lytell bygger then a walnot, and many lytell loues be ioyned togyther, the whiche doth serue for great men, and it is safferonde:[6] I prayse it not. I do loue manchet breade, and great loues the whiche be well mowlded and thorowe [7]baken, *the* brande abstracted and abiected; and that is good for all ages.[8] Mestlyng breade is

[1] AB omit "Of." [2] than AB. [3—3] put in from P.
[4] sign. F .iii. back. [5] chapter 358, leaf 106, &c.
[6] See p. 261, l. 13. [7] F .iv. not signed.
[8] aches AB; and AB insert a fresh chapter, headed ¶ Breade made of Mestlynge or of Rye.

made, halfe of whete and halfe of Rye. And there is also mestlyng made, halfe of rye and halfe of barly. And yll[1] people wyll put whete and barly togyther. breade made of these aforesayde grayne or cornes, thus poched togyther, maye fyll the gutte, but it shall neuer do good to man, no more than horse breade, or breade made of beanes and peason shall do[2]; howbeit this matter doth go moche by *the* educacyon or the bryngyng vp of the people, the which haue ben nourisshed or nutryfyde with suche breade. I do speake nowe in barlyes or maltes, parte to be eaten and also dronken. I suppose it is to moche for one grayne, for barly doth ingender colde humours; and peason and beanes, and the substaunce co*m*mynge from theym, repletyth a man with ventosyte; but and[3] yf a man haue a lust or a sensuall appetyd to eate and drynke of a grayne bysyde malte or barlye, let hym eate and drynke of it the whiche maye be made of otes; for hauer cakes in Scotlande is many a good lorde and lordes dysshe.[4]

Mestlin is half wheat, half rye or barley.

Mixed corn bread may fill the guts, but does men no good.

Barley breeds cold humours; peas and beans fill one with wind.

Scotch oat cake is good,

[1] euyll AB.

[2] "I haue" . . quod Peres . . .
A fewe cruddes and creem · and an hauer cake,
And two loues of *benes and bran* · ybake for my fauntis.
Vision of P. Plowman, Text B, p. 107-8, l. 282-5.

As to *horsebread*, cp.
For þat was bake for Bayarde [the horse · was bote for many hungry, 196
And many a beggere for *benes* · buxome was to swynke,
And eche a pore man wel apayed · to haue *pesen* for his huyre.
ib. p. 103.

Bolde beggeres and bigge · þat mowe her bred biswynke,
With houndes bred and *hors bred* · holde vp her hertis;
Abate hem with benes · for bollyng of her wombe.
ib. p. 104, l. 216-18.

[3] AB omit "and."

[4] The Scotch lords had a different character from Holinshed (1586 A.D.), or Hector Boece (died 1536) if Holinshed follows him here:—"But how far we in these present daies are swarued from the vertues and temperance of our elders, I beleeue there is no man so eloquent, nor indued with such vtterance, as that he is able sufficientlie to expresse. For whereas they gaue their minds to dowghtinesse, we applie our selues to droonkennes: they had plentie with sufficiencie, we haue inordinate excesse with superfluitie: they were temperate,

and, therefore, good drink can be got out of oats.

The Devil sends bad Cooks.

Bad brewers and cheating ale-wives.

And yf it wyll make good hauer cakes, consequently it wyll do[1] make good drynke or euyl; euery thyng as it is handled. [2]For it is a common prouerbe, "God may sende a man good meate, but the deuyll may sende an euyll coke[3] to dystrue[4] it;" [5]whorfore, gentyll bakers, sophystycate not your breade made of pure whete; yf you do, where euyl ale-brewers and ale-wyues, for theyr euyl brewyng & euyl measure, shuld clacke and ryng theyr tankardes at dym myls dale, I wold you shuld

we effeminate; and so is the case now altered with vs, that he which can deuoure and drinke most, is the noblest man and most honest companion; and thereto hath no peere, if he can once find the veine (though with his great trauell), to puruey himself of the plentifullest number of new, fine, and delicate dishes, and best prouoke his stomach to receiue the greatest quantitie of them, though he neuer make due digestion of it. Being thus drowned in our delicate gluttonie, it is a world to see, how we stuffe our selues both daie and night, neuer ceasing to ingorge & powre in, till our bellies be so full that we must needs depart. Certes it is not supposed meet that we should now content our selues with breakefast and supper onelie, as our elders haue doone before vs, nor inough that we haue added our dinners vnto their aforsaid meales, but we must haue thereto our beuerages and reare suppers, so that small time is spared wherein to occupie our selues in any godlie exercise; sith almost the whole daie and night doo scarselie suffice for the filling of our panches. We haue also our merchants, whose charge is not to looke out, and bring home such things as necessarilie perteine to the maintenance of our liues, but vnto the furniture of our kitchen; and these search all the secret corners of our forrests for veneson, of the aire for foules, and of the sea for fish; for wine also they trauell, not only into France, whose wines doo now grow into contempt, but also into Spaine, Italie, and Greece; nay, Affrike is not void of our factors, no, nor Asia, and onelie for fine and delicate wines, if they might be had for moneie."—P. 22, Harrison's *Description of Scotland*, prefixed to Holinshed's *Historie*, edit. 1586.

[1] ABP omit "do" (= cause to). [2] F .iv. back.

[3] sende euyl cokes P. [4] dystroy A; destroye B.

[5–5] P has for the next two paragraphs: "But wyues, & maydes, & other bruers, the whiche dothe dystrue malte the whiche shulde make good ale, And they [D .iv. back] the which that doth nat fyll theyr potes, geuynge false measure,—I woulde they were clackynge theyr pootes and tancardes at dymmynges dale. And euyll bakers the whyche doth nat make good breade of whete, but wyl myngle other corne with whete, or do nat order and seson hit, gyuinge good weyght, I wolde they myght play bo pepe thorowe a pyllery."

shake out the remnaunt of your sackes, standynge in the Te*m*mes vp to the harde chynne, and .iii. ynches aboue, that whan you do come out of the water you myght shake your eares as a spanyell that veryly co*m*meth out of[1] the water.[2] Gentyll bakers, make good breade[5]! for good breade doth comforte, confyrme, and doth stablysshe a ma*n*nes herte, besyde the propertyes rehersed. Hote breade is vnholsome for any man, for it doth lye in *th*e stomacke lyke a sponge, haustyng vndecoct humours; yet the smel of newe breade is co*m*fortable to the heade and to the herte. ¶ Soden breade, as symnels and crackenels, and breade baken vpon a stone, or vpon yron, and breade that saffron is in,[3] is not laudable. Burnt breade, and harde crust*es*, & pasty crust*es*, doth ingendre color, aduste, and melancoly humours; wherfore chyp the vpper crust of your breade.[4] And who so doth [6]vse to eate *th*e seconde cruste after meate, it maketh a man leane. And so doth wheten breade, the which is ful of brande. ¶ Breade, the whiche is nutrytyue, & praysed in physycke, shuld haue these propertes. Fyrste, it must [not][7] be newe, but a daye & a nyght olde, nor it is not good whan it is

I should like to duck rascally bakers.

Good bread comforts a man's heart.

Hot bread is like a sponge.

Symnels and Cracknels are not good.

Chip your upper crusts off.

Bread should be 24 hours old,

[1] B omits "of."

[2] Sir H. Ellis (*Brand*, iii. 53, ed. 1843) says of the Cucking-Stool, "It was a punishment inflicted also anciently upon brewers and bakers transgressing the laws. . . In 'The Regiam Majestatem,' by Sir John Skene, this punishment occurs as having been used anciently in Scotland: under 'Burrow Lawes,' chap. lxix., speaking of Browsters, i. e. '*Women quha brewes aill to be sauld*,' it is said—'gif she makes gude ail, that is sufficient. Bot gif she makes evill ail, contrair to the use and consuetude of the burgh, and is convict thereof, she sall pay ane unlaw of aucht shillinges, or sal suffer the justice of the burgh, that is, *she sall* be put upon the *Cock-stule*, and the aill sall be distributed to the pure folke.' Lysons cites an instance of an alewife at Kingston-on-Thames, being ducked in the river for scolding, under Kingston Bridge, in April 1745, in the presence of 2000 or 3000 people." (Ellis's *Brand*, iii. 52.)

[3] See p. 258, l. 4 from foot.

[4] See *The Babees Book*, p. 200, 266/4.

[6] sign. G .i.

[7] not AB.

not mouldy, well-baked, slightly salt.

past .iiii. or .v. dayes olde, except the loues be great; nor it must not be moldy nor musty; it must be well muldyd[1]; it must be thorowe bake; it muste be lyght, & not heuye, and it must be temporatly salted. Olde breade or stale breade doth drye vp the blode or naturall moyster[2] of man, & it doth ingender euyll humours, and is euyll and tarde of dygestyon; wherfore there is no surfet so euyll as the surfet of eatynge of euyll breade.

Stale bread is slow of digestion.

¶ The .xii. Chapyter treateth of potage, of sewe, of stewpottes, of grewell, of fyrme*n*te, of pease potage, of almon mylke, of ryce potage, of cawdels, of culleses, and of other brothes.

Potage and Broth fill a man with wind.

AL maner of lyquyd thynges, as potage, sewe, & all other brothes, doth replete a man that eateth theym, with ve*n*tosyte. Potage is not so moch vsed in al Crystendom as it is vsed in Englande. Potage is made of the lyquor in the which flesshe is soden[4] in, with puttyng-to chopped herbes, and otemel and salt. The herbes with the whiche potage is made with all, yf they be pure, good, and clene, not worme[5]-eaten, nor infected with the corrupte ayre descendynge vpon them, doth comforte many men, the ventosyte notwithstandyng. But for asmoch as dyuers tymes, many partes of Englande is infected with the pestylence, thorow the corrupcyon of the

Potage is more used in England than anywhere else.

Herbs for potage must be good.

In pestilence time

[1] moulded AB; mylded P. [2] moyst AB.
[3] sign. G .i. back. [4] sod AB. [5] warme, *orig.*; wanne P.

ayre, the which doth infecte the herbes, In such tymes it is not good to make any[1] potage, nor to eate no potage. In certayn plac[e]s beyonde see where as I haue traueyled in, in the pestylence tyme a general commaundment hath ben sent from the superyoryte to the commonalte, that no man shuld eate herbes in suche infeccyous tymes.

don't make potage,

or eat herbs.

[2] ¶ Of[3] sewe and stewpottes.

¶ Sewe and stewpottes, and grewell made with otmell, in all the[4] which no herbes be put in, can do lytel displeasure, except that[5] it doth replete a man with ventosyte; but it relaxeth the belly.

Oatmeal gruel, &c., don't hurt one much.

¶ Of[3] fyrmente.

¶ Fyrmente is made of whete and mylke, [6]in the whiche, yf flesshe be soden, to eate it[7] is not commendable,[6] for it is harde of dygestyon; but whan it is dygested it doth nouryssha, and it doth strength[8] a man.

Frumenty is indigestible, but nourishing.

¶ Of[3] pease potage & beane potage.

¶ Pease potage and beane potage doth replete a man with ventosyte. Pease potage is better than beane potage, for it is sooner dygested, & lesser of ventosyte: they both be abstercyue,[9] and do clense the body. They be compytent of nutryment; but beane potage doth increase grosse humours.

Pease potage is better than bean potage.

¶ Of[3] almon mylke & of[3] ryce potage.

¶ Almon mylke and ryce potage: Almons be hote and moyste; it doth comforte the brest, and it doth mollyfye the bely, and prouoketh vryne. Ryce potage made with almon mylke doth restore and doth comforte nature.

Almonds mollify the belly.

[1] AB omit "any." [2] sign. G .ii. [3] AB omit "Of."
[4] in the P. [5] AB omit "that."
[6–6] P omits this, but adds at the end, after *man*, "but flesshe soded in mylke is nat commendable." [7] it, it AB.
[8] strengthen AB. [9] abstercyue, *orig.*

¶ Of[1] ale-brues, caudelles, & colesses.

[2]¶ Ale-brues, caudelles, and colesses, for weke men and feble[3] stomackes, the whiche can not eate solydate meate, is suffered.[4] But caudels made with hempe-sede, and collesses made of shrympes, doth comforte blode and nature.

Cullisses of Shrimps comfort the blood.

¶ Of[1] honny soppes, and other brothes.

¶ Honny soppes & other brothes, of what kynde or substaunce soeuer they be made of, they doth[5] ingender ventosyte; wherfore they be not good nor holsome for the colycke nor the Illycke,[6] nor other inflatyue impedymentes or syckenesses, specyally yf honny be in it, the sayinges of Plyne, Galene, Auycene, with other Aucthours, notwithstandynge; for in these dayes experyence teacheth vs contrary to theyr sayinges & wrytynges;[7] for althoughe the nature of man be not altered, yet it is weker, and nothynge so stronge nowe as whan they lyued,[a] &c. [[a] & dyd practes & makyng *the* bokes.—P.]

Honey-sops breed wind.

Don't mind old authors, if Experience contradicts them.

¶ The .xiij. Chapitre treateth of whyt meate, as of egges, butter, chese, mylke, crayme,[8] &c.

IN England there is no egges vsed to be eaten but hen-egges; wherfore I wyl fyrst wryte & pertract of hen-egges. The yolkes of [9]hen-egges be cordyalles, for it is temporatly hote. The whyte of an egge is viscus & colde, and slacke of digestyon, and doth not ingender good blode; wherfore, whosoeuer that wyl eate an egge,[10] let the egge be newe, and roste hym reare, and

Hens' eggs only are used in England.

Eggs should be new, and roasted,

[1] AB omit "Of." [2] sign. G .ii. back. [3] fell AB.
[4] sustered, *orig.* [5] do AB. [6] nor Ilyacke AB. [7] wrytynge AB.
[8] and crayme P. [9] sign. G .iii. [10] Henne egge AB.

eate hym; or els poche hym, for poched egges be best at nyght, & newe reare rosted egges be good in the mornynge, so be it they be tyred with a lytell salte and suger; than[1] they be nutry[ty]ue.[2] In Turkey, and other hyghe chrystyan landes anexed to it, they[3] vse to seth two or thre busshels of egges togither harde, and pull of the shels, &[4] sowse them, and kepe them to eate at all tymes; but hard egges be slowe and slacke of dygestyon, and doth nutryfye the body grosly. Rosted egges be better than sodden; fryed egges be nought; Ducke-egges & geese-egges I do not prayse; but fesaunt-egges and partreges egges, physycke syngulerly doth prayse.

or poached, and eaten with salt.

In Turkey, they boil eggs hard, and pickle 'em.

Fried eggs are bad. Pheasant and Partridge eggs are good.

¶ Of[5] butter.

¶ Butter [is][6] made of crayme, and[7] is moyste of operacion; it is good to eate in the mornyng before other meates. Frenche men wyll eate it after meate. But, eaten with other meat*es*, it doth not onely nowrysshe, but it is good for the breste and lunges, and also it [doth][8] relaxe and [9] mollyfye the bely. Douche men doth eate it at all tymes in the daye, the whiche I dyd not prayse when I dyd dwell amonge them / consyderyng that butter is vnctyous,[10] and euery thynge that is vnctyous[10] is noysome to the stomacke, for as moche as it maketh lubryfactyon. And also euery thyng that is vnctious,[10] That is to say, butterysshe,—oyle, grese, or fat,—doth swymme aboue in the brynkes of the stomacke: as the fatnes doth swymme aboue in a boylynge potte, the excesse of suche nawtacyon or superfyce wyll ascende to the oryse[11] of the stomacke, and doth make eructuasyons / wherfore, eatynge of moche butter at one refection is not co*m*me*n*dable, nor it is not good for theym the

Eat butter early, before other food.

It's good for the lungs.

Dutchmen eat butter at all times in the day. (See p. 147, 149.)

Butterish things swim at the top of other drinks in the stomach.

Butter is bad for ague and fever.

[1] that AB. [2] nutritive P; nutryue AB.
[3] AB omit "they." [4] AB omit "&."
[5] AB omit "Of." [6] is AB. [7] Butter made of crayme P.
[8] doth AB. [9] and doth P; sign. G .iii. back.
[10] vncryous B. [11] oryfe AB; orifice P.

whiche be in any ague or feuer, for the vnctuosyte[1] of it dothe auge and[2] augment the heate of the lyuer: a lytell porcyon is good for euery man in the morenynge, yf it be newe made.

Eat fresh butter in the morning.

¶ Of[3] Chese.

Of 5 sorts of cheese: 1. Green Cheese; 2. Soft Cheese; 3. Hard Cheese; 4. Spermyse Cheese, made of curds and the juice of herbs.

¶ Chese is made of mylke; yet there is[4] .iiii. sortes of chese, whiche is to say, grene chese, softe chese, harde chese, and[5] spermyse / Grene chese is not called grene by the reason of colour, but for the newnes of it / for the whey is not halfe pressed out of it; and in operacy[6]on it is colde and moyste. Softe chese, not to new nor to olde, is best, for in operacyon it is hote and moyste. Harde chese is hote and dry, and euyll to dygest. Spermyse is a chese the which is made with curdes and with the iuce of herbes: to tell the nature of it, I can not / consyderynge that euery mylke-wyfe maye put many iuces of herbes of sondry operacyon & vertue, one not agreynge with another. But and yf they dyd knowe what they dyd gomble togyther without trewe compoundynge, and I knowynge the herbes, then I coulde tell the operacyon of spermyse chese. Yet besyde these .iiii. natures of chese, there is a chese called a rewene[7] chese, the whiche, yf it be well orderyd, doth passe all other cheses, none excesse taken. But take the best chese of all these rehersyd, yf a latel[8] do good and pleasur, The ouerplus doth ingendre grose humours; for it is harde of dygestyon; it maketh a man costyfe, and it is not good for the stone. Chese that is good, oughte not be to harde nor to softe, but betwyxt both; it shuld not be towgh nor bruttell; it ought not to be swete nor sowre, nor tarte, nor to salt, nor to fresshe; it must be of good sauour & taledge,

5. Rewene Cheese, the best of all.

The qualities of good Cheese.

[1] ventuosyte *orig.*, and P; vnctuosyte AB.
[2] AB omit "auge and."
[3] AB omit "Of."
[4] mylke there be P.
[5] or AB.
[6] G .iv. not signed.
[7] Irweue AB.
[8] lytell AB; lytel P.

nor full of iyes, nor mytes, nor magottes / yet in Hygh Almen[1] [2] the chese the whiche is full of magotes is called there the best chese, and they wyll eate the great magot*es* as fast as we do eate comfet*es*.

The High-Germans eat cheese-maggots like we do comfits.

¶ Of[3] Mylke.

Mylke of a woman, and the mylke of a gote, is a good restoratyue; wherfore these mylkes be good for them that be in a consumpcyon, and for the great temperaunce the whiche is in them: it doth nowrysshe moche.

Woman's and goat's milk are good for Consumption.

¶ Cowes mylke and ewes mylke, so be it the[4] beestes be yonge, and do go in good pasture, the mylke is nutrytyue, and doth humect and moysteth the membres, and doth mundyfye and clense the entrayles, and doth alleuyat & mytygate the payne of the lunges & the brest; but it is not good for them the whiche haue gurgulacyons in the bely, nor it is not al the best for sanguyne me*n* / but it is very good for melancoly men, & for olde men and chyldren, specyally yf it be soddyn, addynge to it a lytell sugre.

Cow's and ewe's milk are nourishing.

Milk is bad for grumblings in the belly;

but good for old men and children.

¶ Of[3] Crayme.

¶ Crayme the which dothe not stande longe on the mylke, & soddyn with a lytell suger, is nowrysshynge. Clowtyd crayme and rawe crayme put togyther, is eaten more for a sensuall appetyde than for any good nowrysshe[5]ment. Rawe crayme vndecocted, eaten with strawberyes or hurtes, is a rurall mannes banket. I haue knowen such bankettes hath put men in ieoperdy[6] of theyr lyues.

Clotted cream.

Strawberries and cream will endanger a man's life.

¶ Almon-butter.

¶ Almon-butter made with fyne suger and good rose-water, and eaten with the flowers of many

Almond-butter and violets

[1] Almayne AB. See p. 159, above.
[2] G .iv. back.
[3] AB omit "Of."
[4] *that* the P.
[5] sign. H .i.
[6] ieobardy AB.

vyolettes, is a co*m*mendable dysshe, specyallye in Lent, whan the vyolete*s* be fragrant; it reioyseth the herte, it doth comforte *th*e brayne, & doth qualyfye the heate of the lyuer.

rejoice the heart.

¶ Beene-butter.

Bean-butter fills the paunch and raises wind.

¶ Beene-butter is vsed moche in Lent in dyuers countres. it is good for plowmen to fyl the panche; it doth ingender grose humours; it[1] doth replete a man with ventosyte.

¶ The .xiiii. Chapytre treatyth of Fysshe.

England's the best fish country.

OF all nacyons and cou*n*tres, England is beste seruyd of Fysshe, not onely of al maner of see-fysshe, but also of fresshe-water fysshe, and of all maner of sortes of salte-fysshe.

¶ Of[2] See-fysshe.

Seafish is wholesomer than fresh-water fish.

[3]¶ Fysshes of the see, the which haue skales or many fynnes, be more holsomer than *th*e fresshe-water fysshe, the whiche be in standynge waters. The elder[4] a fysshe is, so much he is the better, so be it that the fysshe be softe and not solydat. yf the fysshe be faste and solydat, the yonger the fysshe is, the better it is to dygest; but this is to vndersta*n*de, that yf the fysshe be neuer so solydat, it muste haue age / but not ouergrowen, except it be a yonge porpesse, the which kynde of fysshe is nother praysed in the olde testament nor in physycke.[5]

Porpoise is bad, say the Bible and Physic.

¶ Fresshe-water fysshe.

Fish from running water is better than fish

¶ The fysshe the whiche is in ryuers and brokes be more holsomer than they the which be in pooles,

[1] and AB. [2] AB omit "Of."
[3] sign. H .i. back. [4] older AB.
[5] See *The Babees Book* Index, "Porpoise," and "Purpose."

pondes, or mootes, or any other standynge waters; for they doth laboure, and doth skower them selfe. Fysshe the whiche lyueth & doth feede on the moude, or els do feede in *the* fen or morysshe grounde, doth sauer of *the* moude, whiche is not so good as the fysshe that fedyth and doth skowre them self on the stones, or grauell, or sande.

From standing water.

Mud-fish taste of mud.

¶ Of Salte fysshe.[1]

¶ Salte fysshe,[2] the whiche be powderyd[3] and salted with salte, be not greatly to be praysed, specyally yf a man do make his hoole refecty[4]on with it; the qualyte doth not hurte, but the quantyte, specyally suche salte fysshes as wyll cleue to the fyngers whan a man doth eate it. And *the* skyn of fysshes be vtterly to be abhorryd,[5] for it doth ingender viscus fleume and color adust. Al maner of fysshe is colde of nature, and doth ingender fleume; it doth lytell nowrysshe / Fysshe and flesshe oughte not to be eaten togyther at one meale.

Salt-fish only for a meal is not good.

Don't eat fish and flesh together.

¶ The .xv. Chapitre treateth of wylde fowle, and tame fowle [and][6] byrdes.

OF all wylde foule, the Fesaunt is most beste,[7] Althoughe that a partreche of all fowles is soonest dygested; wherfore it is a restoratyue meate, and dothe comforte the brayne and the stomacke, & doth augment carnall lust. A wood-cocke is a meate of

Pheasant is the best;

Partridge soonest digested.

Woodcock.

[1] Salte fysshes AB. [2] fysshes AB.
[3] sprinkled.—F. [4] sign. H .ii.
[5] See *Babees Book*, p. 154/553; 140/367, &c. [6] and AB.
[7] See *Babees Book*, p. 217, &c.; also p. 218-20, 143-4, &c., for the other wild birds.

Quail. Plover. Lapwing. Turtle-dove. Crane. Heron. Bustard. Bittern. Shoveler. Hen-pheasant. Moorcock. Moorhen.

good temperaunce. Quayles & plouers and lapwynges doth nowrysshe but lytel, for they doth ingender melancoly humours. yonge turtyll-doues dothe ingender good blode. [1] A crane is harde of dygestyon, and doth ingender euyll blode. A yonge herensew is lyghter of dygestyon than a crane. A bustarde well kylled and orderyd is a nutrytyue meate. A byttoure is not so harde of dygestyon as is an herensew. A shoueler is lyghter of dygestyon than a byttoure: all these be noyfull except they be well orderyd and dressyd. A fesaunt-henne, A more-cocke and a more-henne, except they be sutt [2] abrode, they be nutrytyue. All maner of wylde fowle the whiche lyueth by the water, they be of dyscommendable nowrysshement.

¶ Of tame or domestycall fowle.

Capon. Hen. Chicken. Cock. Pigeon. Goose. Duck. Peachick. Peacock.

¶ Of all tame fowle a capon is moste beste,[3] For it is nutrytyue, and is soone dygestyd. A henne in wynter is good and nutrytyue. And so is a chyken [4] in somer, specyallye cockrellys and polettes, the whiche be vntroden. The flesshe of a cocke is harde of dygestyon, but the broth or gely [5] made of a cocke is restoratyue. pygyons be good for coloryke & melancoly [6] men. gose-flesshe and ducke-flesshe is not praysed, except it be a yonge grene goose. yonge peechyken of halfe a yere of age be praysed. olde pecockes be harde of dygestyon.

¶ Of Byrdes.

Sparrow. Colmouse (or Cole Titmouse, *Parus Ater*: Nat. Libr. xxv. 172). Wren.

[7] ¶ All maner of smale Byrdes be good and lyght of dygestyon, excepte sparowes, whiche be harde of dygestyon. Tytmoses, colmoses, and wrens, the whiche doth eate spyders and poyson, be not commendable.[8]

[1] sign. H .ii. back. [2] do syt AB; they sute P.
[3] See *Babees Book*, p. 222, &c. [4] be chycken A; be chyckens B.
[5] a gely AB. [6] melancolycke AB. [7] sign. H .iii.
[8] commestyble AB.

of all smale byrdes the larke is beste: than is[1] praysed *the* blacke byrde & the thrusshe.[2] Rasis and Isaac prayseth yonge staares;[3] but I do thynke, bycause they be bytter in etyng, they shuld inge*n*der colour.

Lark. Blackbird. Thrush. Starling.

¶ The .xvi. Chapytre treatyth of flesshe, of wylde and tame beestes.

BEefe is a good meate for an Englysshe man, so be it the beest be yonge, & that it be not koweflesshe; For olde beefe and koweflesshe doth[4] ingender melancolye and leporouse humoures. yf it be moderatly powderyd,[5] that the groose blode by salte may be exhaustyd, it doth make an Englysshe man stro*n*ge, the educacion of hym with it co*n*syderyd. Martylmas beef, whiche is called "hanged beef" in the rofe of the smoky howse, is not laudable; it maye fyll the bely, and cause a man to drynke, but [6]it is euyll for the stone, and euyll of dygestyon, and maketh no good iuce. If a man haue a peace hangynge by his syde, and another in his bely, that the whiche doth hange by the syde shall do hym[7] more good, yf a showre of rayne do chaunse, than that the which is in his[8] bely, the appetyde of mans sensualyte notwithstandynge.

Young Beef is good for Englishmen.

Salt beef makes 'em strong.

Martilmas or hanged beef is bad.

Use it outside yourself, not inside.

¶ Of[9] Veale.

¶ Veale is [a][10] nutrytyue meate, and doth nowrysshe moche a man, for it is soone dygestyd; wherupon many men doth holde oppynyon that it is the beste flesshe,[11]

Veal is soon digested.

[1] then P. [2] thrusshes B. [3] starlings. [4] do AB.
[5] salted.—F. [6] H .iii. back. [7] a man AB.
[8] within the AB. [9] AB omit "Of." [10] is a AB.
[11] flesse, *orig.*

and the moste nutrytyue meate, that can be for mans sustenaunce.

¶ Of[1] Mutton and lambe.

Mutton I don't like;

¶ Mutton, of Rasis and Aueroyes is praysed for a good meate, but Galen dothe not laude it; and sewrely I do not loue it, consyderyng that there is no beest that is so soone infectyd, nor there doth happen so great murren and syckenes to any quadrypedyd[2] beeste as doth fall to the sheepe. This notwithstandynge, yf the sheepe be brought vp in a good pasture and fatte, and do not flauoure of the wolle, it is good for sycke persones, for it doth ingender good blode.

sheep are so liable to murrain.

But good mutton helps sick folk.

Lamb is not good for old men.

¶ Lambes flesshe is moyste and flumatycke, [3]wherfore it is not all the best for olde men, excepte they be melancolye of complexyon: it[4] is not good for flumatyke men to feade; to moche of it doth hurte.[5]

¶ Of[1] Porke, brawne,[6] bakon, & pygge.

¶ where-as Galen, with other auncyent and approbat doctours, doth prayse porke, I dare not say the contrarye agaynst them; but this I am sure of, I dyd neuer loue it: And in holy scrypture it is not praysed; for a swyne is an vnclene beest, and dothe lye vpon[7] fylthy & stynkynge soyles; and with stercorus matter dyuers tymes doth[8] fede in[9] Englande; yet in[10] Hyghalmen[11] and other hygh countres, (except Spayne & other countres anexed to Spayne), [men] doth kepe theyr swyne clene, and dothe cause them ones or twyse a daye to swymme in great ryuers, lyke the water of Ryne, whiche

Pork I never loved.

A swine is filthy in England;

but is kept clean in Germany,

and has a swim twice a day.

[1] AB omit "Of." On lamb, see *Babees Book*, p. 222.
[2] quatryped AB; quadryped P.
[3] H .iv. not signed.
[4] nor hit P.
[5] ABP omit "doth hurte."—P adds "for the flesshe is waterysshe."
[6] browne, *orig.*
[7] vppon, in AB.
[8] it doth AB.
[9] specyallye in AB.
[10] AB omit "in."
[11] hyghe Almayne AB.

is aboue Coleyne[1]; but Spaynyerdes, with the other regions anexed to them, kepe the swyne more fylthyer than Englysshe[2] persons doth. Further-more, the Ieue, the Sarason, the Turkes, consernynge theyr polytycke wyt and lerenyng in Physycke, hath as moche wyt, wysdom, reason, and knowledge, for the sauyte of theyr body, as any Chrysten man hath ;—and noble physycyons I haue knowen amonges the*m*; yet [3]they all lacked grace, for as moche as they do not knowe or knowledge Iesu Chryste, as the holy scrypture tellyth vs and them.— They louyth[4] not porke nor swynes flesshe,[5] but doth vituperat & abhorre it; yet for all this they wyll eate adders, whiche is a kynde of serpentes, as well as any other Crysten man dwellynge in Rome,[6] & other hyghe countres; for adders flesshe there is called "fysshe of the mountayn." This notwithstandynge, physycke doth approbat adders flesshe good to be eaten, sayinge it doth make an olde man yonge, as it apperyth, by a harte eatyng an adder, maketh hym yonge agayne. But porke doth not so; for yf it be of an olde hogge not clene kepte, it doth ingender grose blode, & doth humect to moche the stomacke; yet yf *the* porke be yonge, it is nutrytyue.

Spanish swine are the filthiest.

I've known noble heathen Physicians.

Jews and Turks hate pork,

but will eat adders like any Christian in Rome. (p. 187.)

Adder's flesh makes a man young.

Young pork nourishes.

¶ Bacon is good for carters and plowmen, the whiche be euer labourynge in the earth or dunge; but & yf they haue the stone, and vse to eate it, they shall synge, "wo be[7] the pye!" wherfore I do say that coloppes and egges is as holsome for them, as a talowe candell is good for a horse mouth, or a peese of powdred[8] beef is good for a blereyed mare; yet sensuall appetyde muste haue a swynge, all[9] these thinges not-[10]withstandynge. [11]porke is conuertyble to mans flesshe.[11]

Bacon is good for ploughmen;

bad for the stone.

Collops and eggs are bad for that too.

[1] See *Introduction*, p. 156. [2] englysse, *orig.* [3] H .iv. back.
[4] loue AB. [5] flesse, *orig.* [6] See *Introduction*, p. 177.
[7] be to AB. [8] salt. [9] at all AB. [10] sign. I .i.
[11—11] P leaves out these words.

¶ Of[1] Brawne.

Brawn is a usual English winter meat.

Keep clear of it.

¶ Brawne is an vsual meate in wynter amonges Englysshe men: it is harde of dygestyon. the brawne of a wylde boore is moche more better than the brawne of a tame boore. yf a man eate nother of them bothe, it shall neuer do hym harme.

¶ Of[1] Pygges.

Pigs in jelly are good.

A young fat pig is good.

But mind; no crackling!

¶ Pygges, specyally sowe pygges, is nutrytyue; and made in a gelye, it is[2] restoratyue, so be it the pygge be fleed,[3] the skyn taken of, and than stewed with restoratyues, as a cocke is stewed to make a gely. A yonge fatte pygge in physeicke is syngulerly praysed, yf it be wel orderyd in the rostynge, the skyn not eaten.

¶ Of[1] Kydde.

Kid's flesh is the best tame animal flesh.

¶ Yonge Kyddes flesshe is praysed aboue all other flesshe, as Auicen, Rasis, & Aueroyes sayth, for it is temperate and nutrytyue, although it be somwhat dry. Olde kydde is not praysed.

¶ Of wylde beestes flesshe.

Nowhere in Christendom are deer so loved as in England.

Give me Venison, though Physic says it's bad.

¶ I haue gone rownde aboute Crystendome, and ouerthwarte Crystendom, & a thousande or two and more myles out of Crystendom, [4]yet there is not so moche pleasure for harte & hynde, bucke, and doo, and for roo bucke and doo, as is in Englande; & although the flesshe be dispraysed in physycke / I pray God to sende me parte of the flesshe to eate, physycke not-withstandyng. The opynyon of all olde physycyons was & is, that venyson is not good to eate, pryncipaylly for two cause[s][5]: the fyrst cause is, that[6] the beest doth lyue in fere[7]; for yf he be a good wood-man, he shall neuer

[1] AB omit "Of." [2] is a AB.
[3] fleyd AB. [4] sign. I .i. back.
[5] causes ABP. [6] that he AB. [7] feare AB.

se no kynde of deere, but at the .x. byt on the grasse, or brosynge on the tree, but he wyll lyfte vp his hed & loke aboute hym, the whiche commeth of tymorysnes; and tymorosyte doth brynge in melancoly humours. wherfore all Physycyons[1] sayth that venyson, which is the seconde cause, doth ingender coloryke humours; & of trueth it doth so: wherfore let them take the skyn, and let me haue the flesshe. I am sure it is a lordes dysshe, and I am sure it is good for an Englysshe man, for it doth anymate hym to be as he is, whiche is, stronge and hardy / but I do aduertyse euery man, for all my wordes, not to kyll, and so to eate of it, excepte it be lefully,[2] for it is a meate for great men. And great men do not set so moch by *the* meate, as they do[3] by the pastyme of kyllyng of it.

The deer is full of fear, and its flesh breeds choleric humours. But I say, let Physicians take the deer's skin: give me its flesh! Venison is a lord's dish, and good for an Englishman. Don't poach for deer. Great men like killing 'em.

[4] ¶ Of[5] Hares flesshe.

¶ A hare doth no harme nor[6] dyspleasure to no man: yf the flesshe be not eaten, it maketh a gentylman good pastyme. And better is for the houndes or dogges to eate *the* hare after they haue kylled it, as I sayd, than man shuld eate it; for it is not praysed, nother in the olde Testament, nother in physycke; for the byble sayth the hare is an vnclene beeste, And physycke sayeth hares flesshe is drye, and doth ingender melancoly humors.

Let hares be hunted; and let the dogs eat 'em; they breed melancholy.

¶ Of[5] Conys flesshe.

¶ Conys flesshe is good, but rabettes[7] flesshe is best of all wylde beestes / for it is temperat, and doth nowrysshe, and [is] syngulerly praysed in physycke; for all thynges the whiche dothe sucke, is nutrytyue.

Rabbit's flesh is the best wild-beast flesh.

[1] Phyon suchons, *orig. and* AB. Physycyons P.
[2] lawfully AB. [3] do AB. [4] sign. I .ii.
[5] AB omit "Of." [6] nor no AB.
[7] Rabbit, the young cony while a sucker. *Babees Book.*

¶ The .xvij. Chapytre doth treate of pertyculer thynges of fysshe and flesshe.

The heads and the fat of fish are bad.

THe heddes of fysshe, and the fatnes of fysshe, specyally of Samon and Conger, is not good for them the whiche be dysposed to haue rewmatycke heddes. And *th*e heddes of lampryes [1] & lamprons,[2] & the strynge the whiche is within theym, is not good to eate. refrayne from etynge of the skynnes[3] of fysshe and flesshe,[4] & bornet[5] meate, and browne meate, for it doth ingender viscus humours, and color, & melancoly, And doth make opylacions.

Don't eat the skin of fish and flesh.

Brains (except a kid's, and some birds') hurt the stomach.

The braynes of any beest is not laudable, excepte the brayne of a kydde; for it is euyl of digestyon, and doth hurte a mans appetyde and the stomacke, for it is colde, and moyste, and viscus. a hote stomacke may eate it, but it doth inge*n*der grose humours. The brayne of a wodcocke, and of a snype, and suche lyke, is co*m*mestyble.

Fore parts better than hind parts.

The foreparte of all maner of beestes & fowles be more hotter, and lyghter of digestyon, than the hynder partes be.

Marrow is nourishing when eaten with pepper.

The marye of all beestes is hote and moyste; it is nutrityue yf it be wel dygestyd, yet it doth mollyfy the stomacke, and doth take away a mans appetyde; wherfore let a man eate peper with it.

Blood, inwards, and entrails, are indigestible.

The blode of all beestes & fowles is not praysed, for it is hard of digestyon. Al *th*e inwarde*s* of beestes and of fowles, as the herte, the lyuer, the lunges, and trypes, and trylybubbes, wyth all the intrayles, is harde of dygestyon, and doth increase grose humoures.

Fat nourishes less than lean.

The fatnes of flesshe is not so moche nutrytyue as [6]the leenes of flesshe; it is best whan leene and fat is

[1] sign. I .ii. back. [2] See *Babees Book*, p. 215, 166, 174, 235.
[3] kynnes, *orig.*; skynnes AB. [4] flesshe and fysshe AB.
[5] burned AB; borned P. [6] sign. I .iii.

myxte one with another. The tunges of beestes be harde of dygestyon, and of lytell nowrysshement. The stones of a cockrell, & the stones of other beestes that hath not done theyr kynde, be nutrytyue.

Tongues.

Testicles

¶ The .xviij. Chapitre treatyth of roste meate, of fryed meate, [of soden or boyled meate, of bruled meate,][1] and of bake meate.

WIth vs at Mou*n*tpylour, and other vnyuersyties, is vsed boyled meate at dyner, and roste meate to supper: why they shulde do so, I cannot tell, onlesse it be for a consuetude. For boyled meate is lyghter of digestyon than rosted meate is. Bruled meate is harde of digestyon, & euyll for the stone. Fryed meate is harder of dygest[y]on[2] than brulyd meate is, and it doth ingender color and melancoly. Bake meate, whiche is called flesshe that is beryd,[3]—for it is buryd i*n* paast,—is not praysed in physycke. All maner of flesshe the whiche is inclyned to humydyte, shulde be rostyd. And all flesshe the whiche is [4] inclyned to drynes shulde be sodde or boyled.

At Montpeller we have boiled meat for dinner and roast for supper.

Broiled meat is indigestible.

Baked meat (buried in paste) is bad.

Roast moist flesh; boil dry flesh.

¶ Fysshe may be sod, rostyd, brulyd, & baken, euery one after theyr kynde, and vse, & fasshyon of the countree, as the coke and the physycyon wyll agre and deuyse. For a good coke is halfe a physycyon. For the chefe physycke (the counceyll of a physycyon excepte) dothe come from the kytchyn; wherfore *the* physycyon and the coke for sycke men muste consult togyther for the preparacion of meate[5] for sycke men.

How to cook fish.

The chief physic comes from the kitchen.

Cook and Doctor must consult.

[1] Put in from AB. [2] dygestyon ABP. [3] buryed AB.
[4] sign. I .iii. back. [5] meates AB.

Physicians are bad cooks.

For yf the physycyon, without the coke, prepare any meate, excepte he be very experte, he wyll make a werysse[1] dysshe of meate, the whiche the sycke can not take.

¶ The .xix. Chapitre treateth of Rootes, and fyrste of the rootes [of][2] borage and of buglosse.

Borage; Bugloss (see p. 280).

THe rootes of Borage and Buglosse soden tender, and made in a succade, doth ingender good blode, and doth set a man in a temporaunce.

¶ The rootes of Alysaunder[3] and Enulacampana.[4]

Alexanders.

¶ The rootes of Alysaunder soden tender and made in [a][5] succade, is good for to dystroye the stone in the Raynes of the backe & blader. [6]The rootes of Enulacampana[4] soden tender, and made in a succade, is good for the breste, and for the lunges, and for all the interyall membres of man.

Elecampane (Scabwort or Horseheal).

¶ The rootes of percelly & of fenell.

Parsley (p. 281).

¶ The Rootes of percelly soden tender, and made in a succade, is good for the stone, and doth make a man to pysse. The rootes of Fenell soden tender, & made in a succade, is good for the lunges and for the syght.

Fennel (p. 281).

¶ The rootes of turnepes & persnepes.[7]

Turnips.

¶ Turnepes boyled and eaten with flesshe, augmentyth the seede of man. yf they be eaten rawe moderatly, it doth prouoke a good apetyde. Persnepes[7] soden & eaten doth increase nature[8]; they be nutrityue, & doth expell vryne.

Parsnips.

[1] veryсshe AB; werysshe P.
[2] of AB.
[3] Fr. *Alexandre* . . the hearb, great Parsley, Alexanders or Alisaunders.—*Cotgrave.*
[4] Elenacampane B.
[5] in a AB; in surcade P.
[6] I .iv. not signed.
[7] Parsneppes AB.
[8] Semen, generative fluid.

¶ Radysshe rootes, and Caretes.

¶ Radysshe rootes doth breke wynde, & dothe prouoke a man to make water, but they be not good for them the whiche hath the gowte. Caretes soden and eaten doth auge & increase nature, & doth cause a man to make water. Radishes. Carrots.

¶ The rootes of Rapes.

¶ Rape rootes, yf they be well boyled, they do[1] nowrysshe, yf they be moderatly eaten: immoderatly eaten, they doth[2] ingender ventosyte, and doth anoye the stomacke. Rapes.

[3]¶ Of[4] Onyons.

¶ Onyons doth prouoke a man to veneryous actes, and to sompnolence; & yf a man drynke sondry drynkes it doth rectyfy and reforme the varyete of the operacyon of them: they maketh a mans apetyde good, and putteth away fastydyousnes. Onions. (See *Babees Book*, p. 150, 214.)

¶ Of[4] Leekes.

¶ Leekes doth open the breste, and doth prouoke a man to make water; but they doth make and increase euyll blode. Leeks.

¶ Of[4] Garlyke.

¶ Garlyke, of all rootes is vsed & most praysed in Lombardy, and other countres anexed to it; for it doth open the breste, & it doth kyll all maner of wormes in a mans bely, whiche be to say, lumbrici, ascarides, and cucurbitini, whiche is to saye, longe wormes, small lytell longe wormes whiche wyll tykle in the foundement, and square wormes; it also hetyth *the* body, and desoluyth grose wyndes. Garlic kills worms in the belly and fundament.

[1] doth AB. [2] do AB.
[3] I .iv. back. [4] AB omit "Of."

¶ The .xx. Chapitre treateth of[1] vsuall Herbes. And fyrste of Borage and Buglosse.

Borage.

BOrage doth comforte the herte, and doth ingender good blode, and [2]causeth a man to be mery, & doth set a man in[3] temporaunce. And so doth buglosse, for he is taken of more vygor, & strength, & effycacye.[4]

Bugloss (see p. 278).

¶ Of Artochockes, and Rokat.[5]

Artichokes.

¶ There is nothynge vsed to be eaten of Artochockes but *th*e hed of them. whan they be almost rype, they must be soden te*n*der in *th*e broth of beef;[6] & after, eate them at dyner: they doth increase nature, and dothe prouoke a man to veneryous actes. Rokat doth increase the seede of man, and doth stumulat the flesshe, and doth helpe to dygestyon.

Rocket.

¶ Of Cykory, and Endyue.

Chicory. Endive.

¶ Cykory doth kepe the stomacke and *th*e heed in temporaunce, and doth qualyfy color. Endyue is good for them the whiche haue hoote stomackes and drye.

¶ Of whyte Beetes, and Purslane.

White Beets. Purslane.

¶ whyte Beetes[7] be good for the lyuer & for the splene, and be abstersyue. Purslane dothe extynct the ardor of lassyuyousnes, and doth mytygate great heate in all the inwarde partes of man.

[1] of certayne A; of certaine B. [2] sign. K .i.
[3] in a AB. [4] efficacytye AB.
[5] Garden Rocket (*Brassica eruca* or *Eruca sativa*) is an annual, of which, when young, the leaves are used as a salad abroad, and were formerly so in Britain. The wild Rocket (*Sisymbrium officinale* or *Erysimum officinale*) is common here, and is sometimes sown and used as a spring pot-herb. *Chambers's Cyclopædia.* [6] AB add "or with beefe."
[7] beeten P.

¶ Of Tyme and Parsley.

¶ Tyme brekyth the stone; it dothe desolue wyndes, And causeth a man to make water. Parsley is good to breke the stone, and cau[1]seth a man to pysse; it is good for the stomacke, & doth cause a man to haue a swete breth. Thyme. Parsley (p. 279).

¶ Of Lettyse, and Sorell.

¶ Lettyse doth extynct veneryous actes, yet it doth increase mylke in a womans breste; it is good for a hote stomacke, and doth prouoke slepe, and doth increase blod, and doth set the blode in a temporaunce. Sorell is good for a hote lyuer, and good for the stomacke. Lettuce. Sorrel.

¶ Of Penyryall and Isope.

¶ Penyryall doth purge mela*n*coly, and doth comforte the stomacke & the spyrites of man. Isope clenseth viscus fleume, & is good for the breste and for the lunges. Pennyroyal. Hyssop.

¶ Of Roosmary, and Roses.

¶ Roosmary is good for palses,[2] and for the fallynge syckenes, and for the cowghe, and good agaynst colde. Roses be a cordyall, and doth comforte the herte & the brayne. Rosemary. Roses.

¶ Of Fenell, and Anys.

¶ These herbes be seldome vsed, but theyr seedes be greatly occupyde. Fenell-sede is vsed to breke wynde,[3] and [is] good agaynst poyson. Anys-sede is good to clense the bladder, and the raynes of the backe, & doth prouoke vryne, and maketh one to haue a soote[4] breth. Fennel-seed (p. 278, 284). Anise-seed.

¶ Of Sawge, and Mandragor.[5]

[6]¶ Sawge is good to helpe a woman to conceyue, and doth prouoke vryne. Mandragor doth helpe a woman to concepcion, and doth prouoke a man to slepe. Sage. Mandragora.

[1] sign. K .i. back. [2] *the* palsey P.
[3] vryde AB. (cp. Glutton going to the ale-house in *Vis. of Piers Plowman.*) [4] swete AB.
[5] Mandragod, *orig. and* P; Mandragor AB. [6] sign. K .ii.

¶ Of all herbes in generall.

No herb or weed without power to help man.

¶ There is no Herbe, nor weede, but God haue[1] gyuen vertue to them, to helpe man. But for as moche as Plyne, Macer, and Diascorides, with many other olde auncyent and approbat Doctours, hath wryten and pertracted of theyr vertues, I therfore nowe wyll wryte no further of herbes, but wyll speke of other matters that shalbe more necessarye.

¶ The .xxi. Chapitre treatyth of Fruytes, and fyrste of Fygges.

Figs are most nourishing,

specially with blancht Almonds,

but provoke venery.

AUicen sayth that Fygges doth nowrysshe more than any other Fruyte: they doth nowrysshe meruelouslye whan they be eaten with blanched Almons. They be also good, rosted, & stued. They do clense the brest & the lunges, & they do open *the* opylacyons of the lyuer & the splene. They doth stere a man to [2]venereyous actes, for they doth auge and increase the sede of generacyon. And also they doth prouoke a man to sweate; wherfore they doth ingender lyce.

¶ Of great Raysyns.

Raisins stir up the appetite.

¶ Great Raysyns be nutrytyue, specyally yf the stones be pullyd out. And they doth make the stomacke fyrme & stable. And they doth prouoke a man to haue a good appetyde, yf a fewe of them be eaten before meate.

¶ Of smale Raysyns of Corans.

Currants are good for the back.

¶ Smale raysyns of Corans be good for *the* raynes of the backe; and they dothe prouoke vryne. Howbeit

[1] hath AB. [2] sign. K .ii. back.

they be not all the best for the splene, for they maketh opylacyon.

¶ Of Grapes.

¶ Grapes, swete and newe, be nutrytyue, & doth stumulat the flesshe; And they doth comforte the stomacke and *the* lyuer, and doth auoyde opylacyons. Howbeit, it doth replete the stomacke with ventosyte. **Fresh Grapes comfort the Liver.**

¶ Of Peches, of Medlers, & Ceruyces.

¶ Peches doeth mollyfy the bely, and be colde. **Peaches.** Medlers, taken superfluous, doth ingender melancolye. **Medlars.** And Ceruyces[1] be in maner of lyke operacyon. **Services.**

¶ Of Strawburyes,[2] Cherys, & Hurtes.

[3]¶ Strawburyes be praysed aboue all buryes, for **Strawberries.** they do qualyfye the heate of the lyuer, & dothe ingender good blode, eaten with suger. Cherys doth **Cherries.** mollyfye the bely, and be colde. Hurtes be of a **Hurtleberries (*Vaccinium*, L. The Whortle-berry).** groser substaunce; wherfore they be not for them the whiche be of a clene dyete.

¶ Of Nuttes, great and smale.

¶ The walnut & the banocke[4] be of one operacyon. **Walnuts.** They be tarde and slowe of digestyon, yet they doth comforte the brayne if the pyth or skyn be pylled of, and than they be nutrytyue. Fylberdes be better **Filberts are best when new.** than hasell Nuttes: yf they be newe, and taken from the tree, and the skyn or the pyth pullyd of, they be

[1] *Pyrus Sorbus*, the True Service. A tree very like the mountain-ash, but bigger, and bearing larger fruit, which, when beginning to decay, is brought to table in France; though it is oftener eaten by the poor than the rich. See *Loudon's Enc. of Trees and Shrubs*, 1842, p. 442-3.

[2] Strawderyes B. [3] sign. K .iii.

[4] and banocke, AB. *Bannut*, a walnut, *West.* [Wilts and Somerset: Stratmann.] The growing tree is called a *bannut* tree, but the converted timber *walnut*. The term occurs as early as 1697 in MS. Lansd. 1033, fol. 2.—*Halliwell's Gloss.*

nutrytyue, & doth increase fatnes; yf they be olde, they shuld be eaten with great raysens. But new nuttes be farre better than olde nuttes, for olde nuttes be colorycke, and they be euyl for the hed, and euyll for olde men. And they dothe ingender the palsey to the tounge, (yet they be good agaynst venym,) And, immoderatly taken or eaten, doth ingender corrupcyons, as byles, blaynes, & suche putryfaction.

Old nuts breed palsy in the tongue.

¶ Of Peason and Beanes.

Peas.

¶ Peason the whiche be yonge, be nutrytyue; Howbeit, they doth replete a man with vento[1]syte. Beanes be not so moche to be praysed as peason, for they be full of ventosyte, althoughe the skynnes or huskes be ablatyd or cast away; yet they be a stronge meate, and doth prouoke veneryous actes.

Beans are strong food.

¶ Of Peares, and Appulles.

Mellow Pears make men fat.

¶ Peares the whiche be melow and doulce, & not stony, doth increase fatnes, ingenderyng waterysshe blod. And they be full of ventosyte. But wardens rosted, stued, or baken, be nutrytyue, and doth comforte the stomacke, specyally yf they be eaten with comfettes. Apples be good, after a frost haue taken them, or [2]whan they be olde, specyally red apples, and[2] they the whiche be of good odor, & melow; they shuld be eaten with suger or co*m*fettes, or with fenell-sede, or anys-sede, bycause of theyr ventosyte; they doth comforte than the stomacke, and doth make good dygestyon, specyally yf they be rostyd or baken.

Roast Wardens comfort the stomach.

Apples should be eaten with comfits or fennel-seed.

¶ Of Pomegranates, & Quynces.

Pomegranates.

¶ Pomegranates be nutrytyue, and good for the stomacke. Quinces baken, the core[3] pulled out, doth mollyfy the bely, and doth helpe dygestyon, and dothe preserue a man from dronkenshyppe.

Baked Quinces soften the belly.

[1] sign. K .iii. back. [2–2] P omits this. [3] gore P.

¶ Of Daates, and Mylons.

[1] ¶ Daates, moderatly eaten, be nutrityue; but they doth cause opylacyons of the lyuer and of the splene. Mylons doth ingender euyl humoures. Dates nourish. Melons.

¶ Of gourdes, of Cucumbres, & pepones.

¶ Gourdes be euyll of nowrysshement. Cucumbers restrayneth veneryousnes, or lassyuyousnes, or luxuryousnes. Pepones[2] be in maner of lyke operacion, but the pepones ingenderyng[3] euyll humours. Gourds. Cucumbers. Pepones.

¶ Of Almondes and Chesteyns.

¶ Almondes causeth a man to pysse; they do[4] mollyfy the bely, and doth purge the lunges. And .vi. or .vii. eate before meate, preserueth a man from dronkenshyp. Chesteynes doth nowrysshe the body strongly, & doth make a man fat, yf they be thorowe rosted, and the huskes abiected; yet they doth replete a man with ventosyte or wynde. Almonds stop drunkenness. Chestnuts fatten.

¶ Of Prunes, and Damysens.

¶ Prunes be nat greatly praysed, but in the way of medysyne, for they be cold & moyste. And Damysens be of *the* sayd nature; for the one is olde and dryed, and the other be taken from the tre. .vi. or .vii.[5] damysens eaten before dyner, be good to prouoke a mans appetyde; they doth mollyfy the bely, and be abstersyue; [6] the skyn and the stones must be ablatyd and caste awaye, and not vsed. Prunes (plums). Damsons: eat 6 or 7 before dinner.

¶ Of Olyues, and Capers.

¶ Olyues condyted, and eaten at the begynnynge of [a][7] refectyon, doth corroborate the stomacke, and prouoketh appetyde. Capers doth purge fleume, and doth make a man to haue an appetyde. Olives. Capers.

[1] K .iv. not signed.
[2] Fr. *Pepon:* m. A Pompion or Melon.—*Cotgrave.*
[3] ingenderythe P. [4] doth AB. [5] Syxe or seuen AB.
[6] K .iv. back. [7] a AB

¶ Of Orenges.

Oranges, and Orange-Marmalade.

¶ Orenges doth make a man to haue a good appetyde, and so doth the ryndes, yf they be in succade, & they doth co*m*forte the stomacke; the Iuce is a good sauce, and dothe prouoke an appetyde.

¶ The .xxii. Chapitre treateth of spyces, and fyrste of Gynger

Ginger. Green ginger.

Gynger doth hete the stomacke, and helpyth dygestyon: grene gynger eaten in the moreni*n*ge, fastynge, doth acuat and quycken the remembraunce.

¶ Of Peper.

Pepper, white, black, and long.

¶ There be .iii. sondry kynd*es* of peper, which be to say, whyte Peper, blacke Peper, & long Peper. All kyndes of pepers doth[1] heate the bo[2]dy, and doth desolue fleume & wynde, & dothe helpe dygestyon, and maketh a man to make water. Blacke peper doth make a man leane.

¶ Of Cloues, and Mace.

Cloves. Mace.

¶ Cloues doth comforte the senewes, & doth dysolue and doth consume superfluous humours, [and][3] restoryth nature. Maces is a cordyall, and doth helpe the colycke, & is good agaynst the blody flyxe and laxes.

¶ Of Graynes, and Safferon.

Cardamons. Saffron.

¶ Graynes be good for the stomake and the head; And be good for women to drynke. Safferon doth comforte the herte & the stomacke, but he is to hote for the lyuer.

[1] to *orig.*; doth AB. [2] sign. L .i. [3] and AB.

¶ Of Nutmeges, & Cynomome.[1]

¶ Nutmeges be good for them the whiche haue colde in theyr hed, and dothe comforte the syght and the brayne, & the mouthe of the stomacke, & is good for the splene. Cynomome is a cordyall, wherfore *the* Hebrecyon[2] doth say, "why doth a man dye, and can gette Cynomome to eate?" yet it doth stop, & is good to restrayne, fluxes or laxes. *Nutmegs. Cinnamon.*

¶ Of Lyqueryce.

¶ Lyqueryce is good to clense and to open the lunges & the brest, & doth loose fleume. *Liquorice.*

¶ The .xxiij. Chapytre sheweth a dyete for Sanguyne men.

[3] SAnguyne me*n* be hoote and moyste of complexion, wherfor they must be cyrcumspect in eatynge of theyr meate, co*n*syderynge that the purer the complex[i]on is, the soner it may be coruptyd, & the blode maye be the sooner infectyd / wherfore they must abstayne to eate inordynatly fruytes and herbes and rotes, as garlyke, onyons, and leekes; they must refrayne from eatyng of olde flesshe, and exchew the vsage of etynge of the braynes of beestes, & from etynge *the* vdders of keyn. They muste vse moderat slepe and moderat dyet, or els they wyl be to fat and grose. Fysshe of muddy waters be not good for them. And yf blode do abou*n*de, clense it with stufes, or by fleubothomye. *Sanguine men mustn't eat fruits, herbs, roots, old flesh, cows' udders, or mud-fish.*

[1] Cynamon B (ed. 1562); Cynamone P. [2] Hebricion ABP.
[3] sign. L .i. back.

¶ The .xxiiij. Chapyter sheweth a dyete for Fleumatycke men.

Phlegmatic men

mustn't eat viscous or white meat,

fish or fruit.

Fleumaticke men be colde and moyste, wherfore they must abstayne from meates the whiche is cold. And also they must refrayne from eatyng viscus meate, specially from [1] all meates the whiche doth ingender fleumatycke humours, as fysshe, fruyte, and whyte meate. Also to exchewe the vsage of eatynge of crude herbes; specyall[y] to refrayne from meate the whiche is harde and slowe of dygestyon, as it appereth in the propertes of meates aboue rehersyd. And to[2] beware not to dwell nyghe to waterysshe and morysshe grounde. These thynges be good for fleumatycke persons, moderatly taken: onyons, garlycke, peper, gynger; And all meates the whiche be hote and drye; And sauces the whiche be sowre. These thynges folowynge doth purge fleume: polypody, netyll, elder, agarycke, yreos, mayden-heere, and stycados.

Phlegmatic men mustn't eat indigestible meats,

but hot and dry ones.

Purgatives of Phlegm.

¶ The .xxv. Chapitre sheweth a dyete for Colorycke men.

Choleric men shouldn't eat hot spices, or drink wine.

Color is hote and dry; wherfore Colorycke men muste abstayne from eatyng hote spyces, and to refrayne from drynkynge of wyne, and eatynge of Colorycke meate: howbeit, Colorycke men may eate groser meate than any other of complexions, except theyr educacion haue ben to the contrary. [3] Colorycke men shulde not be longe fastynge. These thynges folowyng do[4] purge color: Fumytory, Centory, wormewod, wylde hoppes,

Purgatives of Choler.

[1] sign. L .ii.
[2] AB omit "to."
[3] sign. L. ii. back.
[4] doth AB.

vyoletes, Mercury, Manna, Reuberbe, Eupatory, Tamarindes, & the whay of butter.

¶ The .xxvi. Chapitre treateth of a dyetarye for Melancoly men.

Melancoly is colde & drye; wherfore Melancoly me*n* must refrayne from fryed meate, and meate the whiche is ouer salte, And from meate *that* is sowre & harde of dygestyon, and from all meate the whiche is burnet[1] and drye. They must abstayne from immoderat thurste, and from drinkyng of hote wynes, and grose wyne, as red wyne. And vse these thynges, Cowe mylke, Almon mylke, yolkes of rere egges. Boyled meate is better for Melancoly men than rosted meate. All meate the whiche wylbe soone dygestyd, & all meates the which doth ingender good blode, And meates the whiche be temperatly hote, be good for Melancoly men. And so be all herbes the whiche be hote and moyste. These thynges folowynge doth purge Melancoly: quyckbeme, Seene, sticados, hartystounge, mayden heere, pulyall mountane, borage, organum, suger, and whyte wyne.

Melancholy men mustn't eat fried or salt meat.

Melancholy men should drink only light wine,

milk, and egg-yolks.

Purgatives of Melancholy.

¶ The .xxvii. Chapiter treatyth of a dyete and of an ordre to be vsed in the Pestyferous tyme of the[2] pestylence & swetyng sycknes.

Whan the Plages of the Pestylence or the swetynge syckenes is in a towne or cou*n*tree, with vs at Mountpylour, and al other hygh Regyons and countrees *that* I haue dwelt in, the people doth fle from

In Pestilence time in Montpelier,

[1] burned AB. [2] of B.

people flee from the city.

the contagious and infectious ayre; preseruatyues,[1] with other counceyll[2] of Physycke, notwithstandyng. In lower and other baase countres, howses, the which be infectyd in towne or cytie, be closyd vp, both doores & wyndowes; & the inhabytours shall not come a brode, nother to churche, nor to market, nor to any howse or company, for[3] infectyng other, the whiche be clene without infection. A man cannot be to ware, nor can not kepe hym selfe to well from this syckenes, for it is so vehe[4]ment and so parlouse,[5] that the syckenes is taken with *th*e sauour of a mans clothes the whiche hath vysyted the infectious howse, for the infection wyl lye and hange longe in clothes. And I haue knowen that whan the strawe & russhes hath ben cast out of a howse infectyd, the hogges the whiche dyd lye in it, dyed of *th*e pestylence; wherfore in such infectious tyme it is good for euery man *tha*t wyl not flye[6] from the contagyous ayre, to vse dayly—specyally in the mornynge and euenyng—to burne Iuneper, or Rosemary, or Rysshes, or Baye leues, or Maierome, or Franke*n*[se]nce, [or][7] bengauyn. Or els make this powder: Take of storax calamyte half an vnce,[8] of frankensen*c*e an vnce,[8] of the wodde of Aloes the weyghte of .vi. ƀ.[9]; myxe al these togyther; Than cast half a sponefull of this in a chaffyng-dysshe of coles, And set it to fume abrode in the chambers, & the hall, and other howses. And[10] you wyll put to this powder a lytell Lapdanum, it is so moche *th*e better. Or els make a pomemau*n*der[11] vnder this maner. Take of Lapdanum .iii. drammes, of *th*e wodde of Aloes one drame, of amber of grece .ii. drames and a half; of nutmegges, of storax calamite, of eche a drame and a half; confect[12] all those[13]

In low countries, infected houses are shut up, with the men in them.

Infection hangs in clothes,

straw, and rushes.

Burn scented herbs or gums;

or fumigate with Boorde's powder,

or make a Pomander

of spices, &c.,

into a ball.

[1] preseruations B. [2] counsayles AB.
[3] against, for fear of, to prevent. [4] sign. L .iii. back.
[5] peryllous AB. [6] flee AB; fly P. [7] frankensence or AB.
[8] ounce AB. [9] drachms. [10] if.
[11] Pomaunder AB. [12] conferre B. [13] this B.

togyther with Rose-wa[1]ter, & make a ball. And this aforesayd Pomemaunder[2] doth not onely expell contagyous ayre,[3] but also it doth comforte the brayne, as Barthelmew of Montagnaue sayth, & other modernall doctors doth afferme *the* same: whosoeuer *that* is infectyd with the pestylence, let hym loke in my 'breuyary of helth' for a remedy.[4] But let hym vse this dyete: let the Chamber[5] be kept close, And kepe a contynuall fyre in the Chamber, of clere burnynge wodde or chare[6]-cole without smoke; beware of takynge any colde, vse temporat meates and drynke, and beware of wyne, bere, & cyder; vse to eate stued or baken wardens, yf they can be goten; yf not, eate stued or baken peers, with comfettes; vse no grose meates, but those the whiche be lyght of dygestyon.

For remedies for the Pestilence, see my *Breviary*.

Keep a fire in your room.

Don't take cold;

eat stewed wardens, with comfits.

¶ The .xxviij. Chapitre sheweth of a dyete [for them][7] the whiche be in any Feuer or agew.

Fever, Ague.

I Do aduertyse euery man that hath a Feuer or an Agewe, not to eate no meate .vi. howres before his course doth take hym. And [8]in no wyse, as longe as the Agew doth indure, to put of[9] shertte nor dowblet, nor to ryse out of the bedde but whan nede shall requyre; and in any wyse not to go, nor to take any[10] open ayer. For suche prouysyon may be had that at vttermost at the thyrde course he shalbe delyuered of the Feuer, vsynge the medsynes the whiche be in the Breuyary of helthe.[11]

Don't eat meat for 6 hours before the first course.

Don't expose yourself to cold.

You'll be cured at the 3rd course, if you use the medicines in my *Breuyary*.

[1] L .iv. not signed. [2] Pomaunder AB. [3] ayres AB.
[4] Chap. 121, fol. xlv. back, ed. 1552. [5] Chambers AB.
[6] AB omit "chare." [7] for them AB.
[8] L .iv. back. [9] of the AB. [10] the AB.
[11] Chap. 135—150, fol. xlix. back, to fol. lv., ed. 1552.

And let euery man beware of castynge theyr handes & armes at any tyme out of the bed, in or out of theyr agony, or to spraule with *th*e legges out of the bed: good it is for the space of .iii. courses to weare con*tynewelly gloues, and not to wasshe the handes, And to vse suche a dyet in meate & drynke as is rehersyd in the pestylence. [*See above; p.* 291, *lines* 11—15.]

Wear gloves.

¶ The .xxix. Chapitre treatyth of a dyete for them the whiche haue the Iliacke, or the colyck, & the stone.

Iliac and Colic.

THe Iliacke and the Colycke be ingendered of ventosyte, the whiche is intrusyd or inclosed in two guttes; the one is called Ilia, And *th*e other is called Colon. For these two infyrmytes a [1] man muste beware of colde. And good it is not to be longe fastynge. And necessary it is to be laxatyue, and not in no wyse to be constupat. And these thynges folowyng be not good for them the which haue these aforesaid infyrmyt*es*: [2]new bred, stale bred,[2] nor new ale. They must abstayne also from drynkyng of beere, of cyder, and red wyne, and cynamom. Also refrayne from al meates that ho*n*ny is in; exchew eatyng of cold herbes; vse not to eate beanes, peson, nor potage; beware of the vsage of fruytes, And of all thynges the whiche doth ingender wynde. For the stone, abstayn from drynkynge of new ale; beware of beere, and of red wyne and[3] hote wynes; refrayne from eatynge of red herynge, ma[r]tylmas beef and bakon, and salte fysshe, and salt meates. And beware of goynge colde aboute the mydell, specyally aboute the raynes of the backe. And make no restryctyon of wynde and water, nor seege[4] that nature[5] wolde expelle.

Beware of cold.

Don't fast too long,

eat new bread,

cold herbs,

fruit, or anything which raises wind.

For Stone, don't drink wine, or eat red herrings, &c. (See p. 80 above).

[1] sign. M .i. [2—2] hote bread P. [3] and of AB. [4] egestyon P. [5] water AB.

¶ The .xxx. Chapitre treatyth of a dyete for them the whiche haue any kyndes of the gowte.[1]

Gowt.

[2] THey the whiche be infectyd with the gowte, or any kynde of it, I do aduertyse them not to syt long[3] bollynge[4] and bybbynge, dysyng and cardyng, in forgettyng them selfe to exonerat the blader and the bely whan nede shall requyre; and also to beware that[5] the legges hange not without some stay, nor *that* the boot*es* or shoes be not ouer strayte. who soeuer hath *the* gowte, muste refrayne from drynkyng of newe ale; and let hym abstayne from drynkyng of beere and red wyne. Also, he must not eate new brede, egges, fresshe samon, eles, fresshe heryng, pylcherd*es*, oysters, and all shell-fysshe. Also,[6] he muste exchew the eatynge of fresshe beef, of goose, of ducke, & of pygyons. Beware of takyng[7] colde in the legge,[8] or rydyng, or goynge wetshod. Beware of veneryous act*es* after refection, or after or vpon a full stomacke. And refrayne from all thinges that doth ingen*d*er euyll humours, and be inflatyue.

Don't sit bibbing and dicing, and forget to empty yourself. Gowty folk mustn't wear tight boots, or eat salmon, oysters, or ducks; or go wetshod.

¶ The .xxxi. Chapytre treatyth of a dyete for them the whiche haue any of the kyndes of lypored.

[9] HE that is infectyd wyth any of the .iiii.[10] kyndes of the lepored must refrayne from al maner of wynes, & from new drynkes, and stro*n*ge ale; than let hym beware of ryot and

Lepers mustn't drink wine and strong ale.

[1] gowtes AB. [2] sign. M .i. back. [3] to longe AB.
[4] bowlynge AB. [5] AB omit "that." [6] And AB.
[7] takynge of A; takyng of B. [8] legge AB.
[9] sign. M .ii. [10] foure AB.

Lepers mustn't eat spices, tripe, fish, eggs, beef, goose, water-fowl, venison, hare, &c.

surfetynge. And let hym abstayne from[1] etyng of spyces, and daates, and from trypes & podynges, and all inwardes of beestes. Fysshe, and egges, & mylke, is not good for leperous persons: and they must abstayne from eating of fresshe beef, and from eatynge of gose [&] ducke, and from water-fowle and pygions; And in no wyse eate no veneson, nor hare-flesshe, and suche lyke.

¶ The .xxxii. Chapytre treatyth of a dyete for them the whiche haue any of *the* kyndes of the fallyng syckenes.

Epilepsy. (See Breuyary, ch. 122, fol. xlvi.) Folk with Falling Sickness mustn't drink milk or strong ale, or eat fish-fat, viscous fish, garlick, leeks, venison, &c.; or go to meetings of men, or sit too near the fire,

WHo so euer he be, the which haue any of the kyndes of the[2] fallyng syckenes, must abstayn from eatynge of whyte meate, specially of milke: he must [3]refrayne from drynkyng of wyne, newe ale, and stronge ale. Also they shulde not eate the fatnes of fysshe, nor the hedes of fysshe, the whiche dothe ingender rewme. Shell-fysshe, eles, samon, herynge, & viscus fysshes, be not good for Epilentycke men. Also, they muste refrayne from eatynge of garlyke, onyons, leekes, chybbolles, and all vaperous meates, the whiche doth hurte *the* hed: venson, hare-flesshe, beef, beanes, and peason, be not good for Epilentycke men. And yf they knowe that they be infected with this[4] great sycknes, they shulde not resorte where there is great resorte of company, whiche is, in[5] churche, in sessyons, and market-places on market dayes; yf they do, the sycknes wyll infeste[6] them more there than in any other place, or at any other tyme. They must beware they do not syt to nyghe the fyre, for the fyre wyll ouercom them, and

[1] for AB. [2] AB omit "the." [3] sign. M .ii. back.
[4] these AB. [5] in the AB. [6] infecte AB.

wyll induce the sycknes. They must beware of lyeng hote[1] in theyr bed, or to laboure extremely; for suche thynges causyth the grefe to come the ofter.

or work too hard.

¶ The .xxxiii. Chapytre treatyth of a dyete [for them][2] the whiche haue any payne in the[3] hed.

Headache.

[4] Many sycknes, or infyrmytes, and impedymentes, may be in a mans hed, wherfore, who so euer haue any impedyment in the hed, must not kepe the hed to hote nor to colde, but in a temporaunce. And to beware of ingendryng of rewme, whiche is the cause of many infyrmytes. There is nothynge that doth ingender rewme so moche as doth the fatnes of fysshe, and the heddes of fysshe, and surfettes,[5] & takynge colde in the feete, and takynge colde in the nape of the necke or hed. Also, they *the* whiche haue any infyrmyte in *the* hed must refrayne of immoderat slepe, specyally after meate. Also, they must abstayne from drynkynge of wyne; and vse not to drynke ale and beere the whiche is ouer stronge. vocyferacyon, halowynge, cryeng, and hygh synging, is not good for the hed. All thynges the whiche is vaporous or dothe fume, is not good for the hed. And all thynges the which is of euyll sauour, as caryn, synkes, wyddrawghtes,[6] pisse-bolles, snoffe of candellys, dunghylles, stynkynge canellys, and stynkynge standyng waters, & stynkynge marshes, with suche contagyous[7] eyres, doth hurte the hed, and the brayne, and the memory. All odyferous sauours be good for the hed, and the brayne, and the memory.

Keep the head cool. Don't eat things that breed rheum; don't sleep too long, drink wine, or hallo. Keep out of stinks, and smell sweet odours.

[1] to hote AB. [2] for them ABP. [3] theyr AB.
[4] sign. M .iii. [5] surfestes, *orig.* [6] wynkraughtes.
[7] sign. M .iii. back.

Consumption.

¶ The .xxxiiii. Chapitre treateth of a dyete for them the whiche be in a consumpcyon.

Avoid sour things.

Use cordials, nourishing food, sugar, and sweet wines.

Don't eat fried or burnt meat;

but eat stewed pig or cock.

WHo soeuer he be that is in a consumpcyon muste abstayne from all sowre and tarte thynges, as venegre & alceger,[1] & suche lyke. And also he must abstayne from eatynge of grose meates, the whiche be harde and slowe of dygestyon, And vse cordyallys and restoratyues, and nutrytyue meates. All meates and drinkes the which is swete, & that suger is in, be nutrytyue; wherfore swete wynes be good for them the whiche be in consu*m*pcion,[2] moderatly taken. And sowre wyne, sowre ale, and sowre brede,[3] is good for no man; For it doth freate away nature. and let them beware, that be in[4] consumpcion, of fryde meate, of bruled meate, and bronte[5] meate, the whiche is ouer rostyd. And in any wyse let them beware of anger & pencyfulnes. These thynges folowynge be good for them the whiche [6]be in co*n*sumpcions[7]: a pygge or a cocke stewed and made in a gely, cockrellys stewed, gootes mylke and suger, almon mylke in the whiche ryce is soden, and rabettes stewed,[a] &c. [[a] & newe layd egg*es*, & rere yolkes of egg*es*, & ryce sode*n* i*n* almo*n* mylke. P.]

Asthma.

¶ The .xxxv. Chapitre treateth of a dyete for them the whiche be asmatyke men, beyng short wyndyd, or lackynge breth.

[1] aleger AB; alegar P. [2] consumpcions AB. [3] beere AB. [4] in a AB. [5] of burned AB. [6] M .iv. not signed. [7] consumpcion AB.

SHortnes of wynde commeth dyuers tymes of impedymentes in *the* lunges, and straytnes of[1] the brest, opylatyd thorow viscus fleume; and other whyle whan the hed is stuffyd with rewme, called the pose, lettyth the breth of his naturall course. wherfore he *that* hath shortnes of breth muste abstayne from eatyng of nuttes, specyally yf they be olde: chese[2] and mylke is not good for them; no more is fysshe and fruyte, and rawe or crude herbes. Also all maner of meate the whiche is harde of dygestion, is not good for them. They muste refrayne from eatyng of fysshe, specially from eatyng fysshe the which [3]wyll cleue to the fyngers, & be viscus & slyme; & in any wyse beware of the skyns of fysshe, & of all maner of meate the whiche doth ingender fleume. Also they muste beware of colde. And whan any howse is a swepynge, to go out of the howse for a space in to a clene[4] eyre. The dust also that ryseth in the strete thorow the vehemens of the wynde or other wyse, is not good for theym. And smoke is euyll for them; and so is all thynge that is stoppynge: wherfore necessary it is for them to be laxatyue, [& to be in a clene & pure eyre. P.]

Asthma comes from phlegm obstructing the lungs.
The Pose.
Don't eat nuts or cheese, &c.
For Asthma don't eat viscous fish.
Beware of cold and dust,
and smoke.

¶ The .xxxvi. Chapitre treatyth of a dyete for them the whiche haue the palsye.

Palsy

THey the whiche haue the Palsye, vnyuersall or pertyculer, must beware of anger, hastynes, and testynes, & must beware of feare, for thorow anger or feare dyuers tymes the Palsye do come

Don't get testy.

[1] in AB. [2] and chese P. [3] M .iv. back. [4] clere P.

Don't get drunk, or eat nuts, or lie on the ground. Fox-stink is good for palsy.

to a man. Also they must beware of dronkennes, and eatyng of nuttes, whiche thynges be euyll for the palsye of the tonge. coldnes, and contagyous and stynkynge & fylthy ayres be euyll for the palsye. And lette euery [1]man beware on[2] lyeng vpon the bare grounde, or vpon the bare stones; for it is euyll for the Palsye. the sauour of Castory, & the sauour of a fox, is good agaynst the palsye.

¶ The .xxxvii. Chapitre doth shew an order and a dyete for them the whiche be madde, and out of theyr wytte.

Madmen must be kept in safe guard. Mychell, a lunatic, killed 2 people and himself. Keep lunatics in a close dark room, with a keeper whom they fear. Don't put pictures in their rooms. Shave their heads once a month,

THere is no man the whiche haue any of *the* kyndes of madnes but they ought to be kepte in sauegarde, for dyuers i*n*co*n*uenye*n*ce *tha*t may fall, as it apperyd of late dayes of a lunatycke man named Mychell,[3] the whiche went many yeres at lyberte, & at last he dyd kylle his wyfe, and his wyfes suster, & his owne selfe. wherfore I do aduertyse euery man the whiche is madde, or lunatycke, or frantycke, or demonyacke, to be kepte in saue garde in some close howse or chamber, where there is lytell lyght. And that he haue a keper, the whiche the madde man do feare. And so that the madde man haue no knyf, nor sheers, nor other edge toule, nor that he haue no gyr[4]dyll, except it be a week lyste of clothe, for[5] hurtynge or kyllynge hym selfe. Also the chamber or the howse that the madde man is in, let there be no payntcd clothes, nor paynted wallys, nor pyctures of man nor woman, or fowle, or beest; for suche thynges maketh them ful of fantasyes. lette the madde persons hed be shauen ones a moneth: let them

[1] sign. N .i. [2] of AB. [3] Michel P. [4] sign. N .i. back.
[5] against, to prevent.

drinke no wyne, nor stronge ale, nor stronge beere, but moderat drynke; and let them haue .iii. tymes in a daye warme suppynges, and [a][1] lytell warme meat. And vse few wordes to them, excepte it be for reprehensyon, or gentyll reformacyon, yf they haue any wytte or perseueraunce to vnderstande [what reprehensyon or reformacyon is. P.]

and give them no strong drink.

Speak little to them.

¶ The .xxxviii. Chapytre treatyth of a dyete for them the whiche haue any of the kyndes of the Idropyses.

Dropsy.

SAynt Beede sayeth 'the more a man doth drynke *that* hath the Idropise,[2] the more he is a thurst;' for although *the* syckenes doth come by superabundaunce of water, yet the lyuer is drye, whether it be alchy[3]tes, Iposarca, Lencoflegmancia, or the tympany. They that hath any of the .iiii. kyndes of *the* Idropyses /[4] must refrayne from al thynges the whiche be co*n*stupat and costyue, and vse all thynges the which be laxatyue / nuttes, and dry almondes, and harde chese, is[5] poyson to them; [6]A ptysane and posset ale made with colde herbes doth comforte them. who so euer he be, the whiche wyll haue a remedy for any of these foure kyndes of the Idropyses,[7] and wyll knowe a declaracyon of these infyrmytes, and all other sycknesses, let hym loke in a boke of my makyng, named the Breuyary of helth. For in this boke I do speke but of dyetes, and how a man shuld order his mansyon place, And hym self & his howsholde, with suche lyke thyng*es*, for the conseruacion of helth.[6]

Avoid binding food.

Nuts and cheese are poison. Posset ale is good.

For all sicknesses and their treatment, see my Breuyary.

I only speak here of Diet, and of managing a house.

[1] a AB. [2] Idropsye AB; I dropyse P. [3] sign. N .ii.
[4] Idropsyes AB. [5] AB omit "is." [6–6] Not in AB.
[7] See Boorde's *Breuyary*, chap. 179, 38, 17, 345.

A general Diet.

¶ The .xxxix. Chapytre treateth of a generall dyete for all maner of men and women, beynge sycke or hole.

Every one knows best what helps and what hurts him.

THere is no man nor woman the which haue any respect to them selfe, that can be a better Phesycion for theyr [1]owne sauegarde, than theyr owne self can be, to consyder what thynge *the* whiche doth them good, And to refrayne from suche thynges that doth them hurte or harme. And let euery man beware of care, sorowe, thought, pencyfulnes, and of inwarde anger. Beware of surfettes, and vse not to[2] moche veneryouse actes. Breke not the vsuall custome of slepe in the nyght. A mery herte and mynde, the whiche is in reste and quyetnes, without aduersyte [3]and to moche worldly busynes,[3] causeth a man to lyue longe, and to loke yongly, althoughe he be agyd. care and sorowe bryngeth in age and deth, where[fore] [4] let euery man be mery; and yf he can not, let hym resorte to mery company to breke of his perplexatyues.

Don't be anxious. Sleep at night. A merry heart makes a man live, and look young. Care brings age and death.

Wash your hands often, and comb your head.

¶ Furthermore, I do aduertyse euery man to wasshe theyr handes ofte euery daye; And dyuers tymes to keyme theyr hed euery daye, And to plounge the eyes in colde water in the morenyng. Moreouer, I do councell euery man to kepe the breste and the stomacke warme, And to kepe the feete from wet, and other whyle to wasshe them, and that they be not kept to hote nor to colde, but indyfferently. Also to kepe the hed and the necke in a moderat temporaunce, not to hote nor to colde; [5]and in any wyse to beware not to medle to moche with veneryous act*es*; for that wyll cause a man to loke agedly, & also causeth a man to haue a

Keep your chest and stomach warm, your feet dry, and your head cool. Avoid venery;

[1] sign. N .ii. back. [2] so, *orig.* [3–3] Not in P (ed. 1547). [4] wherfore A; wherefore B. [5] sign. N .iii.

breef or a shorte lyfe. All[1] other matters pertaynynge to any pertyculer dyete, you shall haue[2] in the dyetes aboue in this boke rehersyd.

It shortens life.

¶ The .xl. Chapytre doth shewe an order, or a fasshyon, how a sycke man shulde be ordered, And how a sycke man shuld be vsyd that is lykely to dye.

A sick room, and a Death-bed.

WHoo so euer that is sore sycke, it is vncerteyne to man whether he shall lyue or dye; wherfore it is necessarye for hym *that* is sycke to haue two or .iii.[3] good kepers, the whiche at all tymes must be dylyge*n*t, and not slepysshe, sloudgysshe,[4] sluttysshe. And not to wepe and wayle aboute a sycke man, nor to vse many wordes / nor that there be no greate resort to co*m*mon and talke, [5]For it is a busynes [for][6] a whole man to answere many men, specyally women, that shall come to hym. They the which co*m*meth to any sycke person, ought to haue few wordes or none, except certayne persons the whiche be of counseyll of the Testament makynge, the whiche wyse men be not to seke of such matters in theyr syckenes; for wysdom wolde that euery man shulde prepare for suche thyng*es* in helth. And yf any man for charyte wyll vyset any person, lette hym aduertyse the sycke to make euery thynge euyn bytwext God, and the worlde, & his co*n*scyence; And to recéyue the ryght*es* of holy churche, lyke a catholycke

Have 2 or 3 good nurses.

No wailing or talking,

except to make a Will.

A visitor may advise settling matters,

receiving the Rites of the Church,

[1] Also AB; All, ed. 1547. [2] haue it AB.
[3] thre AB. [4] ABP insert "nor."
[5] sign. N .iii. back. [6] for AB *and* ed. 1547.

man; And to folowe the counseyll of both Physycyons, whiche is to say, the physycyon of the soule, & the physycyon of the body, that is to saye, the spyrituall counseyl of his ghostly father, and the bodely counseyll of his physycyon consernyng the receytes of his medsons to recouer helth. For saynt Augustyn saith, "he that doth not the [1] co*m*maun*d*ement of his physycyon, doth kyll hym self." Furthermore, about a sycke persone shuld be redolent sauour[s], and the chamber shuld be replenysshed with herbes & flowers of odyferouse sauour.[2] & certayne tymes it is good, to be vsed a lytell of some perfume[3] [4]to stande in *th*e mydle of the chamber. And in any wyse lette not many men, and specyally women, be togyther at one tyme in the chamber, not onely for bablynge, but specially for theyr brethes.[5] And the kepers shulde se at all tymes that the sycke persons drynke be pure, fresshe, & stale, and that it be a lytell warmed, turned out of the colde. Yf the sycke man wex sycker and sycker, that there is lykle[6] hope of amendment, but sygnes of deth, than no man oughte to moue to hym any worldly matters or busynes; but to speke of ghostly and godly matters, And to rede the passyon of Cryste, & to say the psalmes of the passyon, and to holde a crosse or a pyctour of the passyon of Cryste before *th*e eyes of the sycke person. And let not the kepers forget to gyue the sycke man that is in suche agony, warme drynke with a spone, and a sponefull of a cawdell or a colesse. And than lette euery man do[7] indeuer hym selfe to prayer, that the sycke p*er*son may fynysshe his lyfe Catholyckely in the fayth of Iesu Cryste, And so[8]

attention to Priest and Doctor.

Keep sweet odours in the sick room.

Don't have women babbling there.

Have the drink fresh.

When Death's coming,

read of Christ' sufferings;

give the dying man a little warm drink;

and pray that he may die in the faith of Christ.

[1] not obserue the co*m*maundementes AB.
[2] flauours AB.
[3] good to vse some perfumes P.
[4] N .iv. not signed.
[5] hote breathes AB.
[6] likely AB; lytle P.
[7] P leaves out "do."
[8] so to AB.

departe out of this myserable world. I do beseche the Father, and *the* sone, and the holy ghost, thorow the meryte of Iesu Crystes passyon, that I and all Creatures lyuynge may do [so].[1] A M E N.

[1] so P

[1] ¶ Imprynted by me Robert Wyer / dwellynge in seynt Martyns parysshe besyde charynge Crosse, at the sygne of seynt John Euangelyste.

For John Gowghe, Cum priuilegio regali.

Ad imprimendum solum.[2]

[? *Cut of St John writing his* Revelations *in the Isle of Patmos.*]

[1] N .iv. back.

[2] Robert Wyer's Colophon to the undated edition in the British Museum of ? 1557 A.D., is: ¶ Imprinted by me Robert Wyer. Dwellynge at the Sygne of Seynt Johñ Euangelyst in S. Martyns Parysshe, besyde Charynge Crosse.

Thomas Colwel's Colophon to the edition of 1562 is: ¶ Imprinted by me Thomas Colwel. Dwellynge in the house of Robert Wyer, at the Signe of S. Johñ Euangelyst, besyde Charynge Crosse. ☞

Wyllyam Powell's Colophon to the edition of 1547 is: ¶ Imprynted at London in Fletestrete at the sygne of the George nexte to saynte Dunstones churche by Wyllyam Powell. In the yere of our Lorde god .M. CCCCC. LXVII. ☞ : ☜

¶ The treatyse answerynge the boke of Berdes,

Compyled by Collyn clowte, dedycatyd to Barnarde barber dwellyng in Banbery.

[1]¶ To drynke with me, be not a ferde
For here ye se groweth neuer a berde.

[Coarse woodcut of a man stooping down and exposing himself, with the legend *Testiculos Habet.*
Any member wanting the cut must apply to
MR FURNIVALL.]

¶ I am a Foole of Cocke lorellys bote
Callyng al knaues, to pull therin a rope.

[1] A .i. back.

[1]¶ The preface, or the pystle.

TO the ryght worshypfulle (Barnarde Barber), dwellynge in. Banberye, Collyn Clowte surrendreth gretynge, with immortall thankes.

IT was so, worshypful syr, that at my last beynge in Mountpyllour, I chaunsed to be assocyat with a doctor of Physyke / which at his retorne had set forth .iij. Bokes to be prynted in Fleetstrete, within Temple barre, the whiche Bokes were compyled togyther in one volume named the Introductorie of knowledge / whervpon, there dyd not resort only vnto hym marchauntes, gentylmen, and wymen / but also knyghtes, and other great men, whiche were desyrous to knowe the effycacyte, and the effecte of his aforesayd bokes; and so, amonge many thynges, they desyred to knowe his fansye consernynge the werynge of Berdes / He answeryd by great experyence: "Some wyl weer berdes bycause theyr faces be pocky, maun[2]gy, sausflewme[3], lyporous, & dysfygured / by the whiche many clene men were infected."[4] So, this done, he desyred euery man to be contentyd: Vvherfore I desyre no man to be dyspleasyd with me. And where-as he was anymatyd to wryte his boke to thende, that great men may laugh therat[5]/ I haue deuysed this answere, to the entent, that in the redyng they myght laughe vs bothe to scorne / And for that cause I wrote this boke, as god knoweth my pretence / who euer keape youre maystershyp in helthe.

[1] sign. A .ij. [2] sign. A .ij. back. [3] See *Forewords*, p. 101.

[4] Speaking of *matters trifelyng*, Wilson, in his *Art of Rhetorique*, 1553 (edit. 1584, p. 8), says: "Suche are triflyng causes when there is no weight in them, as if one should phantasie to praise a Goose before any other Beast liuyng (as I knowe who did) or of fruite to commende Nuttes cheefly, as *Ouid* did, or the Feuer quartaine as *Phaciosinus* did, or the Gnat, as *Virgill* did, or the battaile of Frogges, as *Homer* did, or dispraise beardes, or commende shauen hiddes."—W. C. Hazlitt.

[5] See the Preface to the *Dyetary*, p. 228 above.

¶ Here foloweth a treatyse, made Answerynge the treatyse of doctor Borde vpon Berdes.

CAllynge to remembraunce your notable reproche gyue*n* vnto berdes,[1] I was co*n*strayned to render the occasion therof; wherupon, I founde by longe surmyse and studye that ye had red the storye of Hellogobalus, & founde therin greate and stronge auctoryties / which by lykelyhode mouyd you to this [2] Reformacyon of berdes. For ye knowe that Hellogobalus, beynge gyuen moche to the desyre of the body, & that by moche superfluyte, he[3] thought it requysyght to co*mm*yt the fylthy sy*nn*e of leche[r]y, vpon the receypt*es* of delycate meat*es*. For he caused his cokes to make &

[1] Mr Hazlitt says, 'See Grapaldus *de Partibus Ovium*, and Collier's *Extr. Reg. Stat. Co.* ii. 97.' At the latter reference, 22 Sept. 1579, is, 'H. Denham, Lycenced unto him &c, A paradox, provinge by Reason and example that Baldnes is much better then bushie heare . . vj^{d}.' (Written by Synesius, englished by Abraham Fleming.) After this, Mr Collier prints, from a MS of his own, he says, an amusing dialogue between B[aldness] and H[air], entitled the 'Defence of a Bald Head.' B. argues that baldness is no sign of old age, as many young men are bald from too much wenching;

Then, thinke also of this :
  if you no haire have gott,
How pleasantly your haire you misse,
  when weather it is hot.
Let ruffins weare a bushe,
  and sweat till well nigh dead,
In that Ime bald, I care no rush,
  but onely wipe my head.

Hair ends with

Thy reasons may be good,
  that baldnes is no ill ;
But ladies will love lustie blood
  and haire, say what you will.

[2] A .iij. not signed.

[3] *orig.* yo

ordeyne suche hote meates that maye prouoke or stere hym the rather therunto. And *in* ther so doyng, he made them, some of his preuye chambre, some of his hed lordes of his counsell. But yet the chefe and pryncypall preseptes that he gaue vnto his cok*es*, was this, that they shulde not only polle theyr hedes, but also shaue theyr berdes. For this entente, that when he were dronkyn, or vometynge rype by takyng excesse, that he myghte be well assuryd, that it came not by no heer of from his cokes heddes. For his delyght was not onely in the feminyne kynde / but also delyghted in womenly men / yet he and his fyne vnberdyd faces ledde not onely a vycyous lyfe, but also made a shameful ende. Notwithstandynge other, that, or this storye folowynge, was and is the occasyon why ye [1]abore berdes, and that was this: at your laste beynge in Mowntpyllyer, Martyn the surgyen beyng there with you, & dyd accompany dayly with none so moch as with you: yf ye be reme*m*bred, he brought you to dyner vpon a daye to one Hans Smormowthes house, a Duche ma*n*, in whiche house you were cupshote[2], otherwyse called dronkyn, at whiche tyme your berde was lo*n*ge / so then your assocyat Martyn brought you to bed / and with the remouyng, your stomake tornyd, & so ye vometyd in his bosome; howbeit, as moche as your berde myghte holde, vpon youre berde remayned tyll the next daye in the morenyng. And when ye waked, & smelt your owne berde, ye fel to it a fresshe; and callynge for your frende Martyn, shewynge[3] the cause of this laste myschaunce. Wherupon ye desyred to shaue you. And so, when ye sawe your berde, ye sayd that it was a shamfull thynge on any mans face. And so it is in suche cases, I not denye / yet shall ye consyder, that our Englysshe men, beynge in Englande, dothe vse to kepe theyr berdes moche more clen

[*leaf* A .iv. *is lost.*]

[1] A .iii. back. [2] See p. 156, note. [3] ? shewed hym.

[*leaf* A .iv. *is lost.*]

[sign. B .i.]

As longe as any berdes be worne,
Mockynge shall not be forborne;
But yet at length, his is the scorne.
I fere it not.

Andrew Boorde hates bearded men because he once made his own beard stink.

¶ With berdyd men he wyll not drynke,
Bycause it doth in theyr berdes synke;
The cause therof, ye may soone thynke,
His berde in Flaunders ones dyd stynke,
Whiche by dystulacyon
Of a vomytacyon
Made suche dysturbacyon,
That it abored the nacyon.
I fere it not.

Boorde lookt like a fool when he got drunk.

¶ Some berd*es*, he saith, doth grow a pace,
To hyde an euyll coleryd face;
In fayth, his had an homlye grace,
When he was in that dronkyn case.
But sythe he doth this matter stere,
To make that shauynge shuld be dere,
I thynke it doth full well appere,
That foles had neuer lesse wyt in a yere.
I fere it not.

Boorde says a beard will breed care.

¶ A berde, sayth he, wyl breyd moch care,
If that he with his mayster compare.
Here may ye proue a wyt full bare
That iudgeth so a man to fare.
[sign. B .i. back]
What ma*n* lyuyng, I wold fayne knowe,
That for comp*ar*ason let*es* his berde growe?

He's a spiteful shrew.

But yet, though that a spyghtfull shrow
His spyghtfull wordes abrode doth blow,
I fere it not, &c.

¶ Of berdes, he sayth, ther co*m*ms no gaynes,
& berd*es* quycknyth not the braynes.
Lo, how in Physyke he taketh paynes!
He merytes a busshel of brwers[1] graynes!
He warneth also euery estate
To auoyde berdes, for fere of debate.
If men, lyke hym, shuld vse to prate,
His warnyng then shuld come to late,
I fere it not.

Boorde says beards don't quicken the brains, and do raise quarrels.

¶ If berdes, also, a purse doth pycke,
As ye compare them to be lyke,
yet ye haue gotte more in one wycke,
Then berdes in .x. togyther may stryke.
For by castynge of a pyspotte,
ye haue pollyd many a grote;
yea, and moche more, God wotte,
By falshede ye haue gotte.
I fere it not.

Oh, Andrew, you've cheated men of many a groat by looking at their urine, and by falsehood!

¶ Yet one thynge more, I wyll assayle:
The dau*n*ger of drynkyng ye do bewayle[2].
Beleue ye me, yf all do fayle,
In stede of a cup, ye shall haue a payle;
For you haue gyuen warnynge playne,
That berdyd men shall be full fayne
To brynge a cup, for theyr owne gayne,—
The more fole you, so to dysdayne!
I fere it not.

You've warned men against drinking, and told bearded men to bring a cup for themselves.

¶ Note me well, for it is trewe,
Thoughe berdyd men ye wyll eschewe,
There be moche honyster men than you,
That wyl drynke long, or they do spewe

Some bearded men are more worthy than you, and don't spue, like you.

[1] *so.* [2] See Boorde on Drunkenness, p. 90, above.

As you haue done, I knowe, or this.
wherfore I say, though so it is,
I wyll not tell that is amys;
yet wyll I tell some trewyth yewys[1].
I fere it not.

Boorde, you say that a Beard heats a man.

¶ yet of one thynge that ye do treate,
Howe that a berde, in a great swete,
By lyke doth catche a k[n]auysshe[2] hete:
Therby ye do a grete prayse gete,
For trewely vnfayned,

But your honour is stained.

Your honyste is dystayned;
All though ye haue dysdayned,
Men knowe ye haue sustayned.
I fere it not.

You tell men not to drink when their noses run.

¶ Though in the wynter a dew wyl lye,
That dystylleth from the nose pryuelye;
To refrayne your cup ye pray then hartly;

sign. B .ii. back]

And all is for superfluous glotonye.
For glotony is of suche a kynde,
That ende of excesse he can none fynde,

You've lost wit through gluttony.

Tyll past is both the wyt and mynde;
So one of those ye be assynde.
I fere it not.

[1] *gewis*, certainly.
[2] See l. 156.

The seconde parte

of that songe.

I Lytell thought, ye were so wyse,
Berdes to deuyse of the new guyse;
But truely, for your enterpryse,
ye may go cast your wyt at dyse.
At syncke or syse, whiche so doth fall,
Fere ye not to cast at all;
For yf you lose, your lostes be small:
It is to dere, a tenys ball;
I fere it not.

Boorde, with your new-fashioned beards,

your wit's like a tennis-ball.

¶ A berde vpon his ouer lyppe,
ye saye wyll be a proper tryppe,
Wherby ye shall the better skyppe.
Go your wayes, I dare let you slyppe,
Where as be many more,
I thynke, by .xx. score,
In cocke lorelles bote, before
ye maye take an ore.
I fere it not.

Boorde, begone, you poor fool,

and row low down in Cock Lorell's boat! [B .iii. not signed]

¶ Yet though that ye one thing do craue,
Which is, a muster deuyles berde to haue,
ye make me study, so God me saue!
If this peticion came not of a knaue,
Perhapes some other man dyd make it,
And so ye dyd vp take it;
But best ye were forsake it,
For fere of Pears go nakyt.
Nowe fere you that!

You want a kind of Devil's beard, do you?

Beware of Piers Go-naked.

You say beards hide little brains,

¶ ye say some berdes be lyke la*m*bes woll,
With lytell wyt within theyr skull:
'Who goth a myle to sucke a bull,[1]
Comes home a fole, and yet not full.'

and want magpies to pull our hairs out.

And where ye wyshe the*m* pekt with pyes,
That weres a berde, vnto theyr iyes:
Be wyse, take hede! suche homely spyes

You tell crafty lies.

Oftymes can spye your crafty lyes.
I fere it not.

Pray, Andrew, didn't God make Adam a beard?

¶ But, syr, I praye you, yf you tell can,
Declare to me, when God made man,
(I meane by our forefather Adām)
Whyther that he had a berde than;

If He did, who shaved him?

And yf he had, who dyd hym shaue,
Syth that a barber he coulde not haue.

[B .iii. back]

Well, then, ye proue hym there a knaue,
Bycause his berde he dyd so saue.
I fere it not.

Didn't Christ and His Apostles have beards?

¶ Christ & his apostles, ye haue declaryd,
That theyr berdes myght not be sparyd,
Nor to theyr berdes no berdes co*m*paryd:
Trewe it is, yet we repayryd
By his vocacion, to folowe in generall

And we ought to follow them.

His disciples, both great and small;
And folowyng ther vse, we shuld not fal,
Nothynge exceptynge our berdes at all.
I fere it not.

Sampson, and thousands of old philosophers, wouldn't be shaved.

¶ Sa*m*pson, with many thousandes more
Of auncient phylosophers, full great store,
Wolde not be shauen, to dye therfore;
Why shulde you, then, repyne so sore?

We should imitate them.

A[d]myt that men doth Imytate
Thynges of antyquite, and noble state,

[1] Waltom's calf, says the proverb, did this.

Such counterfeat thinges oftymes do mytygate
Moche ernest yre and debate.
I fere it not.

¶ Therfore, to cease, I thynke be best;
For berdyd men wolde lyue in rest.
you proue yourselfe a homly gest,
So folysshely to rayle and iest;
For if I wolde go make in ryme,
Howe new shauyd men loke lyke scraped swyne,
& so rayle forth, from tyme to tyme,
A knauysshe laude then shulde be myne:
I fere it not.

Bearded men like peace. You're a noodle to rail against them.

[B.iv. not signed] I won't tell you how shaved men look like scraped swine.

¶ What shulde auayle me to do so,
yf I shulde teache howe men shulde go,
Thynkynge my wyt moche better, lo,
Then any other, frende or fo?
I myght be imputed trewly
For a foole, that doth gloryfye
In my nowne selfe onelye;
I thynke you wyll it veryfye:
I fere it not.

What'll be the good of it?

I don't want to show off.

And thus farewel, though I do wryght
To answere for berdes, by reason ryght;
yet vnberdyd men I do not spyght,
Though ye on berdes therin delyght.
And in concludynge of this thynge,
I praye God saue our noble kynge!
Berdes & vnberdyd, to heuen vs brynge,
Where as is Ioye euerlastynge!
I fere it not, &c.

Tho' I defend beards, I don't spite unbearded men.

God save the King! and bring us all, beards and no beards, to Heaven!

¶ Finis.

[B iv, back]

Barnes in the De=fence of the Berde.

Galyen prynce,

of Physycke.

BArnes, I say, yf thou be shent,
Bycause thou wantyst eloquence, — *If my rimes are bad,*
Desyre them, that thyne entent
May stonde all tymes for thy defence,
Consyderynge that thy hole pretence — *think that my wish is to stop quarrelling.*
Was more desyrous of vnyte
Then to enuent curyosyte.

R W

¶ Ad imprimendum solum.

HINDWORDS.

THIS term *Hindwords* is Mr David Laing's; and I gladly adopt it, as it's so much better than the *Post-Præfatio* of Mr W. C. Hazlitt in his Handbook, and of divers other folk.

After the extracts in the Forewords, p. 74—104, from Boorde's *Breuyary*, showing his opinions there, it seems to me now that I ought to have stated some of his opinions in his *Introduction* and *Dyetary* before summing up his character on p. 105. I therefore do this here; better late than never.

Boorde believes in 'the noble realme of England' (p. 116, 144), and, though he reproaches his countrymen for their absurd love of new fashions in dress, and for the treason among them (p. 119), he yet holds that 'the people of England be as good as any people in any other lande and nacion *that* euer I haue trauayled in, yea, and much more better in many thynges, specially in maners & manhod. As for the noble fartyle cou*n*trey of England, hath no regyon lyke it.' So also London is the noblest city in any region, and has the fairest bridge: 'in al the worlde there is none lyke' (p. 119). But Cornish ale Boorde thinks very bad (p. 123). In Wales he notices the people's love of toasted cheese, and that their voices and harps are like the buzzing of a bumble-bee (p. 126), the people very rude and beastly, very fond of the devil in their speech, of selling their produce a year before it comes (p. 127), and of lechery (p. 128). The custom of 'bundling' probably prevailed there; and the priests also increased the population.

The wild Irish, Boorde describes as very rude and wrathful, men and women lying together in mantles and straw (p. 132-3); but among those in the English Pale, which is a good country, Boorde found as faithful and good men as ever he knew (p. 133). The Scotch, among whom Boorde had lived, he didn't much like: they bragged and lied; and either naturally, or from a devilish disposition, didn't love Englishmen, though they resembled the latter in being hardy and strong, well-favoured, and good musicians (p. 137). With Boorde's description of Iceland (p. 141) my friend, Mr Guðbrandr Vigfusson, is much amused, but does not believe in it. Boorde liked Calais, and Flanders (p. 147), though the Flemings were—like the Dutch (p. 149)—great drinkers, and also eat frogs' loins, and toadstools (p. 147), and sold brood mares to England. The church-spire and meat-shambles of Antwerp he thought fine (p. 151); and the Julich (or Juliers) custom of plucking their geese yearly, curious (p. 154). Cologne he calls a noble city, the Rhine a fair water, and its wine good; but the people he found very drunken (p. 156), though many were virtuous and full of alms-deeds (p. 157). The Germans were rude and rustical, eat cheese-maggots, gave their maidens only water to drink (p. 160), and had snow on their mountains in summer (p. 161). Denmark, Boorde found such a poor country, that he couldn't make out how it (and little Saxony, p. 164) came to win England (p. 163). The Bohemians he thought heretics, and they didn't eat ducks (p. 167). The Poles were poor, eat honey, and didn't like wax (p. 168). Hungary was partly in the hands of the Turks, and was full of aliens (p. 170). Greece was Turkish; its capital, Constantinople, and its St Sophia's the fairest cathedral in the world, with a wonderful *sight*[1] of priests (p. 172). Of Sicily, the biting flies (or mosquitoes) Boorde noticed (p. 176); of Naples, the laziness and the hot wells (p. 177); of Italy, the fertility, the noble river Tiber, the fallen St Peter's at Rome, and the abominable vices in the city (p. 178). Venice, Boorde thought the beauty of the world; and he saw no poverty there, but all riches (p. 181-5). The Lombards he found crafty, eaters of adders and frogs, and having spiteful cur-dogs that would bite your legs.

[1] The phrase wasn't slang then.

The Lombards also ploughed with only two oxen, which they covered with canvas, against the flies (p. 187). Genoa was a noble city in a fertile land (p. 189). France a noble country, with Paris and four other universities; but the French had no fancy for Englishmen; they set the fashion to all nations (p. 190-1). They alone, and the English, to Boorde's great disgust, were always changing their dress; every other nation kept to its old apparel. Aquitaine was the cheapest country in the world, and Montpelier the noblest medical university (p. 193-4). The Portuguese were seafarers, and their girls cropt their polls (like the Spanish women), but left a rim of it like a barefoot friar's (p. 197). Spain was a sadly poor place; no good food, wine in goat-skins, hogs under your feet at table, and lice in your bed (p. 198-9). In Castille, &c., the people stupidly called on their dead friends to come to life again (p. 200). Boorde's pilgrimage to, and abode in, Compostella we have noticed above (p. 51); thieves, hunger, and cold, were his foes on it (p. 206). At Bordeaux was the greatest pair of organs in the world, with Vices, giants' heads, &c., that wagged their jaws and eyes as the player played (p. 207). Normandy was a pleasant country, and its people gentle: it and all France really belonged to England (p. 208). Latin was spoken over all Europe (p. 210).

From Barbary, slaves were sold to Europe, and left to die unburied (p. 212). Turkey was a cheap and plentiful country, under the law of Mahomet, whose tricks Boorde shows-up (p. 214-16). Judæa is a fertile land; and Boorde gives full instructions to persons intending to make a pilgrimage to Jerusalem, and describes shortly the Holy Sepulchre (p. 219-20).

In his *Dyetary*, Boorde tells his contemporaries how to choose sites for their houses, how to arrange their buildings, spend their incomes, govern their households, manage their bodies; and what flesh, fish, vegetables, and fruits, are good to eat. The two passages that I specially call attention to are those on the site and plan of a Tudor mansion, p. 238-9, and on what a man should do before going to bed and on rising, p. 246-8. They enable you to realize well the surroundings and life of an English gentleman of Henry VIII's time. The bits on Ale and Beer (p. 256); on bad cooks and brewers, and rascally bakers (p. 260-1); and on Venison (p. 274-5), are also very characteristic.

Our good friend at Manchester, Mr John Leigh, Officer of Health to the Corporation of the town, has been kind enough to read through the Forewords and Boorde's Dyetary, and to send me some notes on the former, which will be found further on, and the following high opinion of Boorde and his Dyetary, which will, I hope, give the reader as much pleasure as it has given me:—

"Either the man was far beyond his time, or the men of the time were better informed than we have given them credit for. How a man who wrote so gravely, and exhibited in his writings such clear sound sense, could have been taken for a 'Merrie Andrewe,' passes one's conception.

"I have carefully read through the *Dyetary*. The first ten chapters are admirable; indeed, the third chapter so thoroughly comprehends all that sanitary reformers have been teaching for the last 20 years, that it is difficult to say that we have made any advance upon it. Certainly, until quite recently, the knowledge of Englishmen on all sanitary matters connected with the surroundings of a house, must have retrograded since Boorde wrote. Nothing can be better than the advice he gives as to the situation of a house, the soil on which it should be erected, the placing of the outbuildings, the avoidance of stagnant water, &c., and the means to be taken to secure a pure atmosphere. The advice given throughout the remaining seven chapters, how to procure and to retain good health, is not surpassed in quality in any book of modern times. It is not necessary to select any special passage where all is good.

"The remaining chapters of the book on special diets are all coloured by the peculiar doctrines of Boorde's time; but, setting those aside, the advice he gives is good. He specifies the articles of diet which are, as determined by long experience, difficult of digestion, or which produce flatulence; whilst such elements of diet as are laxative, diuretic, stimulant to special organs, &c., he points out, albeit there is sometimes a little fancy about the latter.

"Like a sensible man, however, he sums up in his thirty-ninth chapter what it is necessary that a man should do to preserve his health, making much of that depend upon his own experience and common sense. The perusal of the *Dyetary* is calculated to give a medical reader a high opinion of Boorde's sound good sense and powers of observation. I think you have done good service in reprinting the *Dyetary*, and that you will thereby have corrected some erroneous impressions as to the knowledge of the time on sanitary matters."

A man must dwell at elbow-room, says Boorde (p. 233), having water and wood annexed to his house; he must have a fair prospect to and from it, or he'd better not build a house at all (p. 234); he must have pure air round it, and nothing stinking near it (p. 235-7), and must provide, before he begins, all things needful to finish it; for 'there goeth to buyldynge many a nayle, many pynnes, many lathes, and many tyles or slates or strawes, besyde tymber, bordes, lyme, sand, stones or brycke,' &c. (p. 237). Don't front your house to the South, but don't be afraid of the East, as 'the Eest wynde is temperate, fryske, and fragraunt,'—witness Charles Kingsley;—arrange your buildings on my plan in pages 238-9, and have a park, a pair of butts, and a bowling-alley, near them. Provide food and necessaries beforehand (p. 240); divide your income into three parts, 1. for food; 2. for dress, wages, and alms; 3. for emergencies (p. 241); fear God, and make your household do so too, specially punishing swearing (p. 243). Sleep moderately (p. 245), and not during the day; be merry before bed-time, sleep on your side, wear a scarlet night-cap, and have a quilt over you (p. 247); air your breeches in the morning; wash, pray, take exercise, and eat two meals a day (p. 248). Wear a lambskin jacket in winter, and a scarlet petycote in summer (p. 249). Don't stuff (p. 250). Abstinence is the best medicine (p. 251). Only sit an hour at dinner: Englishmen sit too long, and stupidly eat heavy dishes first (p. 252). *Don't drink water* (p. 252-3), except it's mixed with wine (p. 254). In Germany, maidens drink water only; prostitutes drink wine. Abroad there's a fountain in everv town (p. 254).

'Ale for an Englysshe man is a naturall drynke. . . Bere is a naturall drynke for a Dutche man; and nowe of late dayes it is moche used in Englande, to the detryment of many Englysshe men' (p. 256). Cider does little harm in harvest-time; metheglin, fined, is better than mead (p. 257). Bread is best when unleavened and without bran. In Rome the loaves are saffroned, and little bigger than a walnut (p. 258). Rascally bakers I should like to stand in the Thames up to their eyes (p. 261). Potage is more used in England than anywhere else in Christendom (p. 262). Almonds comfort the breast, and mollify the belly (p. 263). Don't mind what old authors say, if experience contradicts them (p. 264). No eggs but hen's are used in England (p. 264); in Turkey they pickle hard eggs (p. 265). Dutchmen eat butter at all times in the day, which I think bad (p. 265). In High Almayne the Germans eat cheese-maggots like we do comfits (p. 267). Milk is not good for those who have grumbling in the belly; strawberries and cream may put men in jeopardy of their lives (p. 267). England is supplied better with fish than any other land (p. 268); but you musn't eat fish and flesh at the same meal (p. 269). A pheasant 's the best wild fowl, and a capon the best tame one (p. 269-70). All small birds are good eating (p. 270). Young beef is good for an Englishman (p. 271); mutton and pork I don't like. In England swine eat stercorous matter, and lie in filth, though in Germany and abroad (except in Spain) they have a swim once or twice a day (p. 272). Jews and Turks hate pork, but will eat adders as well as any Christian in Rome will (p. 273). Bacon 's only good for carters and ploughmen. Brawn 's a usual winter meat in England. Nowhere are hart and hind loved as in England. Doctors tell us that venison is bad for us; but I say it 's a lord's dish: let the doctors take the skin! give me the flesh! (p. 274-5). Let dogs eat hares; don't you (p. 275). Rabbits, sucking ones, are the best wild beasts' flesh (p. 275). At Montpelier they have boiled meat for dinner, roast for supper (p. 277). A good Cook is half a physician. Onions make a man's appetite good, and put away fastidiousness (p. 279). Artichokes' heads and sorrel are good (p. 280-1). 'There is no Herb nor Weede, but God haue gyven vertue to them, to helpe man' (p. 282). Strawberries are praised above all berries; filberts are better than hazle-nuts (p. 283); peas and beans fill a man with wind; roast apples comfort the stomach (p. 284). Olives and oranges provoke appetite; black pepper makes a man lean (p. 285-6). Then I give you diets for Sanguine, Phlegmatic, Choleric, and Melancholy folk (p. 287-9), tell you how to treat Pestilence (p. 289-91), Fever or Ague (p. 291-2), the Iliac, Colic, and Stone (p. 292), Gout, Leprosy (p. 293), Epilepsy (p. 294), Pain in the Head (p. 295), Consumption (p. 296), Asthma, Palsy (p. 297), and Lunatics (p. 298). Hardly, these last: keep 'em in the dark, shave their heads once a month, and use few words to them. Lastly, I treat

Dropsy (p. 299); give general directions on Diet to all people (p. 300); and then tell you how to arrange a sick-bed, a death-bed, urging all to make their peace with God (p. 300-1).

Two quaint and jolly books these are; and if readers are not obliged to me for reprinting them, they ought to be.

On the state of England at Boorde's time, I refer the reader to my *Ballads from Manuscripts* for the Ballad Society, Part I, 1868, 'Poems and Ballads on the Condition of England in Henry VIII's and Edward VI's Reigns;' Part II, 1871, these continued, with Poems against Cromwell, on Anne Boleyn, &c. The contemporary complaints give a very different view of the state of affairs to Mr Froude's *couleur-de-rose* picture. Of early books on the countries of Europe, I know only the *Libel of English Policy*, A.D. 1436, in Mr T. Wright's Political Songs, vol. ii. 1861, and the descriptions, not the history, in Thomas's very interesting *Historye of Italye*, 1561. Both of these I have quoted largely. George North's 'Description of Swedland, Gotland, and Finland. Imprinted at London by Jhon Awdeley, 1561, 4to, 28 leaves, with the Lord's Prayer in Swedish at the end' (*Hazlitt's Handbook*), I don't know. The Russia of Fletcher, and Horsey, Boorde does not touch.

Sprüner's Reformation Map of Europe in the middle of the 16th century, No. VII, in his Historical Atlas, is the best to use for Boorde's *Introduction*. In it, Syria is part of the *Osmannisches Reich*, Turkey in Europe and Asia, and that may account for Boorde treating it as in Europe. For the dress of the inhabitants of the different countries, recourse may be had to the *Recueil de la Diversité des Habits*, Paris, 1562, 8vo, from which Upcott had his Scotchman and Frenchman cut on wood for his reprint of Boorde's *Introduction* in 1814, chap. iv. sign. G ii, chap. xxvii., sign. T.

In conclusion, I have to thank Mr John W. Praed for his help (obtained by Miss C. M. Yonge's kind offices) in Boorde's Cornish dialogue; Dr B. Davies for help in the Welsh; Mr F. W. Cosens and Mr H. H. Gibbs for help in the Spanish; Professor Cassal for help in the French; and Prof. Rieu in the Arabic; also a German officer of the Coin Department in the British Museum (with very little time to spare) for explanations of the names of a few coins.

To Mr Henry Bradshaw, Librarian of the University of Cambridge, I am much indebted for help in the bibliography of Boorde's books, and to his friend, Mr Hollingworth, Fellow of King's, and curate of Cuckfield, for a very pleasant day's entertainment and walk near Andrew Boorde's birthplace.

19*th Sept.*, 1870.

One of Andrew Boorde's phrases, "good felowes the whyche wyll *drynke all out*," p. 151, l. 6, receives illustration from an unexpected source, namely, an English translation in 1576 A.D. of the famous *Galateo* of Della Casa, written about 1550 A.D., and so amusingly sketched for us from the original Italian by our good friend Mr W. M. Rossetti, at the end of his essay on Italian Courtesy Books in Part II, p. 66—76, of the Society's *Queene Elizabethes Achademy*, &c., 1869. Neither he nor I knew at that time of the existence of this translation, though it was entered in Bohn's Lowndes, with others in 1703, and 1774:—

"Galateo of Maister *Iohn Della Casa*, Archebishop of Beneuenta. Or rather, A treatise of the manners and behauiours, it behoueth a man to vse and eschewe, in his familiar conuersation. A worke very necessary & profitable for all Gentlemen, or other. First written in the Italian tongue, and now done into English by Robert Peterson, of Lincolnes Inne Gentleman. *Satis, si sapienter.* Imprinted at London for Raufe Newbery dwelling in Fleetestreate a little aboue the Conduit. An. Do. 1576." black letter 4to, leaves, *A* in 4, *g* in 2, B, C, D, E, F, G, H, I, K, L, M, N, O, P, Q, in fours, with a leaf of errata and verse.

On leaf 115 is this passage:

"Now, to *drink all out* euery man—which is a fashion as litle in vse amongst vs, as *the* terme it selfe is barbarous & straunge: I meane, *Ick bring you*:—is sure a foule thing of it selfe, & in our countrie [Italy, ab. 1550 A.D.] so coldly accepted yet, *that* we must not go about to bring it in for a fashion."

The *Swearing*, of which Boorde complains so much in pages 82, 243, was also complained of by Robert of Brunne in 1303 A.D.; but then the gentry were the chief sinners in this way, and 'every gadling not worth a pear taketh example by you to swear.' *Handlyng Synne*, p. 23-7

NOTES.

I. ON THE *FOREWORDS*.

p. 21. *Agues . . . be infectiouse.* Although at this day medical men are disposed to extend the list of communicable diseases, they have not yet come to regard the agues as amongst them.—John Leigh.

p. 25. *Pronosticacions.*—An amusing instance of how some people believed in prognostications and astronomers' prophecies in Boorde's days, is told by Hall:—

"In this yere [1524 A.D.], through bookes of Emphymerydes and Pronostications made and calculate by Astronomers, the people were sore affrayde; for the sayd writers declared that this yere should be suche Eclipses in watery signes, and suche coniunctions, that by waters & fluddes many people should perishe, Insomuche that many persones vitailed them selfes, and went to high groundes for feare of drounyng; and specially, one Bolton, which was Prior of sainct Bartholemewes in Smythfeld, builded him an house vpon Harow of the hill, only for feare of this flud; and thether he went, and made prouision for all thinges necessarye within him, for the space of two monethes: But the faythfull people put their trust and confidence onely in God. And this raine was by the wryters pronosticate to be in February; wherfore, when it began to raine in February, the people wer muche afrayd; & some sayd, 'now it beginneth:' but many wisemen whiche thought that the worlde could not be drouned againe, contrary to Goddes promise, put their trust in him onely; but because they thought that some great raines might fall by enclinacions of the starres, and that water milles might stand styll, and not grinde, they prouided for meale; and yet, God be thanked, there was not a fairer season in many yeres; & at the last, the Astronomers, for their excuse, said that in their computacion they had mistaken and miscounted in their nomber an hundreth yeres."—*Hall's Chronicle*, p. 675, ed. 1809.

p. 28. *Gotham and Nottingham.* Nearer hand [nearer to Nottingham Castle than Belvoir Castle was], within three miles, I saw the ancient Towne of *Gotham*, famous for the seven sages (or Wise men) who are fabulously reported to live there in former ages. (1639. John Taylor, *Part of this Summers Travels*, p. 12.)

p. 59. *Trust yow no Skot!* "As there are many sundry Nations, so are there as many inclinations: the Russian, Polonian, German, Belgian, are excellent in the Art of Drinking; the Spaniard will Wench it; the Italian is revengefull; the French man is for fashions; the Irish man, Usquebaugh makes him light heel'd; the Welsh mans Cowss-boby works (by infusion) to his fingers ends, and translates them into the nature of lime-twigs; and it is said, that *a Scot will prove false to his Father, and dissemble with his Brother;* but for an English man, he is so cleare from any of these Vices, that he is perfectly exquisite, and excellently indued with all those noble abovesaid exercises." 1652. John Taylor, *Christmas in & out*, p. 9.

p. 64. *Boorde holding land.* The statute 31 Henry VIII, chapter vi, (A.D. 1539) enabled "all . . . Religiouse persons . . to purchase to them and their heires . . . landes . . and other hereditaments . . as thoughe they . . had never bene professed nor entred into any suche religion." This Act also enabled them to sue and be sued, but provided that not "anye of the saide religiouse persons, beinge Priestes, or suche as have vowed religion att twenty one yeres or above, and therto then consented, continuynge in the same any while after, not duly provinge . . some unlaufull cohercion or compulsion . . . be enhabled by . . this Acte . . to marie or take any wief or wyves."

p. 71. *Mr J. P. Collier's inaccuracy.* I believe that among persons who have followed Mr Collier, only one opinion prevails as to his accuracy. While I write, comes an unsought testimony on the point from a conscientious editor; "*King Iohan* as edited by Mr Collier so *swarms* with blunders, that I regard it as just so much waste paper. The late J— B— (good man and true) sent me his copy of Mr C.'s *Iohan*, and every page is speckled with his corrections. I'm sorry to say this is no new thing in following and *testing* Mr Collier."

p. 72. *The sycknes of the prisons.* Boorde has anticipated Howard and other samaritans in announcing that "this infirmitie doth come of the corruption of the ayer," &c. As prisons are now kept, medical men have little opportunity of seeing the special forms of disease referred to by Boorde. They do, however, meet with cases simulating *carcinoma*, in badly-ventilated private houses, which recover on removal to more healthy localities.—John Leigh.

p. 75, 256. *Ale.* I call to minde the vigorous spirit of the Buttry, Nappy, Nut-browne, Berry-browne, Ale Abelendo, whose infusion and inspiration was wont to have such Aleaborate operation to elevate & exhillerate the vitals, to put alementall Raptures and Enthusiams in the most capitall Perricranion, in such Plenitude, that the meanest and most illiterate Plowjogger could speedily play the Rhetorician, and speak alequently, as if he were mounted up into the Aletitude. 1652. John Taylor, *Christmas in & out*, p. 14.

p. 75, 255. *Wines.* See a long list of wines in "Colyn Blowbols Testament" (? 1475-1500 A.D.), printed in Halliwell's *Nugæ Poeticæ*, 1844, and Hazlitt's *Early Popular Poetry*, i. 106, lines 324-341 (line 7 or 8 of the poem is left out); and in "The Squyr of Lowe Degre," l. 753-762, *E. Pop. P.* ii. 51. *Alicant wine*, so called from *Alicante*, the chief Town of *Mursia* in *Spain*, where great store of Mulberries grow, the juyce whereof makes the true *Alicant* Wine.—*Blount's Glossographia.*

p. 78-9. An excellent description of Nightmare and of its causes and remedies. Nothing can be better than the advice. It is honestly worth a guinea even now.

Query, Is the use of 'Saynt Iohns worte' (commonly placed by maidens under their pillows on St John's eve in former times, and in some districts even now, that they may dream of their sweethearts,) adopted on the Hahnemanian principle, that what will cure a disease will produce it?—John Leigh.

p. 79. If the general advice for the cure of *Cachexia* be followed, the treatment by 'Confection of Alkengi' may be safely omitted.—J. L.

p. 80, 271. *Martinmas beef.* "In a hole in the same Rock was three Barrels of nappy liquour; thither the Keeper brought a good Red-Deere Pye, cold roast Mutton, and an excellent shooing-horn of hang'd *Martimas* Biefe." (1639. John Taylor, *Part of this Summers Travels*, p. 26.)

p. 80. *Symnelles.* At Bury in Lancashire, 'Symnell Sunday' is a great day; and rich cakes are prepared for it, containing currants, raisins, candied lemon, almonds, and other ingredients.

In the prescription for Stone, the Broom seeds, parsley seeds, saxifrage (*Saxifraga granulata*), and Gromel seed (those of *Lithospermum arvense*) are all excellent diuretics.—J. L.

p. 81. It is rather an exaggeration to say that "touchynge the contentes of vrines, experte physicions maye knowe the infyrmyties of a pacient *unfallybly*"; but certainly, the careful examination of the contents by the "experte phisicions" of modern times has marvellously increased their knowledge of many diseases.—J. L.

p. 82. "*Impetigo*" is now known to be a fungoid growth, and not a worm.—J. L.

p. 94. The farrago of remedies for the treatment of wounds is now all cast aside. The proper treatment is all contained in Boorde's first two lines of "remedy."—J. L.

p. 97. Boorde's treatment of Tertian Fever not unlikely brought the latter into the category of infectious diseases.—J. L.

p. 97. '*Boorde's treatment of Scurf.*' With the omission of the mercury, we have here a very good sulphur ointment, the free application of which would render the cultivation of the nails unnecessary.—J. L.

p. 99. '*Boorde's cure for asthma.*' The treatment consists in the administration of antispasmodics and expectorants, and the avoidance of such articles of diet as produce flatulence.—J. L.

p. 99. '*Loch de pino.*' In the "Niewe Herball or Histōrie ot Plantes, &c., first set foorth in the Doutche or Almaigne tongue by that learned D. Rembert Dodoens, &c., and nowe first translated out of French into English by Henry Lyte, Esquyer, 1578," it is stated in the description of the virtues of the Pine: "The Kernels of the Nuttes which are founde in the Pine apples are good for the lunges, they clense the breast, and cause the fleme to be spet out; also they nourish wel, and ingender good blood, and for this cause they be good for suche as have the cough."—John Leigh.

p. 99. '*Pylles of Agarycke.*' Dodoens also says, "there groweth on the larche tree a kinde of Mushrome or Tadstoole, that is to say, a fungeuse excrescence called *Agaricus* or *Agarick*, the whiche is a precious medicine, and of great vertue. The best *Agarick* is that which is whitest, very light and open or spongious. . . . *Agarick* is good against the shortnesse of breath called Asthma; *the* hard continuall cough or inveterate cough. . . . Taken about the weight of a Dramme, it purgeth the belly from colde slimie fleme, and other grosse and raw humours which charge and stoppe the brayne, the sinewes, the lunges, the breast, the stomach, the liver, the splene, the kidneyes, the matrix, or any other the inwarde partes. . . It also cureth the wamblinges of the stomacke." —J. L.

p. 99. *Wood powder for Excoriation.* The application of wood-powder to an excoriation is analogous treatment to that of flour to a burn or scald. The object in both cases is to exclude atmospheric air, and to effect the absorption of purulent matter.—J. L.

Wood-dust was also used for the 'violet powder' of the present day: compare Florio's '*Carolo*, a moath or timber-worme. Also, a cunt-botch or winchester-goose. Also *dust of rotten wood vsed about yongue children against fleaing.*'

p. 100. '*Agnus castus.*' "Agnus castus, Hempe tree or Chaste tree, is a singular remedie and medicine for such as woulde live chaste . . . whether in powder or in decoction, or the leaues alone layde on the bed to sleepe uppon. . . . The seede of Agnus Castus driveth away and dissolveth all windinesse and blastinges of the stomacke, entrailes &c." *Lyte's Dodoens*—J. L.

p. 110. *Louis Napoleon.* My revises come on Sept. 5; and on Sept. 2 Louis Napoleon and MacMahon's army surrendered almost unconditionally to the King of Prussia, Bazaine and the Army of the Rhine being held captives at Metz! Well-deserved retribution[1]! May it be speedily followed to the end, and France have meted to her the same measure she declared that she would mete to Prussia, at least, the loss of her Rhine provinces! Meantime, as the uprising of the German nation to defend their Fatherland has been the grandest sight that I have ever seen, and one of the most magnificent that I have ever heard of, making one glad to have lived to witness it, I desire to quote here

[1] Notwithstanding Louis Napoleon's friendship for England. If one's friends take to unprovoked murder, they deserve hanging.

the words of a stranger who is not one of the trimmers who have disgraced part of the English Press :—

"History will record no instance of a greater outrage done to humanity, or one accompanied by circumstances of more malicious perfidy, more selfish premeditation, or a display of combined abjectness, effrontery, and vainglorious miscalculation more disgustful to think of, than this war thrust upon the world by Napoleon III. and his official lackeys. There has never been a nobler movement of national indignation and national resolution, undertaken in a temper more magnificent, more gravely and unexultingly heroic, than the rising of the German people to the challenge. These great facts are, and will remain, true concerning the causes of the war, whatever may be its progress and results. I am not speaking of that which has been obscure or ambiguous in the contradictions and recriminations of diplomatists; but of that which has been obvious in the action and speech of a sovereign and a nation. It is perfectly possible to separate the German nation in this case from Herr von Bismarck; and if Herr von Bismarck is convicted of the crime of seriously entertaining rapacious negotiations (which involves, be it remembered, his further conviction of the folly of self-betrayal) in that case to condemn him, without foregoing a jot of the admiration due to the superb attitude of threatened Germany. To what extent it may yet be possible to separate Napoleon III. from the people among whom he has gagged whatever elements he has not been able to demoralize, and to acquit France of anything worse than military and territorial jealousy, must remain uncertain for the present." AN ENGLISH REPUBLICAN, *in the Pall Mall Gazette*, August 10, 1870, p. 3, col. 2.

II. NOTES ON BOORDE'S *INTRODUCTION.*

p. 119. *Bulwarks, &c.*—Compare Hall, under the xxx. yere of Kyng Henry the VIII. "The same tyme [March, 1538-9] the kyng caused all the hauens to be fortefied, and roade to Douer, and caused Bulwarkes to be made on the sea coastes."—*Chronicle*, p. 827, ed. 1809. And on p. 828, "Also he sent dyuers of his nobles and counsaylours to view and searche all the Portes and daungiers on the coastes, where any meete or conuenient landing place might be supposed, as well on the borders of Englande, as also of Wales. And in alle suche doubtfull places his hyghnes caused dyuers & many *Bulwarkes* & fortificacions to be made."

p. 119. *Castles and Blockhouses built by Henry VIII.* "The most prouident prince that euer reigned in this land, for the fortification thereof against all outward enimies, was the late prince of famous memorie, king Henrie the eight, who, beside that he repared most of such as were alreadie standing, builded sundrie out of the ground. For, hauing shaken off the more than seruile yoke of popish tyrannie, and espieng that the emperour was offended for his diuorce from queene Catherine, his aunt, and thereto vnderstanding that the French king had coupled the Dolphin his sonne with the popes neece, and maried his daughter to the king of Scots, . . he determined to stand vpon his owne defense, and therefore with no small sped, and like charge, he builded sundrie blocke-houses, castels, and platformes, vpon diuerse frontiers of his realme, but chieflie the east and southeast parts of England, whereby (no doubt) he did verie much qualifie the conceiued grudges of his aduersaries, and vtterlie put off their hastie purpose of inuasion." *W. Harrison's Descr. of England*, in *Holinshed's Chronicle*, p. 194, col. 2, ed. 1587.

p. 120. *Caernarvon.* "Wednesday the 4. of August, I rode 8 miles from Bangor to *Carnarvan*, where I thought to have seen a Town and a Castle, or a Castle and a Towne ; but I saw both to be one, and one to be both ; for indeed a man can hardly divide them in judgement of apprehension ; and I have seen many gallant Fabricks and Fortifications, but for compactness and compleatness of *Caernarvon*, I never yet saw a parallel. And it is by Art and Nature so sited and seated, that it stands impregnable ; & if it be well mand, victualled, and ammunitioned, it is invincible, except fraud or famine do assault, or conspire against it." (1653. John Taylor, *A short Relation of a long Iourney*, p. 14.)

p. 120. *The Northern tongue.*—Sane tota lingua Nordanimbrorum, et maxime in Eboraco, ita inconditum stridet, ut nichil nos australes intelligere possimus. Quod propter viciniam barbararum gentium, et propter remotionem regum quondam Anglorum modo Normannorum contigit, qui magis ad austrum quam ad aquilonem diversati noscuntur.—*Willelmi Malmesburiensis monachi Gesta Pontificum Anglorum*, lib. iii. p. 209, ed. Hamilton, 1870.

p. 120. *Salt.* And for Salte, there is great plentie made at the Witches [places whose names end in *-wich*] in Cheshire, and in diuers other places : Besides many Salte houses standyng vpon the coaste of Englande that makes Salte, by sething of salte Sea water.—1580, *Robert Hitchcok's Pollitique Platt*, sign. e. iii.

p. 122. *Cornwall.* The Water-Poet gives the county a much better character a hundred years later : "Cornewall is the Cornucopia, the compleate and repleate Horne of Abundance, for high churlish Hills, and affable courteous people : they are loving to requite a kindenesse, placable to remit a wrong, and hardy to retort injuries : the Countrey hath its share of huge stones, mighty Rocks, noble, free, Gentlemen, bountiful housekeepers, strong and stout men, handsome beautifull women ; and (for any that I know) there is not one Cornish Cuckold to be found in the whole County ; In briefe, they are in most plentifull manner happy in the abundance of right and left hand blessings." 1649. *John Taylors Wandering, to see the Wonders of the West*, p. 10. On pages 17, 18, Taylor gives an account of the pilchard fishing at Mevagesey in Cornwall.

p. 126. *The Welsh and* Cawse boby *or Roasted Cheese.*—The 78th Tale in "A Hundred Mery Talys" from the only perfect copy known, printed by John Rastell in 1526, ed. Oesterley, 1866, p. 131, is

"LXXVIII. *Of seynt Peter that cryed 'cause bobe.'*—I fynde wryten amonge olde gestys, how God made Saynte Peter porter of heuen / and that God of his goodnes, soone after his passyon, suffred many men to come to the kyngdome of heuen with small deseruyng / at whiche tyme there was in heuen a grete company of Welchemen / whiche, with theyre krakynge & babelynge, trobelyd all the other. Wherfore God sayd to Saynt Peter *that* he was wery of them / & that he wolde fayne haue them out of heuen. To whome Saynt Peter sayde 'Good Lorde, I warrant you *that* shalbe shortly done /' wherfore Saynt Peter went out of heue*n* gatys, & cryed *with* a loude voyce 'Cause bobe' / *that* is as moche to say as 'rostyd chese' / whiche thynge *the* Welchmen heryng, ran out of heuyn a great pace. And when Saynt Peter sawe them al out, he sodenly went in to heuen, and lokkyd the dore, and so sparryd all the Welchmen out.

"¶ By this ye may se that it is no wysdome for a man to loue or to set his mynde to moche vpon ony delycate or worldly pleasure wherby he shall lose the celestyall & eternall Ioye."

See also the note below, on p. 156.

p. 127. *St Winifrid's Well.* Taylor the Water-Poet describes this in his *Short Relation of a long Iourney* in 1653, p. 10-12. "Saturday, the last of July, I left Flint, and went three miles to *Holy-Well*, of which place I must speak somewhat materially: About the length of a furlong, down a very steep Hill, is a Well (full of wonder and admiration;) it comes from a Spring not far from Radland Castle; it is, and hath been, many hundred yeares knowne by the name of *Holy-Well*, but it is more commonly, and of most Antiquity, called *Saint Winifrids Well* in memory of the pious and chaste Virgin Winifrid, who was there beheaded for refusing to yield her Chastity to the furious lust of a Pagan Prince: in that very place where her bloud was shed, this Spring sprang up; from it doth issue so forceible a stream, that within a hundred yards of it, it drives certain Mils; and some do say that nine Corn Mils and Fulling Mils are driven with the Stream of that Spring: It hath a fair Chappell erected over it called Saint Winifrid's Chappell, which is now much defaced by the injury of these late Wars; The Well is compassed about with a fine Wall of Free stone; the Wall hath eight Angles or Corners, and at every Angle is a fair Stone Piller, whereon the West end of the Chappell is supported. In two severall places of the Wall there are neat stone staires to go into the water that comes from the Well; for it is to be noted that the Well it selfe doth continually work and bubble with extream violence, like a boiling Cauldron or Furnace; and within the Wall, or into the Well, very few do enter: The Water is Christalline, sweet, and medicinable; it is frequented daily by many people of Rich and Poore, of all Diseases; amongst which, great store of folkes are cured, divers are eased, but none made the worse. The Hill descending is plentifully furnished (on both sides of the way) with Beggers of all ages, sexes, conditions, sorts, and sizes; many of them are impotent, but all are impudent, and richly embrodered all over with such Hexameter poudred Ermins (or Vermin) as are called Lice in England."

p. 127-8. *Foolish Customs in Wales.* Taylor the Water-Poet, in 1653 notices that the Welsh were free from the Sabbatarian superstition of one English place. "Of all the places in England and Wales that I have travelled to, this village of Barnsley [in Gloucestershire] doth most strictly observe the Lords day, or Sunday, for little children are not suffered to walke or play: and two Women, who had beene at Church both before and after Noone, did but walke into the fields for their recreation, and they were put to their choice, either to pay sixpence apiece (for prophane walking,) or to be laid one houre in the stocks; and the pievish willfull women (though they were able enough to pay,) to save their money, and jest out the matter, lay both by the heeles merrily one houre.

There is no such zeale in many places and Parishes in Wales; for they have neither Service, Prayer, Sermon, Minister, or Preacher, nor any Church door opened at all, so that people do exercise and edifie in the Church-Yard, at the lawfull and laudable Games of Trap, Catt, Stool-ball, Rocket &c, on Sundayes."

p. 128. *Prestes shal haue no concubynes* (or wives). The 31st of Henry VIII, chapter 14, A.D. 1539, enacted "that if any p*er*son w*hi*ch is or hath byne a Preest, before this present parliament, or during the time of cession of the same, hath maryed, and hath made any contract of matrimony with any woman, or that any man or woman w*hi*ch before the makinge of this acte advisedly hath vowed chastitie or wydowhode before this p*re*sent parliament or during the cession of the same, hath maried or contracted matrimony with any p*er*son, that then every suche mariage & contract of matrimony shalbe utterlie voide and of none effecte: And that the Ordynaries within whose Dioces or Jurisdicc*io*n the p*er*son or p*er*sons so maried or contracted is or be resident or abydynge, shall from tyme to tyme make sep*ar*ac*io*n and devorses of the saide mariage*s* and contracte*s*.

AND further it is enacted by the auctoritie abovesaide, that if any man w*hi*ch is or hathe bene Preest as is aforesaide, at any tyme from and after the saide xij^th^ daye of July next comynge, doe carnally kepe or use any woman, to whom he is or hathe bene maried, or with whome he hathe contracted matrimony, or openly be conversaunt [or] kepe companye and famyliaritie withe any suche woman, to the evell example of other p*er*sons, everie suche carnall use, copulac*io*n, open conv*er*sac*io*n, kepinge of company and famyliarity, be, and shalbe demed and adjudged, felony, aswell against the man as the woman; and that everie such p*er*son soe offendinge shalbe enquired of, tried, punyshed, suffer, and forfeyt, all and everie thinge and thinge*s* as other felons made and declared by this Acte, and as in case of felonye, as is aforesaide."

The death-punishment for Felony was found too severe; and therefore by the 32 Henry VIII, chapter 10, the penalty was altered to: "First offence, Forfeiture of all Benefices but one, &c. Second offence, Forfeiture of all Benefices land, goods & chattels. Third offence, Imprisonment for Life. The Penalty on Single Women offending was; First offence, Forfeiture of Goods. Second offence, Forfeiture of Half the Profits of her Lands. Third offence, Forfeiture of all Goods, chattels, & Profits of land, and Imprisonment for Life. The Penalty on Wives offending was Imprisonment for Life.

p. 131. *Products of Ireland.*—'The Libel of English Policy,' A.D. 1436, speaks of these, and the country itself. The products are

Hydes, and fish, samon, hake, herynge,
Irish wollen, lynyn cloth, faldynge[1]
And marternus gode, bene here marchaundyse;
Hertys hydes, and other of venerye,
Skynnes of otere, squerel and Irysh [h]are,
Of shepe, lambe, and fox is here chaffare,
ffelles of kydde and conyes grete plenté. (ii. 186.)

Then, as to the country, which is a buttress and a post under England, the writer says,

Why speke I thus so muche of Yrelonde?
ffor als muche as I can understonde
It is fertyle for thynge that there do growe
And multiplyen,—loke who-so lust to knowe;—
So large, so gode, and so comodyouse,
That to declare is straunge and merveylouse.
ffor of sylvere and golde there is the oore
Amonge the wylde Yrishe, though they be pore;
ffor they ar rude, and can thereone no skylle;
So that if we had there pese and gode wylle
To myne and fyne, and metalle for to pure,
In wylde Yrishe myght we fynde the cure;
As in London seyth a juellere,
Whych brought from thens gold oore to us here,
Whereof was fyned metalle gode and clene,
As [to] the touche, no bettere coude be sene.
T. Wright's Political Songs, Rolls Series, ii. 186-7.

And welle I wote that frome hens to Rome,
And, as men sey, in alle Cristendome,
Ys no grounde ne lond to Yreland lyche,
So large, so gode, so plenteouse, so riche,
That to this worde *dominus* dothe longe. (*ib.* ii. 188.)

p. 131, line 8. *And good square dyce.*—There is amung them (the Wild Irish) a brotherhood of Karrowes, that profer to play at chartes all *the* yere long, and make it their onely occupation. They play away mantle and all to the bare skin, and then trusse themselues in strawe or in leaues; they wayte for passengers in the high way, invite them

[1] He rood vp on a Rouncy, as he kouthe,
In a gowne of *faldynye* to the knee.
CHAUCER of his Shipman, *Cant. Tales*, group A. § 1, l. 391.

to game upon the grene, & aske them no more but companions to holde them sporte. For default of other stuffe, they paune theyr glibs, the nailes of their fingers and toes, their dimiffaries, which they leese or redeeme at the curtesie of the wynner.—The *Description of Ireland*, by Richard Stanyhurst (chap. 8), in *Holinshed*, ed. 1577.

p. 131, l. 8-7. *Aqua Vitæ, and the Diet of the Wild Irish.*—"Water cresses (which they terme shamrocks), rowtes, and other herbes, they feede upon; otemeale and butter they cramme together; they drinke whey, mylke, and biefe brothe. Fleshe they devour without bread, and that halfe raw: the rest boyleth in their stomackes with Aqua vitæ, which they swill in after such a surfet by quartes & pottels: they let their cowes bloud, which, growen to a gelly, they bake, and ouerspred with butter, and so eate in lumpes. No meat they fancy so much as porke, and the fatter the better. One of Iohn Oneales household demaunded of his fellow whether biefe were better then porke: 'that,' quoth the other, 'is as intricate a question, as to aske whether thou art better then Oneale.'"—*Stanyhurst's Description of Irelande*, chap. 8, Holinshed, ed. 1577.

p. 131. *Natural disposition of the "wyld Irishe."*—"The people are thus enclined: religious, franke, amorous, irefull, sufferable of infinite paynes, very glorious, many sorcerers, excellent horsemen, delighted with wars, great almsgiuers, passing in hospitality. The lewder sort, both clearkes and lay men, are sensuall, & ower loose in liuyng. The same, beyng vertuously bred up or reformed, are such myrors of holynes and austeritie, that other nations retaine but a shadow of deuotion in comparison of them. As for abstinence and fasting, it is to them a familiar kynd of chastisement."—*Stanyhurst's Description of Irelande*, chap. 8, Holinshed, ed. 1577.

p. 132. *The Wild Irish lack manners.*—"The Irishe man standeth so much upon hys gentilitie, that he termeth any one of the English sept, and planted in Ireland, *Bobdeagh Galteagh*, that is, 'English churle': but if he be an Englishman borne, then he nameth hym, *Bobdeagh Saxonnegh*, that is, 'a Saxon churle': so that both are churles, and he the onely gentleman; and therupon, if the basest pesant of them name hymselfe with hys superior, he will be sure to place himselfe first, as 'I and Oneyle, I and you, I and he, I & my maister,' wheras the curtesie of the Englishe language is cleane contrary."—*Stanyhurst's Description of Irelande*, chap. 8, Holinshed, ed. 1577.

p. 132. *The English Pale.*—"Before I attempt the unfoldyng of the maners of the meere Irish, (wild Irish) I thinke it expedient, to forewarne thee, reader, not to impute any barbarous custome that shall be here layde downe, to the citizens, townesmen, and the inhabitants of the english pale, in that they differ little or nothyng from the ancient customes and dispositions of their progenitors, the English and Walshmen, beyng therfore as mortally behated of *the* Irish, as those that are borne in England."—*Stanyhurst's Description of Irelande*, chap. 8, Holinshed, ed. 1577.

p. 133. *Ireland; No Adders, &c., there.*

"'Tis said no Serpent, Adder, Snake, or Toade,
Can live in Ireland, or hath there aboade."
1642. John Taylor, *Mad Fashions*, p. 4.

p. 133. *Men and women lie together in straw.*—In olde tyme they (the Wild Irish) much abused the honourable state of marriage, either in contractes unlawfull, meetyng the degrees of prohibition, or in diuorcementes at pleasure, or in retaynyng concubines or harlots for wyues: yea, euen at this day where the clergy is fainte, they can be content to marry for a yeare and a day of probation, and at the yeres ende, or any tyme after, to returne hir home with hir marriage goodes, or as much in valure, upon light quarels, if the gentlewomans friendes be unable to reuenge the injury. In lyke maner may she forsake hir husband.—The *Description of Ireland*, by Richard Stanyhurst (chap. 8), in *Holinshed*, ed. 1577.

p. 133. *Superstitions of the Irish.*—Stanyhurst says, "In some corner of the land they used a damnable superstition, leauyng the right armes of their infantes unchristened (as they terme it) to the intent it might giue a more ungracious & deadly blowe. Others write that gentlemens children were baptized in mylke, and the infantes of poore folke in water, who had the better, or rather the only, choyce. Diuers other vayne and execrable superstitions they obserue, that for a complete recitall would require a seueral volume. Wherto they are the more stifly wedded, because such single preachers as they haue, reproue not in theyr sermons the pieuishnesse and fondnesse of these friuolous dreamers. But these and the like enormities haue taken so deepe roote in that people, as commonly a preacher is sooner by their naughty lyues corrupted, then their naughty lyues by his preaching amended. . . . . Againe, the very English of birth, conuersant with the sauage sort of that people, become degenerate; &, as though they had tasted of Circes poysoned cup, are quite altered. Such a force hath education to make or marre."—The *Description of Ireland*, by Richard Stanyhurst (chap. 8), in *Holinshed*, ed. 1577.

John Cal. li. 2 Cant. ant.

p. 135. *Scotland.*—The *Libel* of 1436 says the exports of Scotland are skins, hides, and wool, which pass through England to Flanders,—the wool being sold in the towns of Poperynge and Belle. The imports are mercery, haberdashery, cartwheels and barrows.—*T. Wright's Polit. Songs*, ii. 168.

p. 136. "*Scotlande is a baryn and a waste countrey.*"—Certes there is no region in the whole world so barren & unfruteful, through distaunce from the Sunne.—*Description of Scotland*, chap. 13, *Holinshed*, ed. 1577.

p. 137. *The Scotch 'be hardy men.'*—Thereunto we finde them to be couragious and *hardy*, offering themselues often unto the uttermost perils with great assurance, so that a man may pronounce nothing to be ower harde or past their power to perfourme.—*Description of Scotland*, chap. 1, *Holinshed*, ed. 1577.

p. 141. *Iceland and its Stockfish.*—The *Libel* of 1436 says,

Of Yseland to wryte, is lytille nede,
Save of stokfische; yit for sothe, in dede,
Out of Bristow, and costis many one,
Men have practised by nedle and by stone
Thider-wardes wythine a lytel whylle,
Wythine xij. yere, and wythoute perille,
Gone and comen—as men were wonte of olde—
Of Scarborowgh unto the costes colde;
And now so fele shippes thys yere there were,
That moche losse for unfraught they bare;
Yselond myght not make hem to be fraught
Unto the hawys; this moche harme they caught.
T. Wright's Political Songs, ii. 191.

p. 142. *Iceland curs, and Icelanders eating tallow-candles.*—"Besides these also we haue sholts or *curs dailie brought out of Iseland*, and much made of among vs, bicause of their sawcinesse and quarrelling. Moreouer they bite verie sore, and *loue candles exceedinglie, as doo the men and women of their countrie:* but I may saie no more of them, bicause they are not bred with vs. Yet this will I make report of by the waie, for pastimes sake, that when a great man of those parts came of late into one of our ships which went thither for fish, to see the forme and fashion of the same, his wife apparrelled in fine sables, abiding on the decke whilest hir husband was vnder the hatches with the mariners, *espied a pound or two of candles* hanging at the mast, and being loth to stand there idle alone, *she fell to, and eat them vp euerie one, supposing hir selfe to haue beene at a iollie banket*, and shewing verie plesant gesture when hir husband came vp againe vnto hir."—*Harrison's Descr.*, Bk. iii. chap. 7, p. 231, col. 2, ed. 1586-7.

"My lorde is not at lesure:
The pawre man at the dur
Standes lyke an *yslande cur*,
And Darre not ones sture."

Vox Populi Vox Dei, A.D. 1547-8, l. 473-5, p. 137 of my *Ballads from Manuscripts*, vol. i. Ballad Society, 1868, p. 137, where this note from Nares is given, "Iceland Dogs: shaggy, sharp-eared, white dogs, much imported formerly as favourites for ladies etc. 'Pish for thee, *Iceland dog*, thou prick-ear'd *cur of Iceland!*' Henry V, ii. 1."

p. 142. *The newe founde land named Calico.*—? Calicut, a kingdom of India on the coast of Malabar, about 63 miles long, and nearly as many broad. Its capital is also named Calicut, and was the first place where the Portuguese admiral Vasco de Gama landed on May 22, 1498, and whence he returned to Portugal, laden with the first spoils of the eastern world. This was the beginning of European trade with India. Our word *calico* is taken from Calicut.—*Oxford Encyclopædia*, 1828.

p. 145. *Paschal.*—Can this be the PASCAL or PASCHAL, Pierre, described in the *Bibliographie Universelle*, 1823, vol. xxiii. p. 44, col. 2, as a littérateur without talent, but full of vanity and impudence, who was born in 1522 at Sauveterre in the Bazadois, of a noble family, and died at Toulouse on Feb. 16, 1565, at the age of 43? He got praises in plenty, and a pension, for his proposals to continue Paulus Jovius's Eulogiums of Learned Men, and to write a History of France; but he left only 6 leaves of the latter work finisht when he died, though he had before distributed notes with '*P. Paschali liber quartus rerum à Francis gestarum*' on them.

Pope Pascal II died on January 11, 1118; Pope Pascal III was for a time made Anti-Pope in the days of Alexander III, who was elected on Sept. 7, 1159, and died Aug. 30, 1181.

p. 147. *The Flemings' Fish and Beer.*—"the Flemminges . . . with their greene fishe, barreled Cod and Heringes, caryeth out of Englande for the same yearely, both golde, and siluer, and other comodities: and at the leaste tenne thousande tunne of dubble dubble Beare, and hath also all kinde of Frenche commodities, continually both in tyme of warres and peace, by their trade onely of fishyng."—1580, *Robert Hitchcok's Pollitique Platt*, sign. f. ii. (The book shows how great a help the development of the Herring Fishery would be to England.) For the "Butter," see the note on p. 156.

p. 147, &c. *Flemings, their Beer-drinking, Butter, and Products.*—The *Libel* of 1436 says of the Prussians, High-Dutchmen, and Easterlings,

Oute of fflaundres . . . . .
. . . they bringe in the substaunce of the *beere*
That they drynken fele to goode chepe, not dere.
Ye have herde that twoo ffleinynges togedere
Wol undertake, or they goo ony whethere,
Or they rise onys, to drinke a barrelle fulle
Of goode *berkyne.*[1] So sore they hale and pulle,
Undre the borde they pissen, as they sitte:
This cometh of covenant of a worthy witte.
Wythoute Calise in ther *buttere* the[y] cakked;
Whan they flede home, and when they leysere lakked
To holde here sege, they wente lyke as a doo:
Wel was that fflemmynge that myght trusse and goo . . .

After bere and bacon, odre gode commodites usene.

Now bere and bacon bene fro Pruse ibrought
Into ffiaundres, as loved and fere isoughte;
Osmonde,[2] coppre, bow-staffes, stile,[3] and wex,
Peltre-ware, and grey, pych, terre, borde, and flex,
And Coleyne threde, fustiane, and canvase,
Corde, bokeram: of olde tyme thus it wase.
But the fflemmyngis, amonge these thinges dere,
In comen lowen[4] beste, bacon and bere:

[1] barley brew [2] a kind of iron.—Halliwell. [3] steel [4] love

Thus arn they hogges; and drynkyn wele ataunt;
ffare wel, Flemynge! hay, harys, hay, avaunt!
Also Pruse men make here aventure
Of plate of sylvere, of wegges[1] gode and sure
In grete plenté, whiche they bringe and bye
Oute of londes of Bealme and Hungrye;
Whiche is encrese ful grete unto thys londe.
And thei bene laden, I understonde,
Wyth wollen clothe, alle manere of coloures,
By dyers craftes ful dyverse that ben oures.
And they aventure ful gretly unto the Baye,[2]
ffor salte, that is nedefulle wythoute naye.

T. Wright's Political Songs, ii. 169-171.

Again, at p. 161 the Spanish imports from Flanders are said to be

ffyne clothe of Ipre, that named is better than oure-is,
Cloothe of Curtryke, fyne cloothe of alle coloures,
Moche ffustyane, and also lynen clothe.
But, ye ffleinmyngis, yf ye be not wrothe,
The grete substaunce of youre cloothe, at the fulle,
Ye wot ye make hit of youre Englissh wolle.

p. 149. *Dutchmen 'quaf tyl they ben dronk.'*

"'Tis said the *Dutchmen* taught vs drinke and swill;
I'm sure we goe beyond them in that skill;
I wish (as we exceed them in what's bad,)
That we some portion of their goodnesse had."

1632. *Taylor on Thame Isis*, p. 27.

p. 150, l. 5. *Antwerp and Barow.*—If this warre [with the Emperor in 1527] was displeasaunt to many in Englande (as you have hard), surely it was as much or more displeasant to the tounes and people of Flaunders, Brabant, Hollande, and Zelande, and in especiall to the tounes *Andwarpe and Barrow*, where the Martes wer kept, and where the resorte of Englishmen was; for thei saied that their Martes were vndoen if the Englishemen came not there; and if there were no Marte, their Shippes, Hoyes, and Waggons might rest, and all artificers, Hostes, and Brokers might slepe, and so the people should fal into miserie and pouertie.—*Hall's Chronicle*, p. 746, ed. 1809.

p. 150. *Brabant, the Mart of all nations.*—The *Libel* of 1436 says,

And wee to martis of Braban charged bene
Wyth Englyssh clothe, fulle gode and feyre to seyne.
Wee bene ageyne charged wyth mercerye,
Haburdasshere ware, and wyth grocerye.
To whyche martis—that Englisshe men call "feyres"—
Iche nacion ofte maketh here repayeres,

[1] wedges

[2] Into the Rochelle, to fetche the fumose wine,
Nere into Britonnse *bay* for salt so fyne. (*ib.* p. 162.)

Englysshe and Frensh, Lumbardes, Januayes,
Cathalones, theder they take here wayes,
Scottes, Spanyardes, Iresshmen there abydes,
Wythe grete plenté bringinge of salte hydes.
T. Wright's Political Songs, ii. 179.

The English were by far the largest buyers at the Marts, of goods brought thither by land as well as sea; and among the articles are,

Yit marchaundy of Braban and Selande,
The madre and woode that dyers take on hande
To dyne wyth; garleke, and onyons,
And salt fysshe als, for husbond and comons.
But they of Holonde, at Caleyse byene oure felles
And oure wolles, that Englyshe men hem selles. (*ib.* p. 180.)

p. 151. *Antwerp Church and its Spire.*—"The great glory of Antwerp is its cathedral, the finest building in the Low Countries; it is said to be 500 feet long, 240 wide, and has a spire of stone . . 366 feet (high); consequently it is lower than the spire of Salisbury cathedral, if the [generally acknowledged] height of this spire can be depended on." *Penny Cyclopædia.*

p. 151. *Hanawar or Hanago*, or Hainault, is called *Hennigow* in the map of Europe in XII Landtaflen, printed at Zurich by Christoffel Froschower, M.D.LXII., and is placed South (instead of East) of Artois, and north of Paris. The map is turned and lettered with its North, instead of its South point, towards you. 'Lunden' is wholly on the south of the Thames.

p. 156. *Butter and Dutchmen.*—A tale in *The Sack-Full of Newes*, ed. 1673, sign. B., illustrates this: "There was a widow in London that had a Dutchman to her servant, before whom she set a rotten Cheese & butter for his dinner: and he eate of the butter because he liked it, and his Mistresse bad him eat of the cheese. 'No, Mistresse,' *quod* he, 'the butter is good enough.' She, perceiving he would eat none of the bad cheese, said, 'Thou knave, thou art not to dwell with honest folkes!' 'By my troth, Mistresse,' said he, 'had I taken heed ere I came hither, I had never come here.' 'Well, knave,' *quod* she, 'thou shalt go from on whore to another.' 'Then will I go,' *quod* he, 'from you to your sister;' and so departed."

See also in "The Figure of Nine, Containing these Nine Observations, Wits, Fits, and Fancies, Jests, Jibes, and Quiblets, with Mirth, Pastime, and Pleasure.

The Figure of Nine to you I here present,
Hoping thereby to give you all content,"

over a circular device, with the legend *Cor unum via una.* "Printed for J. Deacon, and C. Dennisson, at their Shops at the Angel in Guiltspur-street, and at the Stationers Arms within Aldgate." A in eight.

"*Nine sorts of men love nine sorts of dishes.*—A Dutchman loves butter, an Englishman Beefe, a Scot loves an Oat-cake, the VVelshman loves Couse-bobby [toasted cheese], an Irishman Onions, a Frenchman loves Mutton, the Spaniard tobacco, the Seaman loves Fish, and a Taylor loves cabbage." sign. A. 3, back.

p. 161. *holmes* (fustian). A.D. 1474. "Item, x. elnes of blak *holmess* [printed *holmefs*] fustian to the trumpatis doublats, iij. s. the eln."—Dauney's Extracts from Accounts in his *Ancient Scotish Melodies*, Edinb. 1838 (Bannatyne & Maitland Clubs).

p. 163. *The old warriors and present poverty of Denmark.*—The *Libel*, A.D. 1436, says,

In Denmarke ware fulle noble conquerours
In tyme passed, fulle worthy werriours,
Whiche, when they had here marchaundes destroyde,
To poverte they felle,—thus were they noyede ;—
And so they stonde at myscheffe at this daye ;
This lerned I late, welle wryten, this no naye.
T. Wright's Polit. Songs, ii. 177.

p. 169. *Bugles.*—See Topsell's *History of Four-footed Beasts* : "Of the Vulgar *Bugil.* A Bugil is called in Latine, *Bubalus*, and *Buffalus* ; in French, *Beufle* ; in Spanish *Bufano* ; in German, *Buffel.* . . This vulgar Bugil is of a kinde of wilde Oxen, greater and taller then the ordinary Oxen, and their limbs better compact together. . . They are very fierce, being tamed ; but that is corrected by putting an Iron ring through his Nostrils, whereinto also is put a cord, by which he is led and ruled, as a Horse by a bridle ; (for which cause, in Germany they call a simple man over-ruled by the advise of another to his own hurt, 'a Bugle, led with a ring in his nose.' His feet are cloven, and with the formost he will dig the earth, and with the hindmost fight like a Horse, setting on his blows with great force, and redoubling them again if his object remove not. His voyce is like the voyce of an Oxe ; when he is chased he runneth forth right, seldom winding or turning, and when he is angred, he runneth into the water, wherein he covereth himself all over, except his mouth, to cool the heat of his blood." p. 45, ed. Rowland, 1658.

p. 171. *A gret citie called Malla-vine.*—And Men gon thorghe the Lond of this Lord [the Kyng of Hungarye], thorghe a Cytee that is clept Cypron, and be the *evylle Town*, that sytt toward the ende of Hungarye.—*Mandeville's Voiage and Travaile*, p. 7, ed. 1839.

p. 176. *Naples.*—Thomas speaks thus of the Neapolitans, *Hist. Italye*, lf. 114, "the Neapolitanes are scarcelye trusted on their wordes. Not that I thynke they deserue lesse credyte than other men, but because the wonted general ill opinion of their vnstedfastnesse is not taken oute of men's hertes. Yet is the Neapolitane, for his good enterteinment, reckened to be the veraie courtesie of the worlde, thoughe most men repute him to be a great flatterer, and ful of crafte.

"What wol you more? They are rych, for almost euery gentylman is lorde and kynge within hym selfe ; they haue veray fayre women, and the worlde at wyll; in so muche as Naples contendeth wyth Venice, whether should be preferred for sumptuouse dames. Finallye, the court about the *Vicere* was wont to be very princelye, and greater than that of Myllayne for trayne of gentilmen; but now it is somewhat diminished."

p. 178. *Italy: 'the people be homly and rude.'*—Thomas (leaf 3, back, leaf 4) praises the Italian gentlemen *very* highly: "so honourable, so courteise, so prudente, and so graue withall, that it shoulde seeme eche one of thaim to haue had a princelye bringynge vp. To his superior, obediente; to his equall, humble; and to his inferiour, gentle and courteyse; amyable to a straunger, and desyrous with curtesie to winne his loue.

"I graunte, that in the expense or loue of his money to a straunger, he is waré, and woull be at no more cost than he is sure eyther to saue by, or to haue thanke for: wherein I rather can commende him than otherwyse. But this is out of doubte, a straunger can not be bettor enterteigned, nor moore honourablie entreated, then amongest the Italians." Thomas also praises highly the Italian universities "Padoa, Bononia, Pauia, Ferrara, Pisa, and others"; none of which Andrew Boorde says he saw. But Thomas says the condition of the poor is very bad; they are hardly able to earn bread.

p. 178. *St Peter's fallen to the ground.*—Though Rome was sackt in 1527 by the Emperor's army under the command of the Duke of Bourbon (see the account in *Hall's Chronicle*, p. 726-7, ed. 1809), yet it was Julius II who had the old basilica of St Peter's pulled down, in order to provide a site for his mausoleum, which Michael Angelo had designed. On April 18, 1506, Julius II laid the foundation-stone of the present church. Bramante made designs for it, and four great piers and their arches were completed before he died in 1514. The work stood still for nearly 30 years; Michael Angelo altered the design; and his Cathedral was nearly finisht in 1601, when Paul V and the Cardinals commissioned Carlo Maderno to lengthen the nave, &c. Urban VIII dedicated the church on the 18th of November 1626, a hundred and twenty years after the building began. *Spalding's Italy and the Italian Islands*, iii. 154: see a plan and account of the old Basilica, *ib.* ii. 46-50.

p. 178. *Rome.*—See W. Thomas's chapter "Of tho present astate of Rome," leaf 37, &c., of his *Hist. of Italye*, ed. 1561. Of the new Cathedral of *St Peter's*, he says:—"But aboue all, the newe buildyng, if it were finished, wolde be the goodliest thyng of this worlde, not onelye for the antike pillers that haue ben taken out of the antiquitees, and bestowed there, but also for the greatnesse and excellent good proporcion that it hathe. Neuerthelesse it hath been so many yeres adoing, and is yet so vnperfect, that most men stand in dout whether euer it shalbe finished or no."—1549, *W. Thomas's Hist. of Italye*, leaf 40, back, ed. 1561.

p. 181. *Venice.*—Thomas, in his *Historye of Italye*, 1549, p. 74, ed. 1561, says of Venice, "I thynke no place of all Europe, hable at this daye to compare with that citee for noumber of sumptuouse houses, speciallye for theyr frontes. For he that would rowe through the *Canale grande*, and marke wel the frontes of the houses on bothe sydes, shall see theim more lyke the doynges of prynces then priuate men. And I haue been with good reason persuaded, that in Venice be aboue .200. palaices able to lodge any king."

p. 182. *The Merchandise of Venice* was, according to the *Libel* of 1436, grocery, wines, monkeys, knicknacks, and drugs:

The grete galces of Venees and fflorence
Be wel ladene wyth thynges of complacence,—
Alle spicerye, and of grocers ware,
Wyth swete wynes, alle manere of chaffare,
Apes, and japes, and marmusettes taylede,
Nifles, trifles, that litelle have availede,
And thynges wyth whiche they fetely blere oure eye,
Wyth thynges not enduryng that we bye . .
And . . for infirmitees . . . . skamonye,
Turbit, euforbe, correcte, diagredie,
Rubarde, sené; and yet they bene to nedefulle.

T. Wright's Political Songs, ii. 173.

p. 183. *No Lords in Venice.*—"*Democratia*, a free state or common wealth, hauing no Prince or superior but themselues (as Venice is) except those officers that themselues appoint." *Florio.*

p. 184, note. *Italian Wives, and their Husbands' Jealousy.*—Thys vyce is of property to the Ytaliens, to shytte vp theyr wyues as theyr treasour. And, on my fayth (to my iudgemente) to lytle purpose; for the mooste part of women be of thys sorte, that moost they desyre that [which] moost too them is denyed; and whan thou woldest, they wyl nat; and whan thou woldest nat, they wolde; and yf they haue the brydle at libertye, [the] lesse they offende; so that it is as easy to kepe a woman against her wyll, as a flocke of flies in the hete of the sonne, excepte she be of her selfe chaste. In vayne doth the husband set kepers ouer her; for who shal kepe those kepers? She is crafty; and at them lightely she beginneth; and whan she taketh a fantasy, she is vnreasonable, and lyke an vnbrydeled mule.—*The goodly History of the moste noble and beautyful Ladye Lucres of Scene in Tuskan, & of her louer Eurialus, verye pleasaunt and delectable vnto the reder.* ¶ Anno Domini M.D.LX. [col.] Imprinted at London, by Iohn Kynge. (sign. D .ii.) This is the 2nd edition, and Mr Henry Huth has lent me the copy from which I extract. The book is in Captain Cox's list. Its author, Æn. S. Piccolomini, returns to the husband-&-wife question on leaves F iv, v, vi: "And on the morowe, eyther for that it were necessary to take hede, or for some yl suspecte, Menelaus [the husband] walled vppe the wyndowe [by which Eurialus had got in to Lucres]. I thynke as our Cytezens [of Sienna] be suspectuous and full of coniectures; so dyd hee feare the *com*modyte of the place, & woulde eschewe the occasion; for though he knewe noughte, yet wyste hee well *that* she was much desyred, and daylye prouoked by great requestes, & [he] iudged a womans thought vnstable, whiche hath as many myndes as trees hath leues, & *that* theyr kynde alway is desyrous of newe thynges, and seldom loue they theyr husbands whom they haue obteyned. Therefore dyd he folowe the common opynyon of maried men, too auoyde myshap, thoughe it come wyth good lucke."

The food and ways of Italian servants about 1440 A.D. are shown by a passage in this *Lucres & Eurialus*, written by Pope Pius II in his young days, when he was Æneas Sylvius Piccolomini: "looke that oure supper be redy! We must be meri while our mayster[1] is furth; our maistres[2] is better felowe; shee is merye & liberal; he is angry, full of noyse, couetous, and harde. We are neuer wel wh*en* he is at home. Se, I pray the, what lanke belyes we haue! He is hungry hym selfe, to sterue vs for hunger; hee wyll not suffer one moyste peece of browne breade to be loste; but the fragm*en*tes of one daye he kepeth fyue dayes after, & the gobbets of salte fysh & salt eles of one supper, he kepeth vnto another, and marketh the cut chese, least anye of it shulde be stol*en*. . . . How muche are we better wi*th* our maistres, *that* feedeth vs not onlye with veale & kidde, but with hennes and byrdes, & ple*n*tye of wyne? Go, Dromo, and make the kytch*en* smoke!" "Mary!" quod Dromo, "that shall be my charge; & soner shall I laye the tables thanne rub the horse! I brought my mayster i*n*to the cou*n*tree to-daye, that the Deuyll breke hys necke! and neuer spake hee woorde vnto me, but badde me, wha*n* I brought home my horses, to tell my maystres *that* hee woulde not come home too nyghte. But by God," quod he, "I prayse the, Zosias, *that* at the last hast founde faute at my maysters co*n*dycions. I had forsaken my mayster, yf my maystres had not geue*n* me mi morowe meles as she hath. Lette vs not sleape to-night, Zosia; but lette vs eate & dryncke tyll it bee daye. My mayster shall not winne so muche this moneth, as we shal wast at one supper."

Gladlye dyd Eurialus [Lucres's lover, hiding in the hay till he could get to her] here this, and marked the maners of serua*n*ts, & thought he was serued a lyke. ed. 1560, sign. F .iii., F .iiii. The unique copy of the first edition in the British Museum is more correctly printed than the second, but has lost its last leaf, with the last verse of the Envoy. This has now been supplied by me from Mr Huth's copy of Kynge's edition. The story of the novel is told in the Forewords to my edition of Captain Cox, or Laneham's Letter (Ballad Society, 1871).

p. 185. *The Venetians' timber, &c., in readiness for war.*—"the *Arsenale* in myne eye excedeth all the rest: For there they haue well neere two hundred galeys in such an order, that vpon a very smal warnyng they may be furnyshed out vnto the sea. Besydes that, for euery daye in the yeare (whan they would goe to the coste) they should be able to make a newe galey; hauinge such a staple of timber (whyche in the water wythin Th' *arsenale* hathe lyen a seasoninge, some .20. yeare, some .40. some an .100. and some I wot not how longe) that it is a wonder to see." —*Thomas's Hist. of Italye*, leaf 74, bk. Read the whole chapter.

[1] *orig.* maysters [2] *orig.* maisters.

p. 187. *Lombard's craftiness.*—"The kynge this tyme [Henry VIII in 1511-12] was moche entysed to playe at tennes and at dice; which appetite, certain craftie persons about him perceauynge, brought in Frenchemen and *Lombardes* to make wagers with hym; & so he lost much money: but when he perceyued their craft, he exchuyd their compaignie, and let them go."—*Hall's Chronicle*, p. 520, ed. 1809.

p. 188. *Iene* or *Genoa, and the Genoese.*—See Thomas's interesting description of Genoa, on leaves 160 back, to 163, of his *Historye of Italye.* He was immensely struck by the beauty of their women, and the freedom they had.

"*Of theyr trade and customes.*—All the Genowaies in maner are merchant men, and very great trauailers of strange countreis. For I haue been reasonably persuaded that there be .5. or .6. thousand of them continually abroade, either merchauntes or factours: so that they haue few places of the worlde vnsought, where anye gaine is to be had. For the merchaundise that they bring home hath spedy dispatche, by reason theyr citee is as a keye vnto all the trade of Lumbardy, and to a great part of Italie.

They at home make such a noumber of silkes and veluettes as are hable to serue many countreys: whyche is the chiefe merchaundise that they sende forthe. In deede they are commonly noted to be great vsurers.

¶ One thing I am sure of, that if Ouide were nowe aliue, there be in Genoa that could teache him a dousen poinctes *De Arte Amandi.* For if Semiramis were euer celebrated amongest the Assirians, Venus amongest the Greckes, Circes among the Italians, sure there be dames in Genoa that deserue to be celebrated & chronycled for their excellente practise in loue. And trulye the Genowayes them selfes deserue that their wyfes should be praised; because I saw in no place where women haue so muche lybertee. For it is lawfull there openly to talke of loue, with what wife so euer she bee. Insomuch that I haue seene yonge men of reputacyon, standyng in the strete, talke of loue with yong mistresses beyng in theyr wyndowes aboue; and openlye reherse verses that they had made, one to the other. And in the churches, specially at euensong, they make none other prayers. So that he that is not a louer there, is meete for none honest companye. Many men esteme this as a reproche to the Genowaies; but they vse it as a policie; thinkyng that their wifes, throughe this libertee of open speache, are ridde of the rage that maketh other women to trauaile so much in secret.

¶ In dede, the women there are exceding faire, and best appariled, to my fantasie, of all other. For thoughe their vppermost garments be but plaine clothe, by reason of a law, yet vnderneth they weare the finest silkes that may be had, and are so finely hosed and shoed, as I neuer sawe the like, open faced, and for the moste parte bare headed, with the heare so finely trussed and curled, that it passeth rehearsall. So that, in myne opinion, the supreame court of loue is no where to be sought, out of Genoa" (leaves 161 bk, and 162).

p. 188. *The Genoese, their trading and products.*—The *Libel* of 1436 says,

> The Janueys comyne in sondre wyses
> Into this londe, wyth dyverse marchaundyses,
> In grete karrekkis arrayde, wythouten lake,
> Wyth clothes of golde, silke, and pepir blake
> They bringe wyth hem, and of wood grete plenté,
> Wolle, oyle, wood aschen, by wesshelle [=vessels] in the see
> Coton, roche-alum, and gode golde of Jene.
> And they be charged wyth wolle ageyne, I wene,
> And wollen clothe of owres, of colours alle.
>
> *T. Wright's Political Songs*, ii. 172.

p. 188. The trade of Italy with England, of which Hall speaks, under 1531 A.D., "Merchaunt straungers, and in especiall, *Italians*, Spanyardes, & Portyngales, daily brought Oade, Oyle, Sylke, Clothes of Golde, Veluet, & other Merchaundyse into this Realme, and therefore receiued ready money" (*Hall's Chronicle*, p. 781, ed. 1809), was doubtless carried on by the Genoese, Lombards, Venetians, and Neapolitans, whose merchandisings are noticed by Boorde.

p. 190. *French fashions.*—"With them [the French Ambassadors in 1518] came a great numbre of rascal, & pedlers, & Iuellers, and brought ouer hattes and cappes, and diuerse merchaundise, vncustomed, all vnder the coloure of the trussery of the Ambassadours. . . . The young galantes of Fraunce had coates garded with one colour, cut in .x. or .xii. partes, very richely to beholde. . . The last day of September, the French Ambassadors toke their barge, & came to Grenewiche. The Admyrall [Lord Boneuet] was in a goune of cloth of siluer, raysed, furred with ryche Sables, & al his company almost were in a new fassion garment called a *Shemew*, which was in effect a goune, cut in the middle."—*Hall's Chronicle*, p. 593-4, ed. 1809. The old chronicler didn't think much of the last of French soldiers:

"surely the nature of the Frenchmen is, not to labor long in fightyng, and muche more braggeth then fighteth."—*Hall's Chronicle*, p. 124, at foot, ed. 1809.

p. 196, l. 8-15. *Portuguese products and merchandise.*—The *Libel*, A.D. 1436, says,

> The marchaundy also of Portyngale
> To dyverse londes torne into sale . . .
> Here londe hathe oyle, wyne osey, wex, and grayne,
> ffygues, reysyns, hony, and cordeweyne,
> Dates and salt, hydes, and suche marchaundy.
>
> *T. Wright's Polit. Songs*, ii. 162-3.

p. 196, l. 10. *Portugal poor.*—A.D. 1524. "the Emperor answered: 'The very pouertie of your countrey of Portyngale is suche, that of your selfes you be not able to liue; wherfore of necessitie you were driuen to seke liuyng; for, landes of princes you were not able to purchase, and lande of lordes you were not able to conquere. Wherfore on the sea you were compelled to seke that which was not found.'"—*Hall's Chronicle*, p. 677, ed. 1809.

p. 197. *The fashion of the Spainierdes.*—"after whome came in .vi. ladyes appareled in garmentes of Crymosyn Satyn, embroudered and traucssed with cloth of gold, cut in Pomegranettes and yokes, strynged after *the facion of Spaygne.*"—*Hall's Chronicle*, p. 516, ed. 1809.

p. 198. *The Products of Spain* are stated in the *Libel* of 1436 to be

> . . . fygues, raysyns, wyne bastarde, and dates;
> And lycorys, Syvyle oyle, and grayne,
> Whyte Castelle sope, and wax, is not in vayne;
> Iren, wolle, wadmole; gotefel, kydefel, also,—
> ffor poynt-makers fulle nedefulle be the two;—
> Saffron, quiksilver (wheche arne Spaynes marchandy)
> Is into ffaundres shypped fulle craftyle,
> Unto Bruges, as to here staple fayre,
> The haven of Sluse here havene for here repayre,
> Wheche is cleped Swyn; thaire shyppes gydynge
> Where many wessell and fayre arne abydynge.
>
> *T. Wright's Political Songs*, ii. 160.

p. 202. *The poverty of Navarre (& Spain).*—"The English souldiers, what for sickenes, and what for *miserie of the countrey*, euer desired to returne into England . . . saiyng, that thei would not abide and die of the flixe in *suche a wretched* country."—*Hall's Chronicle*, p. 532, ed. 1809. Navarre was won by the Spaniards under the Duke of Alva, in the 4th year of Henry the 8th, A.D. (22 April, 1512 to 21 April, 1513). See *Hall's Chronicle*, p. 530, ed. 1809.

p. 203. *Hanging long on the Gallows.*—This must have been done also in some cases in England: "the harlot, Wolfes wyfe . . . at the last, she and her husband, as they deserued, were apprehended, arraigned, & hanged at the foresayd turnyng tree [a place on the Thames], where *she hanged still, and was not cut doune*, vntil suche tyme as it was knowen that beastly and filthy wretches had moste shamefully abused hêr, beyng dead."—*Hall's Chronicle*, p. 815, ed. 1809.

p. 205-6. *The Pilgrims to St James of Compostella.*—Contrast the reality with the Court notion of "pilgrims from St James" in February, 1510-11: "Then came nexte the Marques Dorset and syr Thomas Bulleyn, like *two pilgrims from sainct Iames*, in taberdes of blacke Veluet, with palmers hattes on their helmettes, wyth long Iacobs staues in their handes, their horse trappers of blacke Veluet, their taberdes, hattes, & trappers, set with scaloppe schelles of fyne golde, and strippes of blacke Veluet, euery strip set with a scalop shell; their seruauntes all in blacke Satyn, with scalop shelles of gold in their breastes."—*Hall's Chronicle*, p. 518, ed. 1809.

p. 207. *Britanny's products; and its hatred of England.* The *Libel*, A.D. 1436, says,

Commodité therof there is and was,
Salt and wynes, creste clothe, and canvasse ....
And of this Bretayn, who-so trewth[e] levys,
Are the grettest rovers and the grettest thevys
That have bene in the see many oone yere:
That oure marchauntes have bowght full dere;
ffor they have take notable godo of oures
On thys seyde see, these false coloured pelours,
Called of Seynt Malouse, and elles where,
Wheche to there duke none obeysaunce woll bere.
Wyth suche colours we have bene hindred sore,
And fayned pease is called no werre herefore.
Thus they have bene in dyverse costes manye
Of oure England, mo than reherse can I;
In Northfolke coostes, and othere places aboute,
And robbed, and brente, and slayne, by many a routte;
And they have also ransonned toune by toune,
That into the regnes of bost[1] have ronne here soune.

T. Wright's Polit. Songs, ii. 164.

p. 207, line 1. *Bayonne once English.*—It was lost in the 29th year of Henry VI (1 Sept. 1450 to 31 Aug. 1451). Hall says in his *Chronicle*, p. 224, ed. 1809, "When the cities and tounes of Gascoyne wer set in good ordre, the Erle of Dunoys and Foys, with greate preparacion of vitaill, municion and men, came before the citie of Bayon, where, with mynes and battery thei so dismaied the fearful inhabitantes, that neither the capitain nor the souldiors could kepe them from yeldyng: so by force they deliuered the toune; and their capitain, as a prisoner, offred a great some of money for the safegard of their lifes and goodes."

p. 209. *Boulogne.*—"Althoughe this peace [of 1546 A.D.] pleased both the Englysh and the French nacions, yet surely both mistrusted the continuaunce of the same, considering the old Prouerbe, 'that the iye seeth, the harte rueth;' for the French men styll longed for Bulleyn, and the Englyshmen minded not to geue it ouer."—*Hall's Chronicle*, p. 867, ed. 1809.

p. 218. *Jewry or Judæa.*—See, under "Asie," the chapter "Of Jewry, and of the life, maners, and Lawes of the Jewes in *the Fardle of Facions*, conteining the aunciente maners, customes, and Lawes of the peoples enhabiting the two partes of the earth called Affrike and Asie. Printed at London, by Ihon Kingstone and Henry Sutton. 1555, sign. Ii. back." 'Palestina, whiche also is named Judea, beinge a seueralle province of Siria, lieth betwixte Arabia Petrea and the countrie Cœlosiria. So bordering vpon the Egiptian sea on the west, and vpon the floude Jordan on the Easte, that the one with his waues wassheth his clieues, and the other sometime with his streame ouerfloweth his banckes.

(sign. I vii. back.) 'The lande of Siria (whereof we haue named

[1] of the best. MS. Cotton. Vitel. E. x.

Jewrie a parte) is at this daie enhabited of the Grekes called Griphones, of the Jacobites, Nestorians, Saracenes, and of two christian nacions the Sirians and Marouines. . . . The Sarracenes, whiche dwelle aboute Jerusalem (a people valeaunt in warre) delighte muche in housbandrie and tilthe.'

p. 219, 60, 144. *Venice, &c., and Englishmen abroad.*—In the Gentleman's Magazine for October, 1812, reprinted in Fosbroke's British Monachism, ch. vii, p. 337, ed. 1843, are some extracts from a MS Diary of a Pilgrimage to Jerusalem made by a Sir Richard Torkington in 1517. He started on March 20, 1517, from Rye in Sussex, and got back to Dover on April 17, 1518: "We war owt of England in ower sayd pylgrymage the space of an holl yer, v. wekys, and iij. dayes." "We com [29 April, 1517] to the goodly and ffamose Cite of Venys. Ther I was well at ese, ffor ther was no thyng that I desired to have, but I had it shortly. At Venyse, at the fyrst howse that I cam to except oon, the good man of the howse seyd he knew me, by my face, that I was an englyshman. And he spake to me good englyssh. thanne I was jo[yo]us and glade. ffor I saw never englyssh man ffrom the tyme I departed owt of Parys to the tyme I cam to Venys. which ys vij. or viij.C. myles."

p. 220. *Joppa.*—"At Jaffe begynnyth the holy londe; and to every pylgryme, at the ffyrst foote that he sett on the londe, ther ys grauntyd plenary remission *De pena et a culpa.* In Jaff, Seynt Petir reysid from Deth, Tabitam. the sarvaunt of the Appostolis. And fast by ys the place where Seynt Petir usyd to ffysh, And our Savior Crist callyd hym, and seyd *sequere me.*"—Sir Richard Torkington's Diary, 1517; in Fosbroke's *British Monachism*, p. 338, col. 1, ed. 1843.

III. NOTES ON BOORDE'S *DYETARY.*

p. 225. *Sir R. Drewry.*—In Hall's account of the Insurrection in Suffolk, A.D. 1525, he says "the people railed openly on the Duke of Suffolke, and *sir Robert Drurie*, and threatened them with death."—*Chronicle*, p. 699, ed. 1809.

p. 232. Compare "The boke for to lerne a man to be wyse in buylding of his house for the helth of [his] body, and to holde quyetnes for the helth of his soule and body &c." [Coloph.] Imprynted by me Robert Wyer, dwellynge at the sygne of St. Iohn Euangelyst, &c. 8vo, 16 leaves. *Brit. Museum.* (Hazlitt's Handbook, p. 366, col. 2.)

p. 236. *Let nother flaxe nor hempe be watered.*—"Here and there was an artificial flat-bottomed pool of water, formed by damming up one of the many rivulets which ran from their sources in the distant hills to empty themselves into the adjacent Rhine. At the bottom of each pool were bundles of flax undergoing the first process preparatory to their ultimate conversion into linen fabrics. The odour of the decomposed or decomposing flax was the reverse of agreeable. Indeed, the prevalence of bad smells was the chief drawback to the enjoyment of the prospect."

Daily News, Sept. 13, 1870; letter from Achern, Sept. 6, describing the country from Achern to Auenheim, a small village, close to the right bank of the Rhine, near Strasburg, which was then besieged by a German army..

p. 239. *Dovehouse.*—The Norfolk and Suffolk rebels under Kett in 1549 say in their list of Grievances: "We p[r]ay that noman vnder the degre of a knyght or esquyer, kepe a *dowe house*, except it hath byn of an ould aunchyent costome." •Was this because the doves eat the poorer men's grain, as the rich men's pheasants and partridges—and worse, hares and rabbits,—now do? See my *Ballads from Manuscripts*, i. 149.

p. 241. See the 'Proverbys of Howsolde-kepyng' in my ed. of *Political, Religious, and Love Poems*, for the Society, 1866, p. 29.

p. 243. *Instructing the Ignorant.*—Teaching them a Robin-Hood ballad or the Primer, perhaps, after Robert Crowley's exhortation to unlearned curates in his *Voyce of the last Trumpet*, 1550. (E. E. T. Soc. 1871.)

p. 244. *Epilencia*, &c. were generally called Epile*psia*, Anale*psia*, and Catale*psia*. See Boorde's *Breuiary*, ch. 122, Fol. xlvi.

p. 250. *Boarded Chambers.*—Wooden floors were not common in Boorde's days. One of his remedies for a stitch in the side is "take vp the *earth* within a dore, that is *well troden*, and pare it vp with a spade, after [= a piece like] a cake; and cast Vineger on it, and tost it against the fyer; and in a lynnen clothe laye it hote to the syde."—*Breuiary*, Pt. II, *The Extrauagantes*, Fol. xi, back. See too the well-known quotation from Erasmus on the filthy clay floors of England, in the *Babees Book*, Forewords, p. lxvi.

p. 252. *Water.*—*Eau & pain, c'est la viande du chien:* Prov. Bread and water is diet for dogs. *Cotgrave.*

p. 253. *Standing Water.*—*L'eau qui dort est pire que celle qui court:* Pro. So is a sleepie humor worse then a giddie. *Il n'y a pire eau que la quoye:* Prov. The stillest waters (and humors) are euer the worst. *Cotgrave.*

p. 254. *Wyne . . must be . . fayre . . and redolent*, &c.—The compiler of what Mr Dyce, in his Skelton's Works, vol. i. p. xxx, calls 'that tissue of extravagant figments which was put together for the amusement of the vulgar, and entitled the *Merie Tales of Skelton*' (T. Colwell), probably had Boorde's opinion on wine before him when he wrote "all wines must be *strong*, and *fayre*, and well coloured; it must have a *redolent* sauoure; it must be *colde*, and *sprinkclynge in the peece or in the glasse*."—Tale xv. *Skelton's Works*, vol. i. p. lxxiii.

p. 260. *London bakers' trickery.*—A.D. 1522. In this yere the bakers of London came and told the Mayre that corne would be dere; wherupon he and the aldermen made prouision for xv.C. quarters; & when it was come, they [the bakers] would bye none, and made the common people beleue that it was musty, because they would vtter their owne, so that the lord Cardynal was faine to proue it, and found the bakers false, and commaunded them to bye it.—*Hall's Chronicle*, p. 650, ed. 1809.

p. 273. *The Jews love not pork.*—"Swines flesche thei eate none, for that thei holde opinion that this kynde of beaste, of it selfe beinge disposed to be skoruie, might be occasion againe to enfecte them of newe." —*The Fardle of Facions*, 1555. I. iv, not signed.

p. 273. *Adder's flesh eaten, and called "fysshe of the mountayn."*

Now followeth the preparing of Serpents: Take a mountain Serpent, that hath a black back, and a white belly, and cut off his tail, even hard to the place where he sendeth forth his excrements, and take away his head with the breadth of four fingers; then take the residue and squeese out the bloud into some vessel, keeping it in a glass carefully; then fley him as you do an Eele, beginning from the upper and grosser part, and hang the skin upon a stick, and dry it; then divide it in the middle, and reserve all diligently. You must wash the flesh and put it in a pot, boyling it in two parts of Wine; and, being well and throughly boyled, you must season the broth with good Spices, and Aromatical and Cordial powders; and so eat it.

But if you have a minde to rost it, it must be so rosted, as it may not be burnt, and yet that it may be brought into powder; and the powder thereof must be eaten together with other meat, because of the loathing, and dreadful name, and conceit of a Serpent: for being thus burned, it preserveth a man from all fear of any future Lepry, and expelleth that which is present. It keepeth youth, causing a good colour above all other Medicines in the world; it cleareth the eye-sight, gardeth surely from gray hairs, and keepeth from the Falling-sickness. It purgeth the head from all infirmity; and being eaten (as before is said), it expelleth scabbiness, and the like infirmities, with a great number of other diseases. But yet, such a kinde of Serpent as before we have described, and not any other, being also eaten, freeth one from deafness. —*Topsel's History of Four-footed Beasts and Serpents*, ed. J. Rowland, M.D., 1658, p. 616.

Mandeville says that in the land of Mancy, that is, in Ynde the more, and which is also called 'Albanye, because that the folk ben whyte,' "there is gret plentee of Neddres, of whom men maken grete Festes, and eten hem at grete sollempnytees. And he that makethe there a Feste,—be it nevere so costifous,—and he have no Neddres, he hathe no thanke for his travaylle."—*Voiage and Travaile*, p. 208, ed. 1839.

p. 275. *Great Men hunting.*—See, in 1575, G. Gascoigne. *Noble Art of Venerie.* Works, vol. ii. p. 305, ed. 1870.

"The Venson not forgot, moste meete for Princes dyshe:
All these with more could I rehearse, as much as wit could wyshe.
But let these few suffice, it is a *Noble sport*
To recreate the mindes of Men in good and godly sort.
A sport for Noble peeres, a sport for gentle bloods,
The paine I leaue for seruants such as beate the bushie woods,

To make their masters sport. *Then let the Lords reioyce,*
Let gentlemen beholde the glee, and take thereof the choyce.
For my part (being one) I must needes say my minde,
That Hunting was ordeyned first for Men of Noble Kinde.
And vnto them, therefore, I recommend the same,
As exercise that best becommes their worthy noble name."

p. 279. *Garlic* is good for 'longe whyte wormes in the mawe, stomake, and guttes,' says Boorde: "If any man wyll take a Plowemannes medicine, and the beste medicine for these wormes, and al other wormes in mannes body, let hym eate *Gerlyke.*" Breuiary, fol. lxxiii, ch. 212.

p. 279. *Garlic.*—Tharmie this [= thus, in 1512 A.D.] lyngeryng [in Navarre], euer desirous to be at the busines that thei came for, their victaile was muche part *Garlike;* and the Englishemen did eate of the *Garlike* with all meates, and dranke hote wynes in the hote wether, and did eate all the hote frutes that thei could gette, whiche caused their bloudde so to boyle in their belies, that there fell sicke three thousande of the flixe; and thereof died .xviii. hundred men.—*Hall's Chronicle,* p. 529, ed. 1809.

p. 289. *Sweating Sickness.*—After this great triumphe [Henry VIII's jousts in June, 1517] the king appointed his gestes for his pastyme this Sommer; but sodeinly there came a plague of sickenes, called the *Swetyng sickenes,* that turned all his purpose. This malady was so cruell that it killed some within three houres, some within twoo houres, some, mery at diner and dedde at supper. Many died in the kynges Courte, the Lorde Clinton, the Lorde Grey of Wilton, and many knightes, Gentlemen and officiers. For this plague, Mighelmas terme was adiourned; and because that this malady continued from July to the middes of December, the kyng kept hymself euer with a small compaignie, and kept no solempne Christmas, willyng to haue no resort, for feare of infeccion; but muche lamented the nomber of his people, for in some one toune halfe the people died, and in some other toune the thirde parte, the Sweate was so feruent and infeccious.—*Hall's Chronicle,* p. 592, ed. 1809. See the history of this plague in *Chambers's Book of Days,* under April 16; also in my *Ballads from Manuscripts,* Part II, 1871.

INDEX OF SUBJECTS AND WORDS.

122/9 means page 122, line 9; 133 means page 133.

Abarde, 120, ? in Cornwall.
a base, 238, lower down, beyond.
ABC, 20, alphabet.
abiected, 258, 285, thrown away.
ablatyd, 284, 285, thrown away.
Abraham, 233.
abstercyue, 263, abstersive, 285.
abstinence the best medicine, 251.
abstraction, 101, what you draw out?
Acayra, 172, Achaia.
acca, ava, agon; children's cries, 91.
acetose, confection of, 102.
Acobrynge, 197, Alcoutrin?
Acon, 219, Aix-la-Chapelle, Aachen.
acuate, 244, sharpen.
Adam: who shaved him? 314.
adders, none in Ireland, 133; eaten in Lombardy, 187; eaten in Rome, and called 'fish of the mountain,' 273, 350.
Adrian, Pope, 24, 78.
adulterating bakers, 260-1.
adultery of wives, Boorde's remedy for, 68.
affodyl, 102, daffodilly?
afyngered, 122/9, a hungered, hungry.
agarycke, 288; pilles of, 99.
agedly, 300.
Agnus castus, 100.
ague, 21, 325; how to treat, 291; butter is bad for, 266.
Agur, the son of Jakeh, 67.
air, the need of good, 235, 238.
al, 122/1, ale.
alaye, 254, temper.
alchermes, 103.
alchytes, 299.
ale, 256; awfully bad in Cornwall, 122, 123; and in Scotland, 136; John Taylor on, 326.
ale-brewers and ale-wives, bad, to be punisht, 260.
ale-brews, 264; ale-brue, 97.
ale pockes in the face, 95.
ale, posset, 256.
alexanders, 278, the herb Great Parsley.
Alicant wine, 75, 255, 327.
aliens, Boorde dislikes them, 60.
alkemy, 161, 163, tin.

alkengi, the confection of, 79.
all out, drink, 151/6, 324.
all-to-nowght, 62, good-for-nothing.
allygate, 245, allege.
Almanac and *Prognostication*, supposed to be A. Boorde's, 26-7.
Almayne, Low, 155-8; High, 159-162; maidens of, don't drink wine, 254.
Almen, 53, Germany.
almond-butter, 267.
almond-milk, 263.
almonds, 285.
aloes, 290.
Alygaunt, 255, 75, Alicant wine.
amber de grece, 93.
Amsterdam, 149.
amytted, 25 admitted.
an, 246, if.
anacardine, confection of, 95.
analencia, 244, a kind of epilepsy? See Boorde's *Breuyary*, fol. xlvi.
Ancress at St Albans is infested by a spirit, 78.
Andalase, 196, Andalusia.
Angeou (Anjou), white wine of, 75.
anise-seed, 284.
Antwerp described, 151; its church and spire, 339.
Anwarpe, 219, Antwerp, 338.
apples, 284.
appoplesia, 244, apoplexy.
appostata, 62, apostate.
approbat, 273, approve.
approbat, *adj.*, 282, approved.
aqua vitæ, 258, 351; Irish, 131/8; 334.
Aquitaine, 191; described, 193, 206.
Araby, 20, Arabic.
archane, 21, secret, hidden.
Argentyne, 156.
Aristotle, 91.
armipotentt, 53, powerful in arms.
Arragon described, 195, 53.
Arran, Earl of, named Hamilton, 59.
Arras cloth made in Brabant, 151/2; in Liege, 155.
artichokes, 280.
artoures, 101; artures, 91/7, arteries.
Artuse, 176, river Arethusa, in Sicily.
Arundel, 120.
Arundel, Sir John, 55.
aryfye, 247, burn and dry up.
ascarides, 81, 279, little long worms in the anus.
Ascot, 110.
Asia, Boorde never in it, 145.
aspers, 216, Turkish silver coins.
asthma, Boorde's cure for, 99.
asthmatic men, a diet for, 297.
Astronamye, the Pryncyples of, by Andrew Boorde, 16, 22-23.
astronomers or astrologers, the gammon of, 325.
astronomy, importance of the study, 25.
avarice, 86.
Aueroyes quoted, 272, 274.
Augsburg, 161.
aungels, 121, gold coins worth from 6*s*. 8*d*. to 10*s*.
auripigment, 102.
Avycen quoted, 91, 258, 274, 282.
auydous, 252, avidous, greedy.

backehowse, 239, bakehouse.
bacon, good for carters, bad for the stone, 273.

bagantyns, 189, Italian brass coins: *bagatino*, a little coine in Italie. *Florio.*

baked pears, 291.

baken, 284, baked.

bakers, rascally, 260, 349.

Bale, Bp, on A. Boorde, 33.

ballot in Venice, 184-5.

banocke, 283, a kind of walnut.

Barbarossa, 55, 213.

Barbary sleeves, 106.

Barber, Barnarde, 305, 307.

Barberousse, 213, 55, Heyradin Barbarossa.

Barcelona, 55.

Bargen in Hainault, 151, Bergen.

barges, the fair little ones in Venice, 183; '*Gondola*, a little boat or whirry vsed no where but about and in Venice.' 1611, *Florio.*

barley, 259.

barley malt is the best for ale, 256.

Barnes in the Defence of the Berde, 305—316; date of, 19-20.

Barow, 150, 338.

Barsalone, 195, Barcelona.

Barnsley in Gloucestershire; Sabbatarian superstition in, 332.

Barslond, 160, the Tyrol.

Bartholomew of Montagnave, 291.

Base-Almayne, 148, the Netherlands; described, 155-7.

Bastard wine, 75, 255.

Bath, waters at, 120.

Batmanson, Prior, 47, 48, 57, 58.

Batow, 150/5.

Bayonne, 206-7, 347.

bean-butter, 268.

bean-potage, 263.

beans, 284; and peas, 259; and stockfish, Danish food, 163/5.

Beards, Boorde's lost treatise on, 307, 309, 26; Barnes's answer to it, 305—316.

beards, Harrison on, 16, note.

bears, white ones in Norway, 141/18.

beasts, reasonable; men and women are, 91, 93.

bedauer, 122/16, 21, ? father or partner.

bedtime, what to do at, 246.

beef good for Englishmen, 271.

beer, 256.

beets, white, 280.

bekyng, 185, 207, pointing, poking.

Bell, Humfrey, 74.

Belvedere, a fort in Windsor Forest, 110.

benche-whystler, 245-6.

bengauyn, 290, ? gum Benjamin.

Berdes (beards), Boorde's *Treatyse vpon*, 26, 308.

Bergevenny, Lord, frees his villein Andrew Borde, 41-2.

Berwick, 120, 136.

beryd flesshe, 277, meat-pie.

beshromp, 207/8, hate ?

Bindley, Mr, 227, note.

Bion (Bayonne) described, 207-8.

birds, small, 270.

Biscay described, 199, 200; 53.

Bishop must be 30 years old, 44.

Bishops should examine and license Midwives, 84.

Bishops-Waltham in Hampshire, 52, 53, 60; eight miles from Winchester, 145.

blackbird, 271.

blanched almonds, 282.

blaynes, 284, blains, sores: cp. chil*blains*.

bleareyed mare, 273.

blockhouses in England, 119, 329.

blood not good to eat, 276.

boar, the brawn of, 274.

boar's grease, 97, 102.

Board Hill in Sussex, 38-9.

boarded chamber, 250, 349.

boasters, the Scotch are great ones, 137.

Boece, Hector, on Scotchmen's degenerate ways, 259-60, note.

boggery (buggery) in Rome, 77.

Bohemia and the Bohemians, 166-7.

boiled meat, 289; is digestible, 277.

boiling meat in a skin, 132.

Boleyn, Anne, her badge on the dining-room ceiling of Great Fosters, 7.

Boleyn, Bolyn, 209, Boulogne.

bollynge, 293, drinking with a bowl.

Bolton, Prior of St Bartholomews, Smithfield, makes a fool of himself, 325.

bongler, 21, bungler.

Bonn, red Rhenish wine grown about, 75.

Boord's Hill, 23.

BOORDE, Andrew; his Works (list, p. 9), 10—26, 64; his Life (table of facts of, 10), 36—105; his Letters, I, 45; II, 53; III, 55; IV, 57; V, 58; VI, 59; his Will, 73; his opinions and practice, from his *Breuyary*, 74—104; his *Introduction*, 111—222, 317; his purpose in it, 144-6; his *Dyetary*, 223—304, 319; his motives in writing, 20-1; places visited by him, 63; supposed portraits of him, 74; he hates water, but likes ale and wine, 75; dislikes whirlwinds, 75; trusts in God's will, which is his, 75-6; fears that devils may enter into him, 76; is shocked at the vices of Rome, 77-8; has *cachexia*, 79; has the stone, 80; gets a nit or fly down his throat, 81; his urine, 81; has seen worms come out of men, 81; complains of Englishmen's neglect of Fasting, 82, Swearing and Heresies, 82-3; Laziness of young people, 83, want of training for Midwives, 84, Cobblers being Doctors, 84-5, the Mutability of men's minds, 85, the Lust and Avarice of men, 85-6; alludes to the bad food of the poor, 86-7, and early marriages, 87; thinks Lying the worst disease of the Tongue, 88; praises Mirth, 88-9; treats of a man's Spirits, 88-9, of the Heart, 89, of Pain and Adversity, 89, Intemperance, Drunkenness, 90, Man and Woman (which be reasonable Beastes), 91, Marriage, 91, the words of late-speaking Children, 91, the King's Evil, 91-3, men's Five Wits, 93, Wounds, 94, Obliviousness, 94, Dreams, and man's Face, 95; his Medical Treatment of Itch, 96, Tertian Fever, 96, Scurf, 96, curded Milk in Women's Breasts, 96, pregnant Women's unnatural Appetite, 98, Ulcer in the Nose, 98, Asthma, 99, Palsy, 99, Excoriations, 99, Fatness, 100, Priapismus or involuntary Standing of a Man's Yard, 100, Web in the Eye, 100, rupture of the Gut-Caul, a Sauceflewme Face, 101-2; his opinion on the Soul of Man, 102, on Freewill, 103; his Exhortation to his Readers, 103; his Preamble or advice to Sick and Wounded men, 104; his character, 105; was esteemed by his contemporaries and successors, 105-6; sham portraits of him, 108, 143, 305; he loves venison, 274; doesn't like pork, 272; his powder for the Pestilence, 290.

Boorde, Sir Stephen, 39; Stephen, 43.

boots rubd with grease, 99.

borage, 253, 278, 280, 289.

Borde, Andrew (son of John Borde), Lord Bergavenny's villein, 41-3.

Borde, Dr Richard, 43, 65.

Border, the Scotch, 136.

bornet, 276, burnt.

Bostowe, 120, ? Bristol. § .154. In eadem valle est vicus celeberrimus, *Bristou* nomine, in quo est portus navium ab Hibernia et Norregia et ceteris transmarinis terris venientium receptaculum. A. D. 1125-40. *William of Malmesbury*'s *Gesta Pontificum Anglorum*, bk iv, p. 292, ed. Hamilton, 1870. See also "The Childe of *Bristowe*," a poem by Lydgate, in the *Camden Miscellany*, vol. iv, and Hazlitt's *Early Pop. Poetry*, i. 110.

Boulogne, Henry VIII's conquest of, 18, 209, 347.

Boune, 219, Bonn.

bovy, 167, a beast in Bohemia.

Bowker, Agnes, 78.

bowling-alley to be near every mansion, 239.

Bowyer, Magdalen; Dr J. Storie's wench, 69.

boys marrying, 87.

Brabant and the Brabanders, 150, 338.

Bradshaw, H., 11, note 2; 324.

brains bad to eat, 276.

bran of bones, 94.

brande, 258, bran.

brawn, 274.

bread, a pen'orth of, lasted Boorde a week, 51.

bread strengthens the heart, 89.

bread, the kinds and properties of, 258-262.

Breuyary of Health by Andrew Boorde, 20-22; the name explained, 21; extracts from, 74—104; references to, 291, 299, &c., &c.

Brewer, Prof. J. J. S., 43.

brewhouse, place for the, 239.

brewsters, bad; the Scotch punishment for, 261.

Bridlington, 120.

Bright-Hemston, 120, Brighton.

Brindisi, the cathedral of Naples, 177.

Britany, 207; its products, and its hatred of England, 346.

bronte, 296, burnt.

broths, 264.

brount, 245, long spell.

brown paper; wipe your pimply face with, 102.

bruled, 277, broiled.

Brune, Nicholas, 74.

Brussels, 151.

bruttell, 266, brittle.

bryched, 94, last line, ? come to puberty.

bryched, 95, breeched.

buck and doe, 274, fallow deer.

bugle, 167, 340, a kind of ox.

bugloss, 278, 280, 253.

building, the things needed for, 237.

bulwarks put up by Henry VIII, 119, 329.

Bune, 156, 219, Bonn.

bur roots, 102.

Burdiouse, 206, 207, Bordeaux.

Burdyose, 53, Bordeaux.

Burges, 147, 219, Bruges.

Burgos in Spain, 199.

Burgundy, 191.

burial-customs, absurd, in Castille, &c., 200, and in Wales, 128.

burnet, 289, burnt.

Burse, or Bourse, of Antwerp, 151.

Butte, Dr, phisicion to Henry VIII, 49, 226.

butter, 265.

Butter, eaten in Flaunders, 147/4; barrelled, salt, and bad, in Holland, 149/5, 14; salt, in the Netherlands, 156/11; 339.

butterish, or unctuous, 265.

'Buttermouth Fleming,' 147/3.

buttery, the ghost of the, 75.

buttery, &c., to be kept clean, 237; place for it, 238.

butts, a pair to be near every mansion, 239.

'By a bancke as I lay,' a ballad, 71, note.

Byborge, 163, Wiborg in Denmark.

byles, 284, boils.

byokes, 179, baiocchi; It. *Baiócco*, a mite or such like coine: *Florio*. *Bajocco*, a Roman copper coin worth about a halfpenny. *Baretti*.

Byon, 53, 206, 207, Bayonne.

byttoure, 270, bittern.

cachexia, 79, 327.

Caernarvon, 120, 330.

Cagliari in Sardinia, 55.

cakes, 9 for a penny in Aquitaine, 194.

Calabria, 175-6.

Calais, 120; described, 147; 209.

calculus, 80, the stone.

Caldy, 216, Chaldee.

Calvary, Mount of, 220.

Calyco, 142/7, Calicut?, 336.

caliditie, 100, 102, heat.

calles, 91, cauls.

Cambridge, 120; Boorde's letter from, to Cromwell, 62; Boorde's books in the University Library, 11, 12, 16.

Camden Society's Council of 1870, admire Mr J. P. Collier's editing, 71, note.

camel, Mahomet's, 215-16.

camomyll, 99, camomile.

camphor, oil of, 100.

Can, 208, Caen.

Candia, 172, 182, 219.

candle-ends eaten in Iceland, 141/4; 142, 336.

candles, 264.

canelles, 236, 295, channels, drains.

cankers in the face, 95.

Canterbury, 147.

Cantica Canticorum, quoted, 238.

capers, 285.

capon the best fowl, 270.

Caprycke, 255, wine from Capri.

carcinoma, 72, prison-sickness.

Cardinals, Spanish, 204; Italian ones' pages, 77.

cardyng, 293, playing cards.

Carewe, Sir Wymonde, 64-5.

Carlisle, 120.

carrots, 279.

carters, bacon good for, 273.

Carthusian Order; the strictness of it, 46; A. Boorde couldn't abide its 'rugorosite,' 47.

caryn, 236, carrion.

Castel Angelo in Rome, 77.

Castile, 53, 195, 198; described, 200-1.

castors, 141, beavers, in Norway.

castory, 298.

castynge of a pys-potte, 311, looking at the urine in one.

cat, game of, mentioned, 332.

catalencia, 244, 349, catalepsy.

Catalonia, 56; described, 194-5.

caudle or cullis for a dying man, 302.

caudles, 264.

cauterise, 101.
caves, Icelanders lie in, 142.
cawse boby, 126, 330, 340, roasted cheese.
cedar-trees, 218.
Celestynes in Rome, 77.
cellar, place for the, 238.
centory, 288, centaury.
ceruyces, 283, services, a big kind of pear.
chaffyng, 290, warming.
cham, 122/1, am.
chamber of estate, 238.
chapels at Rome defiled, 77.
charcoal, 291.
Charneco wine, 255, note.
Charterhouse, the Head, 55, the Grande Chartreux.
Charterhouse in London, Boorde in it, 42, 43, 45, 47, 49, 51, 52; in Rome, 77.
chartes (cards), the Irish play at, 333.
Chaucer's Reeves Tale, 33; his Somonour's sawcefleem face, 101-2.
che, 122/1, I.
cheese-maggots eaten in Germany, 160.
cheese, the five kinds, and the qualities of, 266-7.
cherries, 283.
Chester, my Lord of, 57, ? the Prior.
chesteynes, 285, chestnuts.
Chichester, 120.
chicken, 270.
chicory, 280.
chierurgy, 20, 21, surgery.
chilblains, 86.
chimneys, don't piss in them, 237.
chip the top-crust off your bread, 261.
choleric men, 245; a diet for, 288.
Christ and his Apostles wore beards, 314/131.
Christ bids men watch, 245.
Christ, the pillar that he was bound to, 76.
Christie-Miller, Mr S., 19, 106-7, 227.
churchmen's courtesans in Italy, 184.
chybbolles, 294.
chyl, 122/14, will.
Ciclades, 172, the Cyclades.
cider made of pears or apples, 256.
cinnamon, 287, 292.
cipres, 218, cypress.
Ciracus, 176, Syracuse.
claret wine, 255.
Clemers gylders, 140, 153.
Cleveland, 142-3.
Clipron, a noble city in Hungary, 171.
clock: the Italians count to 24 o'clock, 178-9.
clockyng in ones bely, 86.
cloves, 286.
clowtyd (clotted) cream, 267.
coactyd, 53, compelled.
cobblers, &c., turn doctors, 85.
cochee, pills of, 99.
Cocke Lorelles bote, a fool of, 306; take an oar in, 313/101.
cockrellys, 270, young cocks stewed, 296.
cockrel's stones good to eat, 277.
cock's flesh, 270.
cognacion, 233, kindred.
Cokermouth, 120.
Cokersend, 120.
cokes come, 185, cock's comb.

colesses, 264, cullisses, broths.

colic, broths bad for, 264; beer bad for, 256; mead bad for, 257.

Collie weston, 106?

Collier, J. P., quoted, 30; his daring invention, 71; his coolness, 72, note; his inaccuracy, 326; has mist two Boorde entries in the Stationers' Register A, 14.

colloppes and egges bad for the stone, 273.

Collyn Clowte's treatyse answerynge the boke of Berdes, 305—316.

colmouse (the bird), 270.

Colyn, (Cologne, 219), the noble city, 75, 156; the thread of, 337.

comb your head often, 300.

comfettes, 284, comfits.

common, 301, chatter.

compacke, 91, compact, constituted.

company, honest, 89. See *mirth.*

Complaynt of Scotland, 1548-9; its opinion of Englishmen, 59, note 3.

Compostella, Boorde's pilgrimage to, 51, 199, 204, 346.

conies, 275, grown-up rabbits.

connexed, 102, 103, bound together.

Constantinople described, 172.

constupat, 292, constipated.

consumption; woman's and goat's milk are good for, 267; a diet for, 296.

Cony[ng]sby, Wm, gives A. Boorde 2 tenements in Lynn, 73.

cook, a good one is half a physician, 277.

Cooper, W. Durrant, his "unpublished correspondence" of Boorde, 45.

Copland, old Robert, 15, 16. (See my Forewords to *Gyl of Breyntford's Testament*, &c., 1871.)

Coplande, Wm; his editions of Boorde's *Introduction*, 14—19; he printed first at the Rose-Garland, second at the Three Cranes, third at Lothbury, 18.

corans, 282, dried currants; raisins of Corinth.

Cordaline Friars at Jerusalem, 220.

cordyallys, 296, cordials.

Corfu, 182.

corn shouldn't be exported from England, 118.

Cornelis of Chelmeresford, 17, note.

Cornish men described, 122-4; language, samples of, 123-4.

Cornwall, 120, 330.

coroborate, 285, strengthen.

Corpus Christi day, 219.

Corser, Mr, 11, 27.

costine, oil of, 95.

cotydyal, 226, col. 2; 241, daily.

Coualence, 219, Coblentz (*Confluentia*).

couetyse, 86, covetousness.

coun, 122/17, grant.

Course, 75, 255, Corsican.

courtesans in Venice, 183. *Cortegiána*, a curtezane, a strumpet, *quasi Cortése áno*, a curteous tale! *Florio.*

cow-flesh, 271.

Cox, Captain, 32.

coyte, 258, water and yeast.

crab-lice, 87.

crache, 97, scratch.

crackling not to be eaten, 274.

cracknelles, 80, 261.

crake, 137, brag: the Scotch do it.

cramp-rings, the hallowing of, 92.

crane, 270.

cream, 267.

croaking in one's belly, 86.

crocherds, 157, Dutch coins worth about ⅜d, ? kreutzers, 161.

Cromwell, Thomas, loses Boorde's Handbook of Europe, 24, 145; Boorde's 5 letters to him, 53, 55, 58, 59, 62; his kindness to Boorde, 52; is made a brother of the Charterhouse, 57.

Cross, Holy, said to be at Constantinople, 173; cross to be held before a dying man, 302.

crowns and half-crowns, 121; Scotch crown of 4s. 8d. is called a Pound, 137; Dutch crown 4s. 8d., 157; French, 191.

crusts are unwholesome, 261.

Cuckfield (Cookfield), Sussex, 39.

Cuckold, a town in Yorkshire, 61.

cucurbiti, 81, worms.

cucurbitini, 279, square worms.

cunables, 208, cradle.

cupboard, lean against it when you sleep in the day, 246.

cupshote, 309; cupshoten, 156/2, drunk.

curding of milk in women's breasts, 97.

cur-dogs in Lombardy, 187.

cursados, 197, crusados, Portuguese gold coins worth 5s. a piece. Sp. *Cruzádo*, m. a peece of money so called, in Portingall, of the value of a French crowne. *Minsheu*.

cycory, 253, chicory.

Cyuel, 196, Seville.

dagswaynes, 139, rough coverlets (see Harrison's *Descr. of England*).

dairy, 239.

Dalmacye, 172, Dalmatia.

damsons, eat 6 or 7 before dinner, 285.

dandruffe, 95, dandriff.

Dansk whyten, 163, Danish tin and brass coins.

Dartmouth, 120.

dates, 285.

daundelyon, 253, dandelion.

deathbed service, 302.

debt, the evils of, 242.

decepered, 103, deciphered ?, separated.

degges, 81, worms in a man's feet.

demoniack, 298.

denares, 179, Italian pence: *Denári*, pence, money, coine. *Florio*.

Denmark and the Danes, 162-3, 339.

Devil, swearers are possest of him, 83.

devilish disposition of Scotchmen, 61.

devils in a German lady, 76.

Devil's nails unpared, 117/30 (a phrase).

Deynshire, 129, Devonshire.

Diascorides, 282.

diaserys, 100.

Dibdin on Boorde's *Introduction*, 36.

dice, Irish, 131/8; the strong and weak man at, 245.

diet, a general one, for all people, 300.

dinner, sit only an hour at, 252; bad English customs at, 252.

dishes, eat only of two or three, 248, 252.

'dispensyd with the relygyon,' 44-5, 57, 58.

disquictness, 89.

Ditchling in Sussex, 41-2.

Dobie's Hist. of St Giles' and St George's, Bloomsbury, quoted, 65, note.

doble, double, 191, a French coin worth 2 brass pence.

doctor and cook must work together. 277-8.

Doge or Duke of Venice, 183-5.

dogs, wounds from, 94.

Dolphemy, 191, Dauphiny.

done theyr kynde, 277, copulated.

dormitary, 95?

dorow, 122/19, through.

dove's-dung in a plaister, 97.

dovehouse, 239, 349.

Dover, 120, 147, 219.

dragagant, 97, gum Tragacanth.

dragges, 87/8, drugs.

drawghtes, 236, drains?

dreams, Boorde on, 95.

Drewry, Sir Robert, 225, 348.

drink: when the drink is in, the wit is out, 94.

drinks, don't mix your, 248.

dronkenshyppe, 284, drunkenness, 285.

dropsy, a diet for the, 299.

drunkards, great, are Flemings, 147, 337; Hollanders, 149; Low-Germans, 156.

drunkards quarrel, 94.

drunkenness, 90.

dry your house before you live in it, 239.

dryn, 122/4, therein.

Dublin, 132.

ducat, 171, 199, a coin coined by any Duke: 'Ducáti, duckets, crownes.' *Florio.*

duckemet, 253, duckmeat, small green water-weed.

duck-flesh, 270.

ducks and mallards not liked in Bohemia, 167.

ducks' eggs, 265.

Duke of Venice, 183-5: '*Doge* a Duke of Venice or Genoua.' *Florio.*

dulcet pears, 256, sweet pears.

dunghills not to be near a house, 236, 239.

dup, 122/7, do up, fasten up.

During, 155, 219, Duren.

Durrant Cooper, W., quoted, 47, 54, 59, 73.

dust bad for asthma, 297.

Dutchman: beer 's a natural drink for one, 256.

Dutchmen eat butter all day, 265; how they drink, 149, 338.

dyasulfur, 99.

dycke, 122/3, thick.

D[yer], E., his list of story-books, &c., 30.

dyery, 239, dairy.

Dyetary of Health, editions of, 11—14; print of, 223—304; described by Boorde, 227, col. 2, 299: sketch of it, 319—323, with Mr Ju. Leigh's opinion on it, p. 320.

dylygentler, 243, diligentlier.

dym myls dale, 260?

dyn, 122/3, thin.

dyng, 122/7, thing.

dyscommodyous, 234, inconvenient, evil.

dystayned, 312, stained.

dysturbacyon, 310, disturbance.

dysyng, 293, playing with dice.

earthen floors, 349.

east wind is good, 238.

easy boots for gowt, 293.

edge-tools, lunatics not to have, 298.

Edinburgh, 136.

educacion, 271, bringing-up, feeding from one's youth, 259; what you've been brought up to.

Edynborow, 61, Edinburgh.

egestion, 248, out-puttings, excrement.

eggs, the kinds and qualities of, 264.

Egypt and the Egyptians, 217.

Eladas, 172, part of Greece, or Turkey in Europe.

elbow-room wanted for a man in the country, 233.

elder, 288.

Ellis, F. S., 12.

Ellis, Sir Hy., first printed Boorde's letters, 45; quoted, 56.

Emperor Charles V, of Austria, 53, 55, 56, 130/4, 151/6, 154/13, 195.

endewtkynge, 153, a brass coin in Brabant. A *deut* (*liard*, farthing) is a small Dutch copper coin; 8 of them to a stiver, and 400 to a Dollar banco (4*s.* 4*d.*). *Weilmeyr's Allgemeines Numismatisches Lexicon.* Salzburg, 1817, i. 113.

endive, 280.

England, no region like it, 118, 144; languages in, 120; wonders in, 120; money of, 121; ought never to be conquered, 164; odible swearing in, 243, 324; Seven Evils in, of which Boorde complains, 82-6; keeps her swine filthy, 273.

England, beer becoming much used in, 256.

England, pestilence in, 262; potage much used in, 262; more sorts of wine in, than anywhere else, 75; better supplied with fish than any other country, 268; deer loved more in, than anywhere else, 274.

English beer liked by Dutchmen, 148/4; by Brabanders, 150/4, 10.

English language, Boorde's opinion of, 122.

Englishman's talk with the Latin man, 210.

Englishmen, Boorde's character of them, 116-8; few of them live abroad, 60, 144; water is bad for them, 252; ale natural to them, 256; beef good for them, 271; they keep their beards clean, 309; few dwell abroad, 60; venison is good for, 274.

Englishwomen, 119.

enulacampana, 99, 278, elecampane, scabwort, or horseheal.

ephialtes, the nightmare, 78.

epilencia, 244, 349, epilepsy.

epilentycke, 294, epileptic.

Epirs, 172, Epirus.

epulacyon, 250, feasting, stuffing.

eructuacyons, 247, 265, belching.

Esdras, 78.

eupatory, 289.

Evil Mayday, 60, note 1.

evil spirits, Boorde on, 75-6.

Evyndale, Lord, namyd Stuerd, 59.

euyt, 133, eft, none in Ireland.

ewes' milk, 267.

eximyous, 21, excellent.

Exmouth, 120.

exonerate, 248, 293, unload, ease of excrements.

Extravagantes, The, by Boorde, 21.

extynct, 280, extinguish.

eye, the: ills that follow if it is not satisfied, 235.

eyes, plunge 'em in cold water every morning, 300.

face of man, Boorde on it, 95.

faldyng, 333, coarse stuff.

falling sickness, 88, 127, 244, epilepsy, &c.; a diet for it, 294-5.
fardynges, 121, farthings.
fasting, neglect of, in England, 82.
fat not so good as lean, 276.
fatness or fogeyness, Boorde's cure for, 100.
feather-beds in Julich, 155/2; lie on one, 247.
fear breeds the palsy, 297.
feet, keep 'em dry, and wash 'em sometimes, 300.
fennel, 99, 278.
fennel-seed, 278, 281, 284.
feryall dayes, 243, festivals, holidays.
fever, butter bad for, 266.
fever, how to treat, 291.
fever, causon and tertian, 97.
fever lurden, 83, laziness.
fifteen substances that Man is made of, 91.
figs, 282, 212.
filberts, 283.
fire, have one in your bedroom, 246.
fish, 268-9; the Scotch boil it best, 136.
fish in Cornwall, 122/13, 123; in Friesland, 139; in Norway, 141; in Iceland, 142; the cooking of, 277; heads and fatness of, bad, 276; bad for epilepsy, 294.
fish and flesh not to be eaten together, 269.
fish of the mountain, 273, adders.
fishpool in a garden, 239.
five wits, 93.
Flanders and the Flemings, 147-8.
flauour, 248, air.
flax, the steeping of, 236, 348.
fleed, 274, flayed, skinned.
Fleet prison, Boorde in, 70, 73.
Fleet prisoners, Boorde's bequest to, 73.
fleg, 122/8, jolly?
Fleming, Abraham, 308, note.
Flemings, the, 148.
Flemish broodmares sold to England, 147/7; Flemish fish and beer, 336-7.
flemytycke, 245, phlegmatic.
flesh-shambles of Antwerp, 151.
fleubothomye, 287, blood-letting.
fleumaticke men, a diet for, 288.
flies, stinging, in Sicily, 176.
flockes, 247, bits of coarse wool.
Florence, 187
Floshing, Flushing, 149.
fools part drunkards, 94.
for, 290/7, for fear of, to prevent.
forepart better to eat than the hindpart of animals, 276.
foul-evil, the, 136/14.
fountain in every town abroad, 254.
fox, the more he's curst the better he fares, 166/4.
fox, boil one, for a bath for a palsied man, 99.
fox, the stink of one is good for the palsy, 99, 298.
fracted, 93, 94 at foot, broken.
France, 53.
France and the French, 190.
frankincense, 290.
frantic, 298.
frayle, 212, basket.
free-will, Boorde on, 103.
Frenchmen have no fancy for Englishmen, 191; eat butter after meat, 265; their fashions, 345; last of their soldiers, 345.

freshwater fish, 268-9.

fried meat, 277.

Friesland, 139.

frogs, guts and all, eaten in Lombardy, 187.

fruits, ch. xxi., p. 282-6.

fumitory, 288; syrup of, 95.

fustian, Genoese, 189; Ulm, 161; white, used for covering quilts, 247

fynger, 122/15, hunger.

fyrmente made of wheat and milk, 263.

fysnomy, 76, physiognomy, likeness, picture of a face.

fystle, 92, 93, 95, boil?

fystuled, 94? festered.

Galateo, Della Casa's, done into English in 1576 A.D.; quoted, 324.

Galen, quoted, 235, 251, 272; cut of, 232.

Galen's Terapentike, 85.

gales, 185, galleys.

galles, 94, galls.

gallows, corpses hanging long on the, 203, 346.

galy, halfpenny, 187. '*Galley-Men*, certain Genoese Merchants, formerly so call'd, because they usually arriv'd in *Galleys*, landed their Goods at a Place in Thames-street, nam'd *Galley-key*, and traded with their own small Silver Coin call'd *Galley-half-pence*.' Kersey's Phillips: p. 105 of my *Ballads from MSS*, vol. i.

galyngale, 89, a spice.

games of trap, cat, &c., 332.

garden of sweet herbs, 239.

gargarices, 79, 98.

garlic, 279, 351.

Garnynham, 225, Sir John Jernegan or Jerningham.

Gascony, 53; described, 207; wine, 255.

Gawnt, 147, Ghent.

geese-eggs, 265.

geese pluckt yearly in Julich, 154-5.

Gelder, 153/2, the chief town of Guelderland.

Gelderlond and the Gelders, 152-3.

gelders arerys, 153/7, gilders worth 23 stivers, or 3*s*. each.

gemmis, electuary of, 103.

Genoa and the Genoese, 188-9; their beautiful women, and what freedom they have, 344; their trade and products, 344-5.

George, Dane (or Dominus), 48.

German lady possest with devils, 76.

Germany, the splendid uprising of, against Louis Napoleon, 110, 328.

Gersey, 120, Guernsey.

Geslyng, 219, Geisslingen in Wurtemburg.

Gestynge in Germany, 161.

gete, 80, jet.

giants' heads that wag their jaws, on organs, 207.

Gibbs, H. Hucks, 12, 109.

gilders, 153, &c., gold coins first made in Gelder, of various names and values.

ginger, 286.

Glasco, 59, Glasgow.

Glasgow, 136.

goatskin gloves to be worn in summer, 249.

goatskins used for wine-bottles, 199.

goats' milk, 296.

gold found in Hungary, 171.

gomble, 266, jumble.

goose-flesh, 270.
goose-pudding, 199.
gos, gosse, 122/7, 14, &c., gossip, mate.
goshawks, 149.
Gotam, Merrie Tales of the Madmen of, 27—30.
Gotham, or Nottingham, 325.
Gowghe, John, his date, 12.
gowt, how to treat, 293.
grains, brewers', after brewing, 311.
Grandpoole, in the suburbs of Oxford, 69.
Granople, 191, Grenoble.
grapes, 283, 212.
Grauelyng in Flanders, 147, 219, Gravelines.
graynes, 286, cardamons.
Great Fosters, a Tudor mansion near Egham, 7.
great men like killing deer, 275, 350.
Greece, 172-3, Turkey in Europe.
Greek, modern, a specimen of, 173-4; wine, 255.
groats and half-groats, 121.
gromel seeds, 80, 327.
grouelynge, 247, face downwards.
ground, don't lie on the, 298.
gruel made with oatmeal, 263.
Grunnyghen, 140, Groningen.
gryfe, 247, ? misprint for 'oryfice.'
Gulyk, 154/1, 9, Julich, or Juliers.
gum Arabic, 97.
gurgulacyons, 267, grumblings (in the belly).
gurgytacyon, 250, 251, swilling.
gut-caul broken, 101.
gylders, 140, 153, gilders, gold coins.
Gynes in Flanders, 147, Guisnes.
Gyppyng in Germany, 161, 219, ? Eppingen in Baden.

halarde, 161, a German coin, ? ¾*d.* or ½*d.*
Halkett, James, Colonel and Baron, 5.
hall of a house, place for the, 238.
halowynge, 295, halooing.
Hammes in Flanders, 147.
Hanago, or Hanawar, 151, Hainault, 339.
Handbook, or *Itinerary of Europe*, Boorde's, 24.
handling or touching women, or others' goods, 85-6.
Handwarp, 151, Antwerp.
hanged beef, 271.
Hardy, Sir T. Duffus, 43.
hare: dogs, not men, should eat it, 275.
harlot, wounds come through one, 94.
harped groats in Ireland, 133.
Harrison on A. Boorde, 106; on Englishman's fantastic dress, 105-6.
Harrow on the Hill, 325.
hart and hind, 274.
harts eat adders to get young again, 273.
harts-tongue fern, 289.
harvest, cider drunk at, 257.
Harwich, 120.
Hastings, 120.
hastynes, 297, hastiness.
hauer cakes, 136, 259, oat cakes.
hawks in Norway, 141; in Holland, 149.
haws, the water of, 80, 253.
Hay (*History of Chichester*) on Boorde, 40-1, note.

Hayden, a town in Scotland, 136.
Hayward's Heath Station, 38.
hazle-nuts, 283.
Hazlitt, W. C., 11, 12, 117, 307; on A. Boorde, 31.
headache, a diet for it, 295.
heart, Boorde on the, 89.
Hebrecyon, 287, Hebrew writer.
Hebrew, modern and ancient, talks in, 221.
Hellespont, names of the, 172.
Hellogabalus, 308, Heliogabalus.
hemp, the steeping of, 236, 348.
hempseed caudles, 264.
hen, 270.
Henry VI crowned in Paris, 208.
Henry VIII fortifies England, 119; won Boulogne, 209.
Henry VIII, the universities for him, 55.
herbs, ch. xx, 280-2.
herensew, 270, heronshaw, heron.
heresies in England, 83.
High and Low Germans, the difference between, 160.
Hindwords, 317.
Hippocrates, 250.
hobby, 131/6, Irish pony.
Holland, 148-9.
holmes, 161, 340, fustian made at Ulm.
Holmsdale, Sussex, 38, 39.
holy days to be kept, 243.
Holy-Well, near Flint, 331.
honey eaten in Poland, 168-9.
honey-sops, 264.
Hooper, W. H., 19, 107, 109.
hops, 256, wild, 288.
Horde, Dr, 53, 54.
horehound, 100.
horne squlyone, 153; a gold coin worth 12 stivers, or 19½*d.*
horripilacio, 75.
horse-bread, 259.
Horsfield's *Hist. of Lewes* quoted, 27.
house of easement, 236, privy (to be far from the house).
house or mansion: how to choose its site, 233-7; how to build and arrange it, 237-9; how to provision and manage it, 240-4.
houses, miserable, in the Scotch borders, 136.
Howghton, Prior, 47, 52, 54, 58, 60.
Hudson, Edward, Boorde's bequest to, 73, 74.
Hull, 120.
humecte, 244, moisten.
Hungary, 170-1.
hurtes, 267, 283; whortleberries.
Huth, Mr Henry, 342.
hydrophobia, Boorde on, 74.
Hygh Almen, swine kept clean in, 272.
Hynton, Prior, 47, 53.
Hyue, 207, a large heath in Bayonne.

Iaffe, 219, 348, Joppa.
Ianuayes, 188, Ianues, 213, Genoese.
Iber, 195, the river Ebro.
Iceland and the Icelanders as brute as a beast, 141; stockfish of, 336; candle-eating in, 336; curs of, 336.
idleness, the deadly dormouse, 83, note; Henry VIII on, 234; Boorde on, 83-4.
idropise, 299, idropyses, 251, dropsy.

iobet, 203, gibbet.

Iene, 188, Genoa.

ignorant, the; instruct them, 243, 349.

ilia, the gut; *iliac*, the disease of it, 292.

iliac, mead bad for the, 257.

imbecyllyte, 245, want of strength.

impetigo, 82, 327.

implementes, 240, furniture and provisions.

incipient, 205, unwise.

Incubus, 78.

incypyently, 60, unwisely.

incypyentt, 56.

indyfferently, 300, moderately warm.

infection, 290.

inferced, 251, ? stuft,—from *farce*, and not 'enforce.'

inflatyue, 293, puffing, blowing-up.

inscipient, 25, unwise, foolish.

intemperance, 90.

interludes, players in, wear long garbs, 207.

Introduction of Knowledge, 111, 112. Wm Copland's first or Rose-Garland edition, 14-18; his second or Lothbury edition, in 1562 or -3, and its changes from the first, 18-19; its pyctures or woodcuts, 15, 107-8; print of it, 111—222; account of it, 317-19.

iochymdalders, 140, Frisian silver coins. '*Iochymdalders* are also Bohemian coins of about the value of 4*s.* 4*d.*, the earliest dollars coined, struck by the Counts of Schlick in the beginning of the 16th century. *Ioachim Thal* is the name of the valley where the silver was found.'

ipocras, 258.

iposarca, 299.

Ireland and the Irish, 131-6, 335; products of Ireland, 333.

Irish, the wild, 334-5.

isope, 99, hyssop.

Italian servants, their food and ways ab. 1440 A.D., 343.

Italian wives, and their husbands' jealousy, 342.

Italians' opinion of England, 119.

Italy, 53.

Italy and the Italians, 178-9, 340, 342-5.

itch, Boorde's treatment of, 96.

Iues, 218, Jews.

Iury, 218, Jewry, Judæa.

jack, 160/8, loose slop ?

jacket, how to line one in winter, 249.

Jeremiah on the North, 238.

Jerningham, Sir John, 225.

Jersey, 120.

Jerusalem, and the pilgrimage thither, 218—220.

Jesus Christ's coat at Constantinople, 173.

Jews, 218; don't like pork, 273, 350.

John, Father, 57.

Joppa, 219, 348.

Judæa and the Jews, 218, 347.

Julia, the courtesans' street in Rome, 77, note.

july, 179, an Italian coin worth 5*d.*: '*giulio*, a coine made by Julius the Pope.' *Florio*; 'a jule, a small Roman silver coin.' *Baretti.*

juniper, 290, oil of, 100.

justices in Friesland 140.

kaig, 204, cage.

kacke, 122/2.

karoll, 191, a Carolus, worth 10 brass pence. *Carolus:* m. A peece of white mony, worth xd. Tour. or a iust English penny. *Carolus de Bezançon.* A siluer coyne; and is worth about ixd. sterl. *Carolus de Flandres.* Another, worth about iijs. sterl. *Cotgrave,* 1611 A.D.

Karre, Boorde calls himself a, 59.

Karrowes, the Irish, 333.

kateryns, 179, 187, Italian coins worth ½d. each.

kayme, 248, comb.

Kempton, 219, Kempten in Bavaria.

kepers, 301, 302, care-takers, nurses.

keyn, 287, kine, cows.

kid, 274.

King's-evil, cured by English kings, 91-3, 121. Span. *Lamparónes,* or *Puércas,* kernels, a swelling in the necke or armepits, the Kings euill. *Minsheu.*

Kingsmill, Sir John, 66.

kitchen-phisick 's best, 277.

knauerynge, 84 ?

kybes, 86, chilblains.

kynde, 277/4, nature, copulation.

Lachar, electuary of, 100.

lamb, 272.

Lambe, Alice, a wench at Oxford, 69.

lampreys and lamprons, 276.

Lane, Martin, 74.

Languedoc, 189, 213; described, 194.

lantern of Antwerp Church, 151.

lapdanum, 290, labdanum.

lapwings, 270.

larder, place for the, 238.

lardes, 59, lairds.

lark, 271.

lassyuyousnes, 280, lasciviousness.

Latin man, 209-11.

Latin miles longer than English, 179.

laury, 99, laurel.

law, Cornishmen go to, for nothing, 122-3.

lax, 287, diarrhæa.

laxative, 292, 297, with open bowels.

laziness of young English folk, 83-4.

learning, neglected in England, 118.

lechery in Rome, 77.

Lee, Roland, Bp of Coventry and Lichfield, 51.

leeks, 279.

leeness, 276, lean-flesh.

legion is 9999, p. 76.

Leigh, John; his opinion of Boorde as a sanitarian, and of his *Dyetary,* 320.

Leith, Boorde at, 61; King James and his French bride, Queen Magdalen, 'landed at the peare of *Lieth* hauen, the 29 of Maie, in the yeare 1537.' *Holinshed's Hist. Scotland,* p. 320, col. 1, ed. 1568.

Leith ale, 136.

lencoflegmancia, 299.

Lent, almond butter and violets are good in, 268.

lepored, 251, 293, leprosy.

Leth, 61, Leith.

lettuce, 281.

letyfycate, 89, make joyful.

leuyn, 258, leaven.

leuyn bread, 80.

Lewke, 154, Liège.

Libel of English Policy, A.D. 1436, quoted in the Notes, 323—346.

lice, the four kinds of, 87 ; Irish, 131/9; Friesic, 139/8; Welsh, 331.

Liège, velvet and arras-cloth made there, 155.

lier, 191, a French coin worth 3 brass pence. *Liard:* m. A brazen coyne worth three *deniers*, or the fourth part of a sol. *Cotgrave.*

lies, the Scotch tell strong ones, 137.

light-witted, 240.

linen socks or hose to be worn next the skin, 248.

liquorice, 100, 287.

Lisle, Lord, 64-5.

literge, 94, litharge.

liver bad to eat, 276.

liver, is the fire under the pot, 250.

lizards, none in Ireland, 133.

loch, 99, lochiscus, lozenge.

Lombardy and the Lombards, 176-7, 343.

Lombardy, garlic used in, 279.

London, the noblest city in the world, 119, 147, 219, 62; Boorde in, 64, 307; its prisons, 72; its godly order against lazy youth, 84, note; its Bridge, none like it in the world, 119.

lords, none in Friesland, 140.

Louvane, 151, Louvain.

Low-Dutch speech, 157.

Low-Germany or the Netherlands, 155-7.

Lower, M. A., quoted, 28, 34, 38-9, 41.

lubberwort, 84.

Lucres and Eurialus, the romance quoted, 342-3.

Luke, 154/1, Liège.

lumbrici, 81, 279, worms in the belly.

lunatics, how to manage them, 298.

lungs bad to eat, 276.

lust and avarice of men, 85-6.

Lustborne, 197, Lisbon. 'This wynter season, on the .xxvi. day of Ianyuer [1531], in the citie of *Luxborne* in Portyngale, was a wonderous Earthquake.' *Hall's Chronicle*, p. 781.

Luther, Martin, 165.

luxuryousnes, 285, lust.

lying, the worst disease of the tongue, 88.

lykle, 302, little.

Lynn, Boorde's property in, 73.

Lynne, 120.

lynsye-woolsye, 249, stuff for petycotes.

Lyons, 191.

lyporous, 307, leperous.

lyste, 298, list, strip.

Lythko, 136, Lithgow.

Lytle Brytane, 207, Britany.

Macadam, Major, 38.

mace, 286.

Macer referred to, 282.

Macomite, 213, Mahomet; his tricks exposed, 215-6.

Macydony, 172, Macedonia.

madmen, how to diet and manage, 298-9.

maggoty cheese liked best in Germany, 267.

Maid of Kent, the, 216.

maidenhair fern, 288, 289.

maidens, German, may only drink water, 160.

maierome, 290, marjoram.

Maligo, 255, Malaga wine.

Malla vina, 171, Mostelavina in Hungary, 340.

malmyse, 254, malmsey
maltworm, 256.
malyfycyousnes, 79, maleficence, influence of evil spirits.
man and woman be reasonable Beastes, 91, 93.
man made of 15 substances, 91
manchet bread, 258.
Mancy, the land of, 350.
mandilion, 106, a short cloak.
mandragora, 281.
mania, 79, madness.
Manley, Wm, 74.
manna, 289.
manners and manhood, Englishmen the best people for, 118.
manyken, 157, a Dutch farthing.
Mare, the Night-, 78.
marivade, 197, 199; $\frac{1}{1250}d.$, 200 maravedies are worth $\frac{4}{25}d.$; '*Marauedis, picciola monéta in spágna*, foure and thirtie of them make sixe-pence sterling.' *Florio*. Sp. *Maravedí*, m. a peece of plate, being of the value of the thirtie and fourth part of a ryall of plate, *id est*, 34 of them to an English six-pence.' *Minsheu*.
marketes, 187, small Italian silver coins. 'Marchétto, a little coine in Italie.' *Florio*.
marketplaces, 294.
Marlyn, 78, Merlin.
marrow, 276.
Martin, Dr, his *Apologie*, 66.
Martylmas beef, 271, 292, 327.
Martyn the surgeon, Boorde's friend at Montpelier, 309.
Mary, Princess (afterwards 'bloody Queen Mary'); Boorde's Dedication of his *Introduction of Knowledge* to her, 122, 14.
Mastryt, 219, Maestricht.
Mathew, Richard, Boorde's devisee and residuary legatee, 73.
Mawghlyn, 151, Mechlin.
mazer, 132, drinking cup with a long stem.
mead, 257.
meals, two a day are enough, 251.
medlars, 283.
Medon, the isle of, 182.
melancholy complexion, 132; men, 245.
melancholy men, a diet for, 289; milk is good for 'em, 267.
melons, 285.
Memmyng, 161, 219, Memmingen in Bavaria.
Mendicant Friars in Rome, 77.
Mense, 156, 160, 219, Maintz, Mayence.
mercury, 289.
Merlin built Stonehenge, 121.
merry heart, keep a, 300.
merry, who is, now-a-days, 88.
mesele, 95, measles-spots in the face.
mestlyng bread, made of mixt grain, 258-9.
metheglin, 257.
Metropolitan of England, 119.
mice, rats, and snails, in rooms, 249.
midwives, evils of untrained ones in England, 84.
Might-of-Constantinople, 172, the Hellespont.
milk, 267; water of, 253.
mind of man, its changeableness, 85.
minors made monks and friars, 43, and note 4.
mirth, one of the chiefest things of Phisic, 88, 89, 228, 244, 249.
mithridatum, 99.

moat to be scowred, 239.
modernall, 291, modern.
moles on the face, 95.
money makes a man's thought merry, 88.
monks, canons, &c., in Rome, 77.
monks' hatred of friars, 34.
Montanus, 67, note.
Montpelier, 50, 63, 307; Boorde's *Introduction* and *Dyetary* dated from, 122, 223, 227, 228, 191; his praise of it, 194, 226; dinner and supper at, 277; pestilence time at, 289.
moorcock and moorhen, 270.
Mores, 212, Moors, white and black.
Morisco gowns, 106.
morkyns, 161, ?misprint for 'Norkyns,' 157, hapence.
Morles, 208, Morlaix in Brittany.
morning, what to do in the, when you rise, 248.
morphewe, 95, morfew.
mortified, mercury, 102.
Moryske, 216, Moorish.
morysshe, 288, moory, swampy.
moude, 269, mud.
Moulton's Glasse of Health, 12.
Mountgrace, the Prior of, 54.
moustache, called 'a berde vpon his ouer lyppe,' 313/95.
mowlded, 258; muldyd, 262, moulded.
mundyfyed, 236, purified.
munited, 119, fortified.
Muscadell wine, 255.
mushrooms in Lombardy, 177.
musical instruments make mirth, 88.
musicians, the Scotch and English are good, 137.
musk, a confection of, 99.
muster, 313/105, kind.
mutton, 272.
Mychell, a lunatic, 298.
myd, 122/18, 123/5, with.
Mydilborow, 149, Middleburgh (in Zealand).
Mylner of Abyngton is not by A. Boorde, 32-3.
myrtles, powder of, 94.
mytes, 267, cheese-mites.
mytes, Dutch, 26 to 1*d*., 157; Hungarian, 171; Turkish, 20 to 1*d*., 173; French, brass farthings, 191. *Mite*, f. A Mite, the smallest of coynes. *Cotgrave*.

nails, tear yourself with a pair if you have the itch, 97.
Naples described, 176-7; Naples, 219; the people of, 340.
Napoleon, Louis, 110.
Nature, leave slight ailments to, 96.
nature, 61, semen.
Navarre described, 202-6; the poverty of, 346.
Navarre, the king of, 56.
nawtacyon, 265, grease floating at the top.
Negyn manykens, 157, Dutch coin worth ¼*d*.
Nemigyn, 153/3, Niemeguen.
nese, 98, sneeze.
Neselburgh, a castle in Hungary, 171.
nettle, 288.
nettles in the cod-peece, a cure for venery, 100.
Neuer, 200, Navarre.
Newcastle, 120.
Newe Cartage, 195, Cartagena.
Newgate, 84.

Newman-brydge in Flanders, 147.
Newport in Flanders, 147.
Nichol Forest, 136.
Nicholas, Dr, 49.
nightcap to be scarlet, 247.
nightmare, 78-9, 327.
nightingales won't sing in St Leonard's Forest, Sussex, 121.
nine sorts of dishes loved by nine sorts of men, 339.
nit, 87, a kind of louse.
nobles, gold, 121.
Norfolk, Duke of, 12, 13, 48, 49, 223, 225.
norkyn, 157, a Dutch coin worth ½*d.*
norkyns and half-norkyns, 153, brass hapence and farthings in Brabant.
Normandy described, 208, 53.
Northern English tongue, 120, 330.
Norway described, 141.
nose, ulcer of the, 98.
nottyd, 212, polled, clipt.
noyfull, 270, harmful.
nurses, 2 or 3 for a sick man, 301.
nutmegs, 287, 290.
nuts bad for the palsy, 298.
nuts, fresh and old, 284.
nym, 122/12, take, hand, give.
nys, 122/10, have not.

obfuske, 244, darken.
obliviousness, 94.
obnebulate, 244, 250, cloud over.
obpressed, 251, prest down.
occult matters, study of, forbidden by statute, 25
oculus Christi, 100, a herb.
odyferous, 295, 302, odoriferous.
old men's lechery, 69, note.
olives, 285.
O'Neale, John, 334.
onions, 279.
opylacyons, 251, 276, 282, 283. L. *oppilo*, stop up, shut up.
opylatyd, 297, stuft up.
oranges, 286.
orchard, have one, 239.
organs, the finest pair in the world are in St Andrew's church in Bordeaux, 207.
organum, 289.
Orleans wine, 75.
Orlyance, 55, 191, 205, Orleans.
Osay, 255, 75, wine from Alsace.
otemel, 262, oatmeal.
oten, 256, oaten.
otters' skins, 333.
Otto, Marquis, shape of a beard, 17, note.
overplus, 266.
ouerthwarte, 274, across.
Oxburdg, 161, Augsburg.
oxen covered with canvas at plough in Italy, 187.
Oxford, 44, 120; Boorde probably brought up at, 40-1, 210.
oxymel, 258.
oyster-shells burnt, 97.
oysters eaten, 255.

paast, 277, paste, piecrust.
pain or dolour, 89.
painted clothes and pictures bad for lunatics, 298.
Pale, the English, in Ireland, 132.
Palphans, 200, ? who.
palpyble, 103, palpable, touchable.
palsy, Boorde's treatment of, 99; a diet for the, 297.

palsy of the tongue bred by old nuts, 284.
Pampilona, 202, Pampeluna.
Pannell, John, 74.
pannicle, 101, little pane or covering: cp. counterpane.
pantry, place for the, 238.
Pardon's Account of St Giles's, Bloomsbury, quoted, 65.
parents' indulgence, evils of, 83, note.
Paris, 191, 208; the University of, 55.
park with deer and conies, 239.
Parker's *Defence of Priests' Marriages* is not Ponet's, 67.
parks, many, in England, 106.
parsley, 281; great, 278, note 3.
parsnips, 278.
parsons, 220, persons.
partridge is easily digested, 269.
partridges' eggs, 265.
Pascall the Playn, 145, 336, 384, ? who.
pastryhouse, place for the, 238.
Patriarchs of England, Jerusalem, &c., 119; of Constantinople, 172.
paysyng wayghtes, 248, poising weights.
peaches, 283.
peachick and peacock, 270.
pears, 284.
Pears Go-nakyt, 313/111.
peas potage, 263.
peason, 284, peas.
pediculus, 87, louse.
pelfry, 142/10.
Pemsey, 73, 120, Pevensey in Sussex.
pence and halfpence, 121; Scotch, are almost ¼*d.* and ⅛*d.*; brass, in the Netherlands, worth 2½*d.*
pencyfulnes, 300, pensiveness.
pendiculus, 207/10, lice.
penurite, 163, poverty.
peny, 242, income.
penyroyal, 281.
pepone, 285, a kind of melon.
pepper, 3 sorts of, 286.
percelly, 278, parsley.
percilles, 80, parsley.
Peregrination of England, by A. Boorde, 23-4.
perlustratyd, 53, travelled through.
perplexatives, 300.
perpondentt, 53, most weighty.
pertract, 264, treat of.
pestilence, 262; a diet, &c., for the, 289-291.
Peter pence, 78.
Petragorysensis, 56, the chief school of the University of Toulouse.
petycote of skarlet, wear one over your shirt in winter, 249.
pheasant is the best wild fowl, 269; pheasant-hen, 270; pheasants' eggs, 265.
phenyngs, 161, German pence.
philosophers' oil, 99.
phlegmatic men, 272.
phylyp, 83, fillip, cut with a club.
pibles, 253, pebbles.
Picardy described, 208-9.
pigeons, 270.
pigs, 274.
pilchards bad for gowt, 293.
pilgrimage to Compostella, 205-6, 346; to Jerusalem, 219.
pissbowls, 295.
pissing and piss-pots, 236.
pitch, tar, and flax, in Poland, 168.

pitch-plaister, 97.
pith (yolk) of eggs, 80.
Pius II, Pope: his *Lucres and Eurialus*, 342.
placable, 234, pleasing.
plack, a Scotch, 137, almost 1*d*.
Pliny referred to, 282.
plomettes, 248, plummets.
Plommoth, 120, Plymouth.
ploughmen eat bean-butter, 268; bacon good for, 273.
plovers, 270.
poacht eggs comfort the heart, 89, 265.
poched, 259, poacht.
pocky faced, 307.
Poland and the Poles, 168-9.
polettes, 270, pullets, young fowls.
pollyd, 311, bagged.
polypody, 288.
pome Garnade, 94, pomegranate.
pomegranates, 284.
pomemaunder, 290-1, pomander, scented ball.
Ponet, John, Bp of Winchester, charges Boorde with keeping three whores, 65, 66.
poor in England, Boorde's allusions to their state, 86-7.
poores, 248, 251, pores.
Pope, the, 53; is disregarded by the Saxons, 165; Bohemians, 166; and Grecians, 173.
Porche mouth, 120, Portsmouth.
pork, 272, 350.
porpoise is bad food, 268.
portingalus, 197, Portuguese coins worth 10 crowns each.
ports and havens of England, 120.
Portugal, 53; described, 197; products of, 345.
pose, 297, rheum in the head.
posset ale, 97, 257, 299.
potage more used in England than anywhere else, 262.
poudganades, 195, pomegranates.
pound Scotch is 4*s*. 8*d*., 137.
powderyd, 271, salted.
Powell's edition of Boorde's *Dyetary*, 13.
Prague, 167.
precordyall, 57, most hearty.
pregnance, 93, pregnancy.
preservatives, 296.
Preston, Mrs, 38.
pretende, 61, intend.
priapismus, 100.
priest at the bedside of the sick, 302.
priests, how they should avoid erections, 100; forbidden to have wives or concubines in 1539, p. 332.
priests, Icelandic, though beggars, keep concubines, 142.
prisons, sickness of the, 72, 326.
privy chamber, 238.
privy to be far from a house, 263.
prognostications of great floods in 1524 A.D., 325.
Pronostycacyon for the yere 1545, Boorde's, 25.
prospect of a house, 234.
Provence, 189, 191, 213.
proverbs and proverbial phrases, 94, 240, 260, 273, 314/114.
provide all necessaries before you begin building, 237, 240
prunes, 285, plums.
Prussian products, 337.
pryncypalles, 233, principal things, chief needs.

ptysane, 258, 299 ; how to make one, 99.
pudibunde places, 253, secret parts.
pulcritudness, 119, beauty.
pulcruse, 234, beautiful.
pulyall mountane, 289.
purslane, 280.
puruyd, 237, purveyed, provided.
Puttyors, 191, Poictiers.
Pyctanensis, 55, Poitou.
pyctures, 16, 18, woodcuts.
pyes, 133, magpies, none in Ireland.
pyking, 217, picking and stealing.

Qoorse, 75, Corsican.
quadryuyall, 238, quadrangular.
quadrypedyd, 272, four-footed.
quails, 270.
quarel, 299, diamond-shaped bit of glass.
quartron, 99, quarter.
quickbeam, 289
quilt of cotton or wool, covered with fustian, used as a counterpane, 247.
quinces, 284.

rabbits when young, sucking; 'conies' when grown up, 275.
rader, 161/13, with a wheel stampt on them : Germ. *rad*, wheel; *rader albus*, a wheel-penny silvered over.
radish roots, 279.
raisins and currants, 282.
rape, 279, a kind of turnip.
Rasis quoted, 271, 272, 274.
raspyce, 75, raspberry wine.
ratty rooms, 249.
ravener, 194, a glutton.
reare, 264, soft (egg).
red-herrings, 292.
redolent, 302, sweet-smelling.
Redshanks in Ireland, 132.
Reene, 139, the Rhine, 156.
refrayne of (= from), 295.
relics at Rome hardly protected from the rain, 76-7.
religious, or persons having taken monastic vows, enabled to hold land, 326.
relygyon, 57, 58, religious order, or vows of a monk.
rents and income, divide yours into three parts, 241.
repercussives, 97, ? drivers inwards of disease.
repletion, 250.
respect, 172, 235, view.
restoratives, 89, 296.
resurrection, the general, 103/3.
reume, things that breed it, 295.
rewene cheese, 266.
Rhenish wine, 75, 156, 255.
Rhodes, 182, 219.
Rhododendron Walk in Windsor Park, 110.
rhubarb, 289.
rhubarb seeds from Barbary sent to Thos. Cromwell in 1535 A.D., 56.
rice pottage, 263.
Rimbault, Dr E. F., 34.
ringworm, 81-2.
Ritson and J. P. Collier, 71.
roasted eggs, 265.
Roberdany wine, 255.
Rochelle, 208.
Rochester, 147.
rock alum, 99.
rocket, 280.
roe buck and doe, 274.

Roman loaves a little bigger than a walnut, 258.
Romans curse the Greeks, 172.
Romanysk wine, 75, 255.
Rome, 53, 219, 341 ; vicious state of, 77-8, 178.
Rome, Bp of, his bulls, 58.
Rome, harlots in, 77, note.
Rome, lechery and buggery in, 77.
Romny wine, 75, 255 (from the Romagna, *Babees Book*, 205).
roots, ch. xix, 278-90.
ropy ale, 123, 256.
rosemary, 281, 290.
roses, 281.
roudges, 139/5, rugs, 142/5.
Runnymede, or Runemede, 110.
rural man's banquet, 267.
rusty armour, sick folk are like it, 104.
ryals, 121, royals, coins worth 5½*d.* in Spain, 199. Sp. *reál*, a riall or six pence. *Minsheu.*
ryders, 140, Frisian coins. '*Ryders* are gold Coins of Guelder, &c., of different sizes and values stampt with a *rider*, an armed man on horseback.'
Rye, J. Brenchley, 12.
ryghtes, 301, rites.
Ryne, 272, river Rhine; swine swim in it.
rynes, 94, rinds, skins.
Rysbanke in Flaunders, 147.
rysshes, 290, sweet-smelling rush.

Sabbatarian superstition, 332.
sables, 249.
saffron, 286 ; it spoils bread, 261.
saffroned bread, 80 ; in Rome, &c., 258.
saffron shirt, 131.
sage (the herb), 281.
Saint Ambrose, 243.
St Andrew's in Scotland, 136.
St Augustine, 105 ; quoted, 302.
St Bartholomew, 205.
St Bede on the dropsy, 299.
St Blase, 182.
St David's in Wales, 120.
St Domingo in Navarre, 202.
St George, 205.
St George's Arm, 172, the Hellespont.
St Giles's Hospital, London ; Dr Borde tenant of one of its houses, 64-5.
St James the More and Less, 205.
St James's in Compostella, story of, 203-4 ; Boorde's pilgrimage to, 205-6.
St John Erisemon's bones at Constantinople, 172-3.
St John's Town in Scotland, 136.
St John's wort, 79, 327.
St Kateryn, 182.
St Leonard's forest in Sussex, 121.
St Loye, 182.
St Luke's bones in Constantinople, 172.
St Malo, 347.
St Mark's, Venice, 185-6.
St Patrick's Purgatory in Ireland, 133.
St Peter and Paul, shrines of, in Rome, 77.
St Peter's Chapel, Rome, 76.
St Peter's Church at Rome in ruins, 77, 178, 341.
St Philip, 205.
St Severin's church in Toulouse, 205.

St Simon, 205.
St Sophia's the finest cathedral in the world, 172.
St Thomas, a town in Hainault, 151.
St Thomas of Alquyne, 78.
St Winifrid's Well, 127, 331.
Salerne, the University of, is near Constantinople, 173.
salet, 240, piece of armour?
Salisbury Plain, 120.
salmon, 102.
salt beef for a blear-eyed mare, 273.
salt fish, 269.
salt wells in England, 120, 330; in Saxony, 165.
Sampson wore a beard, 314/140.
Sandwich, 120, 147, 219.
sanguine men, 245; a diet for, 287.
sanitary matters; value of Boorde's opinion on, 320.
Saracens don't like pork, 273.
sarafes, 171, gold coins worth 5*s.* each, 173, 216.
sarcenet, 96.
sardines, 202/3; Sp. *sardina*, a little pilchard or sardine. *Minsheu.*
Sarragossa, 195.
saucefleme, 95.
Savoy, 191.
Sawsfleme, 251.
sawsflewme face, 101, 307.
saxifrage, 80.
Saxony and the Saxons, 164-5.
Scamemanger, 171, Steinamanger in Hungary.
Scarborough, 336.
scarlet cloth, wipe your scabby face with one, 95.
Schildburg, the German Gotham, 29.
Scio, 185.
Scogin's Jests, 31-2.
Scot, trust you no, 59, 326.
Scotch, disliked by the Dutch, 149.
Scotchmen, with whom Boorde went a pilgrimage to Compostella, 205-6.
Scotland and the Scotch, 135-8, 335; degenerate and luxurious ways of, 259-60.
Scotland, Boorde practises medicine in, 59.
Scotland, oat cakes of, 259.
scrofula, 50.
Sculwelyng, 171, Stuhlweissenberg in Hungary.
scurf, Boorde's treatment of, 97, 327.
scurf and scabs, 95.
sea-fish better than fresh-water ones, 268.
secke, 255, sack (wine).
seege, 292, excrement.
seene, 289, senna.
segge, 122/6, say.
Selond, 149, Zealand, west of Holland.
Semar, Sir Henry, 66.
Seno in Normandy, 208.
Sepulchre, the Holy, described, 220.
Sermons, Boorde's *Boke of*, 24.
servants, Italian, the food and ways of, ab. 1440 A.D., 343.
set a good example, 244.
Seven Kirkes, 219, Siebenkirchen.
sewe, 262, broth.
shave lunatics' heads, 298.
shaved men look like scraped swine, 315/154.
shaving, the foolishness of, 26.
shefte, 240, shift.

shell-fish bad for gowt, 293; bad for epilepsy, 294.

shemew, a new-fashioned garment in 1518, p. 345.

Sherborne, Bp of Chichester, 44.

shoes, the smell of, good for pregnant women's unnatural appetite, 98.

Shotland, 139, Shetland?

shoueler, 270, shoveller, a water-bird.

shrimp cullisses or broths, 264.

shroving time in Rome, 77, note.

siccative, 94, drying.

sick and wounded, Boorde's advice to, 104.

sick man, how to arrange for one, 301.

sick men and women like a bit of rusty harness (armour), 104.

sider, 200, cyder.

singing, mirth in, 88.

sinistral, 53/11.

sinks, 295.

sirones, 81, worms in a man's hands.

situation of a house, the fit, 232-4.

skin, bad to eat, 276.

skin, meat boiled in a, 132; wine kept in a, 199.

skin of fish is bad, 269.

skyn, 99, cause skin to heal.

slaughter-house, place for it, 239.

slaves, 212.

sleep, how to, 244-7.

slepysshe, 301, sleepy.

sloudgysshe, 301, sluggish.

sluttyshe, 236, 301, sluttish.

slyme, 297, slimy.

smatterers in phisic, 104.

smoke bad for asthma, 297.

Smormowth, Hans, Boorde gets drunk at his house, 309.

Smythe's Hist. of the Charterhouse, 52, 54, 59.

snaily rooms, 249.

snakes, none in Ireland, 133.

snappan, 153, a silver coin worth six steuers, or 9½*d.*

sneeze, how to make yourself, 98.

snipe's brain is good, 276.

snoffe of candellys, 295, candle-snuff.

snow on the German mountains in summer, 160-1.

snuft, 98.

soda, 244, ? a disease of the head.

sodde, 277, boiled.

soldes, 171, 173, brass coins worth ⅓*d.* each, 216.

solydat, 268, solydate, 264, solid.

sompnolence, 279, sleepiness.

soocke or Soken of Lynn, 73.

soole, 122/10, soul, flavouring, meat.

Sophy, the, 214.

sopytyd, 250, stupified?

sorrel, 281.

soul of man, Boorde on the, 102-3.

soul, how to care for the, 243, 301-2.

sour things are bad, 296.

Southampton, 120.

southystell, 253, sowthistle.

sovereigns, gold, 121.

sow-pigs, 274.

sowese, sowse, 191, a French sous worth twelve brass pence, 1½*d.* English.

sowse, 265, pickle in brine.

Spain, 53; has dirty swine, 272, 273; products of, 346.

Spain and the Spaniards, 198-9

Spanish girls are cropt like friars, 199.

Spanish imports, 338.

sparrows, 270.

spelunke, 77, cave, shrine.

spermyse cheese, 266.

spirits of men, 88.

spiritual phisician, 104.

sprawl, 292.

spyght, 315/169, spite.

spyghtfull, 310, spiteful.

Spyres, 161, 219, Spiers.

squlyone, a horne, 153, a gold coin worth 19½*d.*

staares, 271, starlings.

stables, place for the, 239.

stamele, 249, fine worsted stuff.

standing-up of a man's hair with fright, 75.

standing water, 253, 349.

standyng of a man's yerde, 100.

Stanyhurst's *Description of Ireland* quoted, 334-5.

starre, 122/18, quarrel.

Stationers' Register A, extracts from, as to Boorde's *Dyetary*, 14; *Introduction*, 19.

Stations of the Holy Land, 220, praying places where you get remission of sins.

stavesacre, 87.

stercorus, 272, dungy.

sterke, 247, stiff.

Sterling, 136.

sternutacion, 79, 98, sneezing.

stewpottes, 263.

sticados, 289.

stick, the, for lazy backs, 84.

stinks, things that make, 295.

stiver, 161, German coin worth 1⅗*d.*

stockfish, 141/5, eaten raw in Iceland, 336.

stomach, the pot, and the liver the fire under it, 250.

stomach, keep it warm, 300.

stone, don't sit or stand on, 249.

stone in the bladder, 80; Martilmas beef is bad for it, 271; elecampane good for it, 278.

Stonehenge, 120-21.

stones of virgin beasts are nutritious, 277.

stool, go to, every morning, 248.

storax calamyte, 290.

strangulion, 256, strangulation or suffocation?

straw and rushes on floors of houses, 290.

strawberries, 267; the water of, 253.

strawberries and cream may endanger your life, 267.

Straytes, 213/6, Straits of Gibraltar, or the Mediterranean.

Stubbs, Prof., 42.

stufes, 95, 287; It. *stúfa*, a stoue, a hot-house; *stufáre*, to bath in a hot-house or stoue. *Florio.*

stuphes, 97; dry, 99.

sturgeon in Brabant, 150/7, 16.

Stuyvers, 2½ make 4*d.*, 157, 199. Dutch *een Stuyver*, a Stiver, a Low-countrie peece of coine of the value of an English Penny. *Hexham*, A.D. 1660.

stycados, 288.

subieckit, 59, subjected, subdued.

subpressed, 250, prest down.

succade, 278, 286, sucket, sugar-stick.

Succubus, 78.

sucking animals, all good to eat, 275.

Suffragan of Chichester, Boorde appointed, 44, 59.

sugar 's nutritive, 296.

sunshine, don't lie or stand in it, 249.

superstitions of the Irish, 335.

supper, make a light one, 249.

suppynges, 299, drinks to sip.

surfeiting, evils of, 250-2.

Sussex, A. Boorde in, 106; St Leonard's Forest in, 121.

sustencyon, 241, sustentation, support.

sutt, 270, set ?

swart, 81, dark-coloured.

Swavelond or Swechlond, 160, Switzerland.

swearing in England, 82-3, 118/37; 243, 324.

sweating sickness, 289, a fever-plague, 351.

sweeping a house, 236, 297.

sweet breath, eat anise-seed for, 281.

sweet wines for consumption, 296.

swing, 273, fling, range, desire.

swyne, a, 272, pig.

Sycel, 175-6, Sicily.

syght, 172, number; a wonderful sight of priests.

symnels, 80, 261, 327.

syncke or syse, 313, cinq or sise, 5 or 6 on the dice.

Synesius on baldness, 308.

synkes, 236, sinks.

taale, 122/11, deal ?

taledge, 266, ? firmness or texture.

tallow candle for a horse's mouth, 273.

tallow eaten in Iceland, 141/5.

tamarinds, 289.

Tarragon, 195.

Tatianus, 67, note.

Taylor, John, the Water-Poet, quoted, 326, 330-2, &c.

temperance, 90.

Temple-Bar, 307.

temporaunce, 300, temperature.

tennis, play at, 248.

tertian fever, 97, 327.

Tessalus, 85.

testons, 191, French coins worth 2*s.* 4*d.* *Teston* . . a Testoone, a piece of siluer coyne worth xviijd. sterling. *Cotgrave.*

testynes, 297, testiness.

Thames, 110; rascally bakers ought to be duckt in it, 261.

thirty the highest number in Cornish, 123.

Thomas's *Historye of Italye*, 1561, quoted, 183-5, 340-4.

Thomas, Walter, of Writtle, 62.

throte-bol, 80, the weasand.

thrush, 271.

thyme, 281.

Tiber, river, 77, 177-8.

tin in Cornwall, 123, 122/13.

titmouse, 270.

Titus, 219.

Tolet, 200, Toledo.

Tolosa, 55, Toulouse.

tongue, and its diseases, 87.

tongues bad to eat, 277.

toome, 122/13, home ?

Torkington, Sir Richard, his pilgrimage to Jerusalem in 1517, p. 348.

torneys, 216, brass coins. Fr. *Tournois:* m. A French penie; the tenth part of a penie Sterling; which rate it holds in all other words (as the *Sol* or *Livre*) whereunto it is ioyned. *Cotgrave.*

Toulouse, 191, 205; its University, 194.
Tower of London, Prior Howghton in it, 52.
trachea, 80, windpipe.
trade, 243, trodden way, path, custom.
Trafford, Prior, 45, 59.
Tre Poll Pen, 122/22, names of Cornish men.
treacle, 188, antidote against poison, 99.
Trent, 160, 219, Trente in the Tyrol.
triangle-wise, 249.
tributor, 181/7, payer of tribute for.
tripe 's bad to eat, 276.
truss your points, 248, tie up, or button your breeches and coat.
trussery, 345, luggage.
trust in God, 75-6.
trylybubbes, 276, tripe.
tunicle, 101.
tunny (fish) in Brabant, 150/16.
turf and dung for fire in Friesland, 140.
Turk, the Great, 171, 214, the Sultan.
Turkey and the Turks, 214-216.
Turkey, hard eggs are pickled in, 265.
Turks don't like pork, 273.
Turner, Rev. E., quoted, 41.
turnips, 278.
turtle-doves, 270.
tuyssyon, 243, tuition, charge, care.
tyme, 281, thyme.
tymorysnes, tymorosyte, 275, fear.
tynt, 255, tent wine.
Tyre, 255, wine from Syria or Sicily.

udders, cows', 287.
ulcer of the nose, 98.
ulcerated wounds, 94.
Ulm, fustian made there, 161.
Umarys, 120, ?
vnberdyd, 309, 315/169, unbearded.
unchristened, 212, not christened.
unctuosyte, 266, oiliness, greasiness.
vndyscouered, 247, uncovered.
unexpert midwives, 84.
unguentum baculinum, 84, 95.
universities mentioned by Boorde, 49-50.
upright, 247, lying face upwards.
urine, 81, 327; is a strumpet, 32, 34.

veal, 271.
velvet made at Liège, 155.
venery, do none after dinner or before your first sleep, 246; or after meals, 293.
veneryous acts, don't go to excess in, 300.
Venetian women, 184.
Venice and the Venetians described, 181-6, 341; Venice, 219, 348; the merchandise of, 342; its Arsenal, and store of timber, 343.
venison, 274-5; is bad for epileptic men, 294.
ventosyte, 248, wind on the stomach.
Vespacian, 219.
villeins, Coke on, 41, note 2.
violets, 289; oil of, 97.
viscous fish, 297.
Visitation of our Lady, July 2, 55.
Vitas Patrum, 217, 'Lives of the Fathers.'

vivifycate, 89, give life to.
vocyferacyon, 295.
voluis, 59, wolves.
Volunteer Review, Easter Monday, 1871, 38.
vomit, how to make yourself, 90.
voven, 171, towns so called in Hungary.
vyces, 207, devices, ? or like Vices in plays.

wa, an infant's cry, 91.
wadmole, 346/12, coarse woollen cloth.
'wait on,' 49.
Wales and the Welsh described, 125-130; free from Sabbatarianism, 232.
walnut, 283.
walnuts in Germany, 160.
Warden, 171, Groswardein, or Peterwardein in Hungary.
warden, 284, a big apple for roasting.
wardens, stewed, 291.
Warton on Boorde's *Dyetary*, 106.
wash your hands, face, and teeth, every morning, 248.
wash your hands often, 300.
wasp in one's nose, 156/8.
water, Boorde hates, 75; 349.
water alone isn't wholesome, 252; the kinds of, 253.
water-drinking and fruit-eating kill 9 English and Scotchmen in Spain, 206.
water the first need for a house, 233-4.
watered, 236, steept, soakt.
Waterford, 132.
watysh, 122/15, what.
web in the eye, 100.
weft in ale, 256.
wells that turn wood into iron, 141.
wertes, 95, warts.
werysse, 278, tasteless ?
wesande, 80, weasand, windpipe.
Weschester, 120, Chester ?
wetshod, gowty men not to go, 293.
whey, 257, 289.
whirlwinds, Boorde dislikes, 75.
white meat, 264; is bad for epilepsy, 294.
wiches in Cheshire, 330.
Wilberforce, Bp Sam., his clergy's 'hindrance,' 34.
wild fowl, 269-70.
wild Irish, 132.
will, duty of making one, 104; making of a sick man's, 301.
Wilson's *Arte of Rhetorique*, quotations from, 116, note, 307.
Winchester, Boorde in, 64, 66; his property in, 73.
wind, things that breed, 292.
Windsor-Park, 110.
wine, the qualities and sorts of, 254-5, 349.
wines don't grow in England, 119.
Wise man, the, 251.
wits of man, the five, 93.
Witzeburg, 165, Wittenberg. 'In the 15th and 16th centuries, Wittenberg was the capital of the electoral circle of Saxony, and the residence of the court.' *Penny Cyclopædia*.
wo be the pye! 273.
wolf- and bear-skins worn in Iceland, 141/12.
Wolsey ordered to York, 225, 49.
woman, Boorde's chapter on, 68.
woman's waistcoat, 97.

women, Boorde accused of conversance with, 62; curding of milk in their breasts, 97; not to marry priests, 332.

women, pregnant, unnatural appetite of, 98.

women, the Dutch, lay their heads in priests' laps, 149.

women, freedom of the Genoese, 344; disposition of the Italian, 342, 184.

women's babbling round a sick man, 301, 302.

wood that turns into stone, 121.

Wood, Anthony a, on Boorde, quoted, 28, 31, 33, 69, 70.

woodcock, 269; its brain is good to eat, 276.

woodcut, the same, used for different men, in the *Introduction, &c.*, 107.

wood-powder for excoriations, 99, 328.

worms, 160/12, 219.

worms in men, 81.

Wosenham, Thomas, 74.

wounds, Boorde on, 94, 327.

wrens eat spiders and poison, 270.

Wrettyll, 62, Writtle in Essex?

Wright, T., on the Gotham Tales, 29.

Wyclif, 166/5, 7.

wyddrawghtes, 295, withdrawers?, drains.

Wyer, Robert, his date, 12; his undated edition of Boorde's *Dyetary*, 12, 13; his device, 304, 224, 316.

wyesephenyngs, 161, white pennies, worth about 1½*d.*

Wynkyn de Worde; his cuts in *Hyckescorner* and *Robert the Deuyll* used by W. Copland in Boorde's *Introduction*, 108.

Yarmouth, 120.

ȝe, 59, yea.

yll, 122/9, badly, extremely.

yongly, 300.

yonker, 160/3, fine fellow, in Germany.

young folks' laziness, 83.

yreos, 94, 288.

Ytale, 53, Italy.

---

drawghtes, p. 236, l. 4 from foot, must mean 'privies'. 'A draught or priuie, *latrina*': Withals, in *Babees Book*, p. 179, note 2.

On *dagswaynes*, p. 139, see Way's note 1 in *Promptorium*, p. 112. He quotes from Horman, "my bed is covered with a daggeswaine and a quylte (*gausape et centone*): some dagswaynys haue longe thrumys (*fractillos*) and iaggz on bothe sydes, some but on one." 'So likewise Elyot gives *Gausape*, a mantell to caste on a bed, also a carpet to lay on a table; some cal it a dagswayne'.

Fuller's

ACCOUNT OF *ANDREW BOORDE*

IN HIS *HISTORY OF THE WORTHIES OF ENGLAND*, 1672.

"ANDREW BORDE, Doctor of Physick, was (I conceive) bred in *Oxford*, because I find his book called the *Breviary* of *Health* examined by that *University*. He was *Physician* to King *Henry* the eighth, and was esteemed a great Scholar in that age. I am confident his book was the *first* written of that faculty in English[1], and dedicated to the *Colledge of Physicians* in London. Take a test out of the beginning of his *Dedicatory Epistle*,

'*Egregious Doctors and Masters of the Eximious and Arcane Science of Physick, of your Urbanity exasperate not your selves against me for making this little volume of Physick, &c.*'

"Indeed his book contains plain matter under hard words, and was accounted such a *Jewel* in that age, (things whilst the *first* are esteemed the best in all kinds,) that it was Printed, *Cum privilegio ad imprimendum solum*, for William Midleton, Anno 1548. He died, as I collect, in the raign of Queen *Mary*." (Part I, p. 215-216.)

PASCHAL *the playn*, p. 145. Fuller explains who this man was. Under *Suffolk*, in his *Worthies of England*, Part III, p. 59, Fuller gives in his list of Prelates :—

"JOHN PASCHAL, was born in this *County (where his name still continueth) of Gentle Parentage, bred a *Carthusian*, and D.D. in Cambridge. A great Scholar and popular Preacher. *Bateman*, Bishop of *Norwich*, procured the Pope to make him the umbratile Bishop of *Scutari*, whence he received as much profit as one may get heat from a Glow-worm. It was not long before, by the favour of King *Edward* the Third, he was removed from a very shadow to a slender substance, the Bishoprick of *Landaffe*; wherein he died Anno Domini 1361."

* *Bale de scrip. Brit. centur.* 5. num. 85.

[1] This is a mistake.

Supplement

TO

Andrew Boorde's Introduction and Dietary.

NOTE ON THE DISCOVERY IN THE BRITISH MUSEUM OF A BOOK WITHOUT AUTHOR'S NAME OR INITIAL, BUT UNDOUBTEDLY THE WORK OF ANDREW BOORDE.

By Charles Faulke-Watling.

This very interesting little volume from the press of Robert Wyer was entered in the Catalogue under the general heading "Book," there being nothing to show until now by whose hand it was written. The writer of this note, while searching for something else, was so struck with the title "The Boke for to lerne a Man to be wyse in building of his house", that he sent for it, thinking that it might supply material for an interesting article commenting on Dr Richardson's recent lectures on the same subject, after a lapse of more than three centuries. This expectation was amply justified, and the subject having been mentioned to Mr Ponsonby Lyons, that gentleman suggested the name of Andrew Boorde as a writer on sanitary matters in the 16th century, whose works might supply additional material for the purpose in view. But when Boorde's works were obtained, it was found that the interest was by no means confined to the subject matter, but that the first eight of the forty chapters contained in his Dietary were as nearly as possible identical with the eight chapters of which the volume now to be described consists.

The book is quite perfect, and in as good condition as when it first came from the press. It is a small quarto of sixteen leaves (A. B. C. D. in fours). There are twenty-five lines to each page, and every chapter has a woodcut initial letter, which is not the case with any of the editions previously known, except that belonging to

Mr Henry H. Gibbs, which has ornamental initials throughout. The attention of Mr Furnivall was called to the book, and he at once pronounced in favour of its being the work of Boorde. It may be that it was his first attempt at authorship, and that after he had acquired some degree of reputation, and was engaged in writing the more comprehensive work which he published under the title "A compendyous regyment or a Dyetary of helth," he prefixed the little treatise now under consideration to the later work instead of republishing it in a separate form. No edition of the Dietary is known which does not contain these eight chapters, but, as will be seen hereafter, the title is not so applicable to them as it is to the succeeding thirty-two chapters, which relate exclusively to questions of regimen and diet, and there appeared at first sight to be some reason for supposing that the break in the continuity of the subject was recognized by several of the printers, who have concluded the eighth chapter with lines gradually decreasing in length. This is the case in all the editions, except Powell's and that in the possession of Mr Gibbs, in both of which Chapter VIII. ends evenly; the irregularity, however, occurs in one or more places in every edition of the Dietary, so that in all probability it should be attributed rather to accident than to design.

The Title-page, Table of Contents, and Colophon of the newly-discovered work are here given in full, and the notes appended will show that they have been carefully collated with those of five editions of the Dietary; attention is also directed to a circumstance of some interest at the end of the third chapter. The other differences between the work described and any one of the editions of the Dietary are not greater than those between that one and each of the others. There is no dedication to the Duke of Norfolk, but that is also the case with the undated edition of the Dietary (A.), as well as with Colwell's edition of 1562 (B.), both in the British Museum. No allusion whatever is made in the dedication printed in the 1542 edition (E.) to any portion of the book having been in existence previous to that date, and this is, of course, an argument against the supposition that the first eight chapters were published in a separate form *before* the appearance of the Dietary, and would tend rather to show that they were really published as an extract from a book previously known. Which of the two hypotheses is the true explanation is the question now submitted for consideration, and the following extracts are given to aid in the solution of the difficulty. The title-page is as follows:

The boke for to
Lerne a man to be wyse in
buyldyng of his howse for
the helth of body & to holde quyetnes for the helth
of his soule, and body.
¶ The boke for a good
husbande to lerne.

¶ We
May-
sters of
Astro-
nomye,
And do-
ctoures
in Phe-
sycke cō-
ſyrmeth
this say-

Woodcut of an
Astronomer.

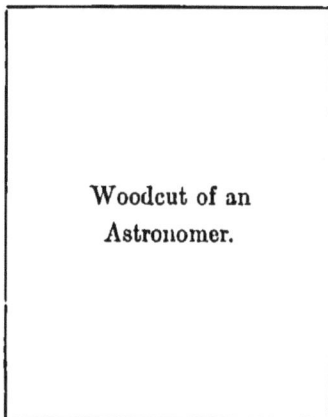

enge to
be good
& trewe
both for
the bo-
dy, and
also for
the sou-
le. ☞

A 1

The woodcut is not the same as that in the copy belonging to Mr Henry Hucks Gibbs, from which Mr Furnivall printed his edition of the Dietary for the Society, nor is it the same as that printed in the undated copy in the British Museum, and in the 1562 edition, which has also been recently acquired by the trustees of the National Library. The double-dated Edition, and that of 1576, have no woodcut on their title-pages. It is noteworthy that the woodcut of the 1542 edition represents St John *without* the eagle. Robert Wyer used as his device a cut of the Saint writing the Revelations, and attended in most cases by an eagle. Herbert makes a special note of the rarity of Wyer's use of the device in which the attendant eagle is omitted.

Another peculiarity to be observed is, that in the tract now described the title-page itself is signed, A. 1.

The next point for description is the table of contents. This has been carefully collated with those of the five editions of the Dietary, and all the various readings are supplied in the foot-notes, chapter by chapter, the heading being numbered 1, and the eight chapters 2 to 9.

[1] ¶ The table of this Boke.

[2] The fyrste chapter doth shewe where a / man shulde buylde or set his howse,/ or place, for the helthe of his body./

[3] ¶ The seconde chapter doth shewe a man,/ howe he shulde buylde his howse, that the / prospect be good for yͤ cōseruacion of helth./

[1] A. ¶ The Table of the Chapters / foloweth; B. The Table / ¶ The Table of the chapters / foloweth; C. ☞ Here foloweth the Table / of the Chapiters; D. ¶ The Table; E. ¶ Here foloweth the Table / of the Chapytres.

[2] D. first; A. B. Chapter (throughout); C. Chapyter; E. Chapytre; B. doeth; D. shew; C. E. shuld; D. should; in A. B. D. "cytuate" for "buylde"; C. E. cytuat; A. B. C. D. E. "set his mansyon place or howse," instead of "howse or place"; except that D. has "mansion," E. "mancyon," and B. C. D. have "house"; C. yͤ.

[3] B. omits ¶ (throughout); D. secōd; C. chapiter; E. Chapytre; C. dothe; D. shew; D. how; C. shuld; D. should; B. D. build; B. C. D. house; A. B. C. D. E. here insert "and"; A. B. prospecte; C 'pspect; A. B. D. the; A. B. C. E. conseruacion; D. conseruation; A. B. C. D. health.

[4] ¶ The thyrde chapter doth shewe a man to / buylde his howse in a pure and fresh / ayre for to length his lyfe./

[5] ¶ The fourth chapt' doth shew vnder what / maner a man shuld buylde his howse in ex/chewyng thynges y^t shuld shorten his lyfe.

[6] ¶ The .V. chapter doth shewe howe a man / shulde ordre his howse, consernynge the im-/plementes, to cōfort the spyrites of man./

[7] ¶ The .VI, chapter doth shewe a man howe / he shulde ordre his howse and howsholde, to / lyue in quyetnes.

[8] ¶ The VII. chapter doth shewe how the hed / of the howse, or howseholder shulde exercy/se hymself, for the helth of his soule & body

[9] ¶ The .VIII. chapter doth shewe how a man / shuld ordre hym self in slepynge & watche,/ and in his apparell werynge.

¶ Explicit tabula.*

[4] C. has ☞ for ¶. D. third; C. Chapyter; E. Chapitre; B. doeth; C. dothe; D. shew; A. mā; B. D. build; A. B. C. D. house; C. ī; C. inserts "a" before "fresshe"; A. B. C. E. fresshe; A. B. C. D. E. lengthen; B. D. life.

[5] A. IIIJ; B. E. IIII; A. B. D. Chapter; C. Chapiter; E. Chapytre; B. doeth; C. dothe; D. shew; A. B. C. shulde; D. should; D. build; B. hys; B. C. D. house; here A. B. C. D. E. all insert the words "or mansyon" (D. spells mansion); A. B. D. omit "in"; C. E. eschewynge; D. eschewing; D. thinges; A. B. D. E. that; A. B. C. shulde; D. should; A. B. D. "the" for "his".

[6] D. fift; C. Chapiter; E. Chapytre; B. doeth; D. shew; C. E. shuld; D. shold; B. C. D. order; B. hys; B. C. D. house; A. B. concernynge; C. E. concernyng; D. concerning; A. B. Implementes; A. B. C. D. E. comforte; A. B. C. E. spyrytes; D. spirites.

[7] C. has ☞ for ¶. D. sixte; C. Chapiter; E. Chapytre; D. shew; C. a mā; B. shoulde; D. should; B. C. D. order; B. C. D. house; B. has "houshold" as a catchword, but at the top of the next page the word is spelt "housholde"; D. quietnesse.

[8] A. VIJ; D. seuēth; C. chapiter; E. Chapytre; D. E. shew; C. E. howe; C. y^e; A. hed of house; B. hed of the house; C. hed of a house; D. head of the house; E. hed of a howse; A. B. C. D. E. insert "a" after "or"; A. B. D. housholder; C. householde; A. B. shuld; D. should; C. excercyse; D. exercise; A. E. C. hym selfe; B. D. himselfe; A. B. C. health; C. E. the soule; A. B. and bodye; D. E. and body.

[9] A. VIIJ; D. eyght; C. chapiter; E. Chapytre; E. show; C. howe; C. mā; A. C. E. shulde; B. shoulde; D. should; B. C. D. E. order; A. hymselfe; B. E. hym selfe; C. him selfe; D. himselfe; D. sleeping; A. B. C. D. E. and; C. E. watchynge; B. apparel; A. B. C. E. wearynge; D. wearing.

* Wyer's undated edition, A. Colwel's of 1562, B. Powell's double-dated edition, 1547-67, C. H. Jackson's of 1576, D. (the table not in black letter). Mr Furnivall's reprint of the 1542 edition, E.

The words "explicit tabula" at the end of the eighth chapter are, of course, peculiar to the treatise which is brought to a conclusion at that point. In all the enlarged editions published under the title "Dietary of Health," the table of contents proceeds, without any break whatever, to give the headings of the remaining thirty-two chapters. The various readings of the concluding words in the different editions will be found at page 231 of Mr Furnivall's reprint.

The next point to be observed is, that in the Dietary there occurs, at the end of the third chapter, a reference to the 27th chapter, but in the book under examination there is no such reference for obvious reasons, but the information referred to appears as a separate paragraph on the *same page*. The extracts are given here, for the sake of comparison, in parallel columns, partly with a view to directing attention to the differences between them, and partly because the circumstance appears, at first sight, to afford some additional ground for believing that the larger work was first published, and the smaller one brought out afterwards in a separate form.

Paragraph at the foot of Chapter III. in the book described.

¶ For whan the plaages of the Pestylence or the swetynge syckenes is in a trowne or countre, at Mountpylour, and in all other hyghe regyons and countres, that I haue ben in, the people doth flye from the contagyous and infectyous ayer, preseruatiues with other councell of Physycke, notwithstandynge. In lower and other baase countres, howses the whiche be infectyd in towne or cytie, be closed vp, both dores & wyndowes, and the inhabytours shal not come abrode, nother to churche nor market, for infectynge other, with that syckenes.

Opening sentences of Chapter XXVII. (Mr Furnivall's reprint.)

Whan the Plages of the Pestylence, or the swetynge syckenes is in a towne or coûtree, with vs at Mountpylour, and all other hygh Regyons and countrees y^{t} I haue dwelt in, the people doth fle from the contagious and infectious ayre preseruatyues, with other counceyll of Physycke, notwithstandyng. In lower and other baase countres, howses the which be infectyd in towne or cytie, be closyd vp both doores & wyndowes: & the inhabytours shall not come a brode, nother to churche: nor to market, nor to any howse or cōpany, for infectyng other, the whiche be clene without infection.

It will be seen that in the tract the author does not use the words "with us" when speaking of Montpelier. Can it be that he wrote the treatise on house-building elsewhere? and, if so, are we to suppose that it was written before or after 1542, the date of his dedication of the Dietary to the Duke of Norfolk, which Mr Furnivall believes to be the date at which the first edition was published? And, speaking of this dedication, does the text afford sufficient ground for believing that it was actually *written* in Montpelier? It is dated from there, but it would be hard to prove that it was not written in London. The author in the body of the dedicatory letter calls attention to a book "the which I *dyd* make in Mountpyller," and which he says "*is* a pryntynge besyde Saynt Dunston's churche." The dedication, as prefixed to the 1542 edition, and the version in Powell's edition of 1547, are printed by Mr Furnivall in parallel columns (page 225 *et seq.*), and we see at once that Powell kept both the original place, Montpelier, and the original day and month, 5th of May, but altered the year, 1542, to the date of his own edition, 1547, to make it look like a new book.

1542 *Edition.*	*Powell's Edition.*
From Mountpyllier. The .v. day of May. The yere of our lorde Iesu Chryste M.v.C.xlij.	From Mountpyller. The fyft daye of Maye. The yere of our Lord Iesu Chryste M.ccccc xlvii.

It is at least possible that the principal object of Boorde, as well as Powell, was to show, not that the dedication was *written* in Montpelier, but that the author had studied in the medical school of that city, which he himself describes as "the hed vniversitie in al Europe for the practes of physycke & surgery or chyrming."

There is nothing more in the book here described that requires any special consideration until the eighth and last chapter is brought to a conclusion, with a caution against travelling in boisterous weather. "¶ Explicit" is printed at the foot of the chapter, and thereafter are inserted the following verses, which do not occur anywhere in the various editions of the Dietary. The last verse is followed by the word "Finis", and beneath that is the Colophon as printed below

¶ Of folyshe Physycyons.

Who that useth the arte of medycyne
Takynge his knowlege in the feelde
He is a foole full of ruyne
So to take herbes for his sheelde
wenynge theyr vertue for to weelde
whiche is not possyble for to knowe
All theyr vertues, both hye and lowe.

¶ Of dolorous departynge.

¶ Neuer man yet was so puyssant
Of gooddes or of parentage
But that mortall death dyd hym daunt
By processe at some strayght passage
yea, were he neuer of suche an age
For he spareth neyther yonge nor olde
Fayre nor fowle, fyerse nor also bolde.

¶ Of the true descripcion.

¶ The wyse man whiche is prudent
Doth moche good where euer he go
Gyuynge examples excellent
Unto them the whiche are in wo
Teachynge them in all vertues so
That they may not in to synne fall
If that they hertely on God call.

¶ Of Phylosophye.

¶ At this tyme doctryne is decayed
And nought set by in no place
For euery man is well appayed
To get good with great solace
Not carynge howe nor in what place
Puttynge the fayre and dygnesophye
Under feete with Phylosophye.

¶ Finis. ¶

Imprynted by me Robert
Wyer,[1] dwellynge at the signe of :S:.
John Euangelyst, in s. Martyns
parysshe in the felde besyde the
Duke of Suffolkes pla-
ce, at Charynge
Crosse.

¶ Cum priueligio, Ad
impremendum
solum.

It now remains to say a few words about the relative ages of the tract described and of the first edition of the Dietary, regarding the question from a purely typographical point of view. All the evidence appears to be in favour of the tract having been printed at an earlier period than the "Dietary." It is well known that the printers of the day allowed the quality of the paper they used to deteriorate as time went on. Now there is a marked difference in the texture and finish of the paper on which the tract is printed and that of the paper which is used for the Dietary, and the superiority belongs entirely to the former. The type used in the tract is, in the opinion of experts, of an earlier character than that used in the Dietary, many of the letters (l, v, &c.) bearing a closer resemblance to the forms used in manuscript, while a careful comparison of those of the woodcut initial letters, which are common to both books, seems to show that if the same blocks were used in both cases they were less worn and in better condition when the tract was printed than when they were used for the Dietary; but, of course, it is quite possible that

[1] Wyer's undated edition says nothing about "the Duke of Suffolk's place," but reads "Dwellynge at the / signe of seynt John E/uangelyst, in S Mar/tyns Parysshe, besy/de Charynge / Crosse /

¶ Cum priuilegio Ad impremendum solum.

For the colophons of the other editions noticed by Mr Furnivall, see page 304 of his reprint. In H. Jackson's edition of 1576 an imprint is given at the foot of the title-page, but the colophon merely consists of the word Finis over the woodcut reproduced by Mr Furnivall from Mr Gibbs's copy, that is, Wyer's ordinary device, St John *attended* by the eagle: it will thus be seen that Mr Gibbs's copy affords examples of two out of the three devices used by that printer, one of them being very rare.

the initials in the two books were printed from different blocks, cut to the same pattern; and if that were the case the argument, based upon the superior clearness of the impressions in the tract, falls to the ground. However, taking all the facts of the case together, the writer, as far as he can venture to form an opinion on such a subject, is inclined to believe that "The boke for to lerne a man to be wyse in the buyldyng of his howse" was printed, if not actually written, at an earlier period than the earliest known edition of the "Compendyous Regyment or Dyetary of Helth," with which it was incorporated; and the supposition that the Dietary, in its complete form, was *first* published, and then that the first eight chapters were extracted and published separately under another title, he believes to be untenable and against the weight of the evidence.

Richard Clay & Sons, Limited, London and Bungay.

www.ingramcontent.com/pod-product-compliance
Lightning Source LLC
Chambersburg PA
CBHW021354210326
41599CB00011B/870
9783337221171